Engineering Geology for Society and Territory – Volume 7

Giorgio Lollino • Massimo Arattano
Marco Giardino • Ricardo Oliveira
Silvia Peppoloni
Editors

Engineering Geology for Society and Territory – Volume 7

Education, Professional Ethics and Public Recognition of Engineering Geology

Springer

Editors
Giorgio Lollino
Massimo Arattano
Institute for Geo-Hydrological Protection
National Research Council (CNR)
Turin
Italy

Marco Giardino
Department of Earth Sciences
Turin University
Turin
Italy

Ricardo Oliveira
COBA
Lisbon
Portugal

Silvia Peppoloni
National Institute of Geophysics and Volcanology
Rome
Italy

ISBN 978-3-319-09302-4 ISBN 978-3-319-09303-1 (eBook)
DOI 10.1007/978-3-319-09303-1
Springer Cham Heidelberg New York Dordrecht London

Library of Congress Control Number: 2014946956

Cover Illustration: San Felice sul Panaro, Modena, northern Italy. The San Felice Vescovo Church, built 1499, was completely destroyed by an earthquake, which struck a vast area of the Po Plain on 20 May 2014. As visible, many near ancient buildings did not suffer similar damages. This proves that the effect of the seismic shock is strongly dependent also from structural features of the building. *Photo*: Giovanni Bertolini.

Printed on acid-free paper

Springer is part of Springer Science+Business Media (www.springer.com)

Foreword

It is our pleasure to present this volume as part of the book series of the Proceedings of the XII International IAEG Congress, Torino 2014.

For the 50th Anniversary, the Congress collected contributions relevant to all themes where the IAEG members have been involved, both in the research field and in professional activities.

Each volume is related to a specific topic, including:

1. Climate Change and Engineering Geology;
2. Landslide Processes;
3. River Basins, Reservoir Sedimentation, and Water Resources;
4. Marine and Coastal Processes;
5. Urban Geology, Sustainable Planning, and Landscape Exploitation;
6. Applied Geology for Major Engineering Projects;
7. Education, Professional Ethics, and Public Recognition of Engineering Geology;
8. Preservation of Cultural Heritage.

The book series aims at constituting a milestone for our association, and a bridge for the development and challenges of Engineering Geology toward the future.

This ambition stimulated numerous conveners, who committed themselves to collect a large number of contributions from all parts of the world, and to select the best papers through two review stages. To highlight the work done by the conveners, the table of contents of the volumes maintains the structure of the sessions of the Congress.

The lectures delivered by prominent scientists, as well as the contributions of authors, have explored several questions ranging from scientific to economic aspects, from professional applications to ethical issues, which all have a possible impact on society and territory.

This volume testifies the evolution of engineering geology during the last 50 years and summarizes the recent results. We hope that you will be able to find stimulating contributions which will support your research or professional activities.

Giorgio Lollino

Carlos Delgado

Preface

In the age of human activities, Engineering Geology plays a key role in the sustainable development of our societies: scientists, regulators, and practitioners of Engineering Geology are called to confront themselves with the purposes, methods, limitations, and findings of their works.

In this perspective, topic seven of the XII Congress of IAEG in Torino on 2014 was an opportunity to illustrate a wide-angle vision on several inter-related issues: the role of Engineering Geologists within the geoengineering profession; the best practice in professional ethics and communication in a changing world; the education and modern development of Engineering Geology profession and its professionals; resource use and reuse in managing risk prevention and impaction a complex framework; engineering our geological responsibility in an uncertain environment; Engineering Geology at tertiary level.

Five part topics were activated, presenting a total of 54 chapters, contributing to:

- stimulating the debate on professional responsibilities of engineering geologists,
- analyzing the interactions of engineering geologists with other professionals,
- evaluating the recognition of the engineering geological profession and its peculiar contribution to society, culture, and economy, and
- reporting examples of the empowerment of research groups and management activities by using web 2.0/3.0 technologies, thus enabling cooperation, knowledge sharing, and collaboration at all levels.

They highlighted implications for the use of the education of engineering geologists at tertiary level and in further education schemes. They also highlighted the importance of having the professionals organized into national groups which stimulate advances in Engineering Geology in their countries.

"Engineering Geological Models" (Part I) discussed the use of engineering geological models within the framework of the total geological approach (Fookes et al. 2000; Baynes et al. 2005; IAEG Commission 25). Such models allow the understanding and prediction of engineering geological conditions and processes, aiming at reducing uncertainties and their impact on our societies. The authors presented examples on innovative use of engineering geological models for different engineering projects, and for different geological and geomorphological environments, envisaging new perspectives and operational outcomes.

"Fifty-Year-Long History of IAEG in Events and Personalities" (Part V) focused on relevant facts and events (congresses, conferences, symposia) of the 50-year-long IAEG history, where many outstanding personalities played a fundamental role as founders of our association. Amongst those who participated in the IAEG work, since its early beginning, some gave great and acknowledged contribution to the development of engineering geology on a world scale. Many witnesses of the events that took place during 50 years, and there are still colleagues and disciples of the remarkable founders of IAEG, keep their historical memory. This part highlighted our duty to share this heritage, passing up the baton to the new young generation of geoscientists. In the 50th Anniversary Book which will be distributed to all participants in the Congress, parts are devoted to the birth of the IAEG and the relevant role of its founders, to the main events organized along the 50 years, to its outstanding activity all over the world, and to the awards that have been established to pay a tribute to those who most contributed to the development of our discipline. The book also includes a History of Engineering Geology which starts with its heritage and reports its evolution and the main achievements until today.

"Geoethics and Natural Hazards: Communication, Education and the Science-Policy-Practice Interface" (Part II) analyzed the critical ethical issues faced by Geoscientists and Engineers in relation to natural hazards (e.g., earthquake, volcano, landslide, and flood events) and risks, and their increasing death toll and social costs owing to population growth, occupation of marginal/unsafe land, and abandon or misuse of land. Sharing and communicating our knowledge more effectively, involving private and public stakeholders, could contribute to a sustainable development of human society and economic activities. In the Anthropocene, Geosciences represent the "connective tissue" of a wider multidisciplinary approach, to build a shared responsibility on the effects of human actions, and to better cope with uncertainties. This part highlighted many natural disasters could be prevented and/or their impact reduced, raising awareness and fostering a true interdisciplinary collaboration that could fulfill ethical obligations of the scientific community as a whole. This shows the growing importance of environment in the practice of engineering geology and also the need for its cooperation with other engineering and social subjects and professionals.

"Resilience Two Citizens and Citizens Four Resilience" (Part III) focused on how engineering geology could benefit from knowledge sharing of natural hazards and collaborative risk management. As natural risks are part of our reality, the authors highlighted how preparedness, as an interdisciplinary issue, could envisage a more effective disaster resilience. The "common and shared knowledge" approach empowered by web 2.0/3.0 technologies, embodies the idea of citizen sciences and the purpose to build a new people-centred resilience: Crowdsourcing and VGI, citizens engagement and participatory practices are a new frontier and a matter of fact. Despite any critics, they have the merit to arouse a debate on cooperation, knowledge sharing, and collaboration at all levels. This part faces, out from the crowd, applicability, opportunity, and constraints of these new approaches, procedures, and technologies for preparedness actions: (A) The "web 2.0 wave": threat or opportunity for disaster resilience? (B) Two-way emergency communications: empower or menace for governmental organizations. (C) ICT laws and regulations: dinosaurs in a glass store? (D) Is research ready for Open Data and Open Knowledge (E) Cultural vs. technological challenges in disaster resilience (E) Web and mobile technologies: experiences and tools.

"Standards, Guidelines and Best Practices for Engineering Geology" (Part IV) offered to professionals an overview of specialized documentation on Engineering Geology: the best practice case studies and compilations, recommended technical procedures in more formal guidelines, rigid regulatory, or prescriptive standards that are legally binding. Such documentation resulted appropriate for a variety of topics relevant to the engineering geology community, and for a suite of topics, including construction materials studies, landslide risk management and land planning, subsurface mining, infrastructure construction, and groundwater extraction. An international open exchange of ideas and knowledge was gained by this part, where authors illustrated their personal, national, or specialized experiences, lessons learned, successes, and failures with fellow professionals. The authors provided much needed guidance and structure to practicing engineering geologists and they underlay our professional obligations to ensure the health, safety, and well-being of society. In the IAEG, this has been best achieved through publication in the Bulletin which was created in 1970 and is today a reference journal in the area, as well as the work produced by the IAEG Commissions.

Interesting points emerged from the IAEG 2014-Topic 7 on "Education, Professional Ethics and Public Recognition of Engineering Geology." A comprehensive view of the proposed contributions fosters the idea of engineering geologists playing the role of acknowledged "interface" between man and nature. They are not only scientists and professionals able to "interpret" both the environmental and the territorial processes, but they also have attitudes and capabilities to communicate information to the general public and to develop guidelines for the correct and safe use of land, namely for the social welfare and economic development of society. The issues proposed by the Topic's sessions, and the way they were discussed within the proposed contributions also highlighted the important role Engineering Geologists can play in disaster resilience. As a conclusion, interesting discussions have been stimulated on the relationships between ethic, science, politics, and citizenship.

Reference

Fookes PG, Baynes FJ, Hutchinson JN (2000) Total geological history: A model approach to the anticipation, observation and understanding of site conditions, Invited Paper, Geoeng 2000 Conference, Melbourne Australia, published in Ground Engineering 34(3): 42–47

Contents

1 Optimization of Large Civil Engineering Projects from an Environmental Point of View 1
Ricardo Oliveira

Part I Engineering Geological Models

2 Engineering Geological Models: Some Examples of Use for Landslide Assessments 11
Jan Novotný

3 Engineering Geological Models IAEG Commission 25 17
Steve Parry, Fred J. Baynes, Martin G. Culshaw, Mark Eggers, Jeffrey R. Keaton, Kori Lentfer, Jan Novotný and Daren Paul

4 The Use of Ground Models for Probabilistic Assessment of Multiple Geohazards Along Lifeline and Transportation Corridors 21
Clark Fenton, Fergus Cheng, Andrew Davie, Andria Loppas, Scott Loudon, Rebekah Marsh, Matthew Pendlebury, Yeldos Sultanbek, Leonidas Tatarakis, Michael Teo, Charlene Ting and Xue Yao

5 Educational GIS-Project "World Experience in Site Investigation and Construction Under Different Engineering-Geological Conditions" . . . 27
Tatyana Averkina, Natalia Pravikova and Natalia Kosevich

Part II Geoethics and Natural Hazards: Communication, Education and the Science-Policy-Practice Interface

6 Geoethics: Perspective Bangladesh 33
Mohammad Omer Faruk Khan

7 Geoethics and Communication 2: Ethics, the Notably Absent from the Internet 39
Stefano Solarino

8 A Road Map for a Deontological Code for Geoscientists Dealing with Natural Hazards 45
Eduardo Marone and Luis Marone

9 Reducing the Gap Between Science, Policy and Practice: The Role of Civil Protection . . . 49
Appiotti Federica, Eleonora Gioia, Gabriella Speranza, Maurizio Ferretti and Fausto Marincioni

10 Geoethics and Communication 1. Scientists, Politicians, Journalists, Media Entertainers, Sociologists: Can Ethics Help to Establish Their Exact Role in Communicating Geoscience? . . . 55
Stefano Solarino

11 Geoethical Aspects in the Natural Hazards Management . . . 59
Giuseppe Di Capua and Silvia Peppoloni

12 Need to Harness Geodiversity to Combat Disaster from Natural Hazards . . . 63
Partha Sarathi Datta

13 What's the Seismic Risk Perception in Italy? . . . 69
Massimo Crescimbene, Federica La Longa, Romano Camassi, Nicola Alessandro Pino and Laura Peruzza

14 Long-Term Disaster-Prevention Strategies Based on Education . . . 77
G. Musacchio, A.E. Bernhardsdóttir, M.A. Ferreira, S. Falsaperla and UPStrat-MAFA Outreach Working Group

15 Geoscientific Communication Problem with Communities for Disaster Prevention and Land Planning in Peru . . . 81
Luisa Macedo, Sandra Villacorta, Shianny Vasquez, Jersy Mariño and Giuseppe Di Capua

16 Promoting Geo-awareness to Make Citizens the First Watchers of the Territory . . . 85
Tiziana Lanza

17 Laboratory Activities for Seismic and Volcanic Hazards Education: A Challenge for Scientists . . . 89
Gemma Musacchio and Nicola Alessandro Pino

18 Geological Engineers from the University of Salamanca: The Experience of Ten Years from Education to the Professional Life . . . 93
J. Nespereira, M. Yenes and J.A. Cabezas

19 Rumours Related to the 2012 Emilia Seismic Sequence . . . 97
Milena Bertacchini, Doriano Castaldini and Giovanni Tosatti

20 The Experience of ScienzAperta, a Week of Scientific Information and Dissemination . . . 103
Giuliana D'Addezio, Giuliana Rubbia, Antonella Marsili and The Laboratorio Didattica e Divulgazione Scientifica Team

21 **Hazardscape, Territorial and Individual Resilience in an Interdisciplinary Study: The Case of Pollino, Southern Italy** 109
Francesco De Pascale, Marcello Bernardo and Francesco Muto

22 **The Role of Earth Science and Landscape Approach in the Ethic Geology: Communication and Divulgation for the Prevention and Reduction of Geological Hazard** 115
Piero Farabollini, Francesca Romana Lugeri, Barbara Aldighieri and Vittorio Amadio

23 ***Piovono idee! (Cloudy with a Chance of Ideas!)*: An Interactive Learning Experience on Hydrogeological Risk and Climate Change** 121
Giovanna Lucia Piangiamore, Gemma Musacchio and Maurizio Bocchia

24 **Interactive, 3D Simulation of Natural Instability Processes for Civil Protection Purposes** 125
Furio Dutto, Massimo Arattano, Marco Bacenetti, Marta Chiarle, Concetta Contrafatto, Marco Giardino, Fabrizio Longo, Luigi Perotti, Filippo Racca, Lorenzo Rocci and Laura Turconi

25 **The Roles and Responsibilities of Engineering Geologists and Other Geoscientists in Serving Society and Protecting the Public—an Overview of International Approaches to Ensuring Effective and Ethical Professional Practice** 131
Ruth Allington and Isabel Fernandez-Fuentes

26 **Natural Hazards, Disasters and Their Mitigation: An Overview with Reference to Indian Scenario** 135
Vasudevan Desikachari

27 **Geohazards in the East Siang District of Arunachal Pradesh, India: Need for Geoethical Approach for Integrated Flood Mitigation Strategies** 141
Swapna Acharjee

Part III Resilience to Citizens and Citizens for Resilience. From Collaborative Risk Management to Knowledge Sharing of Natural Hazards

28 **GeoMedia-web: Multimedia and Networks for Dissemination of Knowledge on Geoheritage and Natural Risk** 147
Marco Giardino, Vincenzo Lombardo, Francesca Lozar, Alessandra Magagna and Luigi Perotti

29 **Reason and Being of Engineering Geology**. 151
J. Javier Diez, Efren M. Veiga, Fernando Rodriguez and Rosa M. Paz

30 **Human Factor in Resilience Systems**. 157
Iria Paz, Rosa M. Paz, Elvira Fernández and J. Javier Diez

31 **Recording the Onset of the Anthropocene** 161
Martin Bohle

32 **Searching for Connections to the Raison d'Être of Engineering Geology in Codes of Ethics** 165
Robert E. Tepel

33 **Web Participatory Framework for Disaster Resilience: Coping with Information Deluge** 169
Elena Rapisardi, Sabina Di Franco and Marco Giardino

34 **Italian Civil Protection 2.0: Towards a Resilient Information Management Framework** 175
Elena Rapisardi, Sabina Di Franco and Marco Giardino

35 **Geomathics Tools and Web-Based Open-Source GIScience Public Participation Field Approach for Field Data Collection: Supports for a Civil Protection Exercise in a Hydrological Risk Scenario** 181
Bacenetti Marco, Bertotto Stefania, Ghiraldi Luca, Giardino Marco, Magagna Alessandra, Palomba Mauro and Perotti Luigi

36 **A Missed Window of Opportunity** 185
Massimo Lanfranco, Elena Rapisardi and Marco Giardino

37 **Pyramids, Toxic Wastes and Nuclear Reactors Containments: A Lesson Drawn from History with a Risk Manager Perspective** 191
Franco Oboni and Cesar Oboni

Part IV Standards, Guidelines and Best Practices for Engineering Geology

38 **Working with Engineers, as an Engineering Geological Consultant** 197
Robert Goldsmith

39 **A Case Study of Intelligent Compaction Used in Road Upgrades** 201
Robert Leyland

40 **Unified Qualification Requirements for Ground Engineering and Engineering Geology Professionals** 207
Helmut Bock, Markus Herten, Reinhard Schwerter and Kurosch Thuro

41 **Guidelines for Site Investigation and Analysis of Nuclear Facilities: A Consultants Perspective** 213
William Godwin, Ramon Secanell and Christophe Martin

42 **Guidelines to Identify and Quantify Expansive Soils in Civil Infrastructure** 217
Marcos Musso

43 **Legislation as a Tool to Mitigate Mass Movements in Spain** 221
Jesús Garrido, José Delgado, Alejandro L. Grindlay, Ignacio Jiménez Soto and M. Isabel Rodríguez

44 Design of a Database for Documentation and Analysis of Lab Data and Rock-Properties Derived in Field Measurements 225
Florian Menschik, Kurosch Thuro, Heiko Käsling and Michael Bayerl

45 Landslide Guidelines and Best Practices for Professional Engineers and Geoscientists 229
Peter Bobrowsky, Doug VanDine and Réjean Couture

46 Engineering Geology at Post-graduated Level at the NOVA University of Lisbon 233
Ana Paula F. da Silva, Pedro C. Lamas, M. Graça Brito and José A. Almeida

47 Dangers of Foundations and Below Ground Construction in the Coastal Zone of the Rhine Delta Sediments, The Netherlands 237
Pieter Michiel Maurenbrecher

Part V Fifty-Year-Long History of IAEG in Events and Personalities

48 Evgenii Mikhailovich Sergeev (1914–1997) as One of the Founders and Leaders of IAEG 243
Victor Osipov, Olga Eremina and Nadezda Rumyantseva

49 The 50th Anniversary of the First International Conference on Permafrost 247
Jerry Brown and Julia Stanilovskaya

50 Presidents and Leaders of the IAEG Visited the Medeo Dam in the Soviet Union 251
Shibakova Valentina and Koshkin Andrey

51 A Short History of Engineering Geology and Geophysics at the British Geological Survey 257
M.G. Culshaw, K.J. Northmore and D.M. McCann

52 ASAGAI, the IAEG National Group of Argentina: Its History, Process and Present 261
Norberto Jorge Bejerman and Maria Beatriz Ponce

53 A Brief History of the Past 65 Years of Romanian Engineering Geology Experience 265
Mihaela Stãnciucu, Cristian Mãrunţeanu, Emilia Elena Milutinovici, Paul Nedelcu and Leonin Bogãţeanu

54 The Activity of Uzbek Scientist in IAEG 269
N.G. Mavlyanova and K.L. Rakhmatullaev

Author Index 273

Consiglio Nazionale delle Ricerche
Istituto di Ricerca per la Protezione Idrogeologica

irpi

The Istituto di Ricerca per la Protezione Idrogeologica (IRPI), of the Italian Consiglio Nazionale delle Ricerche (CNR), designs and executes research, technical and development activities in the vast and variegated field of natural hazards, vulnerability assessment and geo-risk mitigation. We study all geo-hydrological hazards, including floods, landslides, erosion processes, subsidence, droughts, and hazards in coastal and mountain areas. We investigate the availability and quality of water, the exploitation of geo-resources, and the disposal of wastes. We research the expected impact of climatic and environmental changes on geo-hazards and geo-resources, and we contribute to the design of sustainable adaptation strategies. Our outreach activities contribute to educate and inform on geo-hazards and their consequences in Italy.

We conduct our research and technical activities at various geographical and temporal scales, and in different physiographic and climatic regions, in Italy, in Europe, and in the World. Our scientific objective is the production of new knowledge about potentially dangerous natural phenomena, and their interactions with the natural and the human environment. We develop products, services, technologies and tools for the advanced, timely and accurate detection and monitoring of geo-hazards, for the assessment of geo-risks, and for the design and the implementation of sustainable strategies for risk reduction and adaptation. We are 100 dedicated scientists, technicians and administrative staff operating in five centres located in Perugia (headquarter), Bari, Cosenza, Padova and Torino. Our network of labs and expertizes is a recognized Centre of Competence on geo-hydrological hazards and risks for the Italian Civil Protection Department, an Office of the Prime Minister.

Optimization of Large Civil Engineering Projects from an Environmental Point of View

1

Ricardo Oliveira

Abstract

Without exception, the construction and operation of civil engineering projects have resultant environmental impacts. However, in most cases the projects are essential to the economic and social development of the regions where they are located and for some, their sole purpose is to protect people and goods from natural hazards such as floods and landslides. In general, the media and environmentalists tend to enhance the negative impacts of the projects and very seldom make reference to their positive impacts. In this context, the need for high quality studies and designs is assuming increasing relevance for engineering projects, to ensure solutions with the least negative impacts are selected and subsequently constructed and operated by suitably qualified staff. The role of geotechnics in the optimization of civil engineering projects is therefore as important as is the efficient and early intervention of specialists, and their decisions, on the technical, economic, social, environmental and operational aspects of the works. To illustrate that it is often possible to optimize projects from an environmental point of view, several examples are presented in relation to construction materials, hydraulic developments, linear works (roads, railways, airways, and waterways), underground works, maritime works, bridges and viaducts, and natural and excavation slopes. In each case, emphasis is placed on the environmental concerns that require optimization of the design in order to minimise the negative impacts without diminishing the economic and social benefits of the works.

Keywords

Design improvement • Engineering projects • Environment

1.1 Introduction

Until the 70s/80s the design of large engineering projects had to be satisfactory from both technical and economic points of view. This meant that, until that time, the best design solutions were selected by only taking into consideration those points of view. However, quite often those responsible for the design did study different alternatives, some of which had less interference with the environment, but generally only selected them if they were the least cost alternative.

Large engineering projects were constructed all over the world; however people in more developed countries tended to be more sensitive to their environmental impacts.

The primary environmental impacts were related to the archeological, biological, paleontological and physical aspects of the affected region and concerned communities started to contest the projects. In spite of the negative impacts, the construction of those projects was considered essential to the

R. Oliveira (✉)
COBA, S.A., Engineering and Environmental Consultants
and New University of Lisbon, Lisbon, Portugal
e-mail: ricardo.oliveira@coba.pt

G. Lollino et al. (eds.), *Engineering Geology for Society and Territory – Volume 7*,
DOI: 10.1007/978-3-319-09303-1_1, © Springer International Publishing Switzerland 2014

economic and social development of the population affected. Examples include the construction of infrastructure like dams (for water supply, irrigation, clean and renewable hydroelectric energy, leisure, and flood control), canals, motorways, railroads, bridges, tunnels and other underground works, airports and ports. Moreover, some of these works had the objective of protecting people and goods from natural hazards such as landslides, floods and earthquakes.

The media and environmentalist organizations tended to enhance the negative impacts of the projects and very seldom made reference to their positive effects.

Taking into consideration the fact that the implementation of all large engineering projects will generate some negative impacts, it is fundamentally important to develop, both at the feasibility and design stages, solutions that minimize negative environmental impacts and to propose compensatory measures.

In the past, studies and designs conducted by qualified professionals have taken those aspects into consideration, trying to balance the technical and economic feasibility of projects with environmental preservation.

However, we must acknowledge that, in the past, when faced with more than one feasible alternative engineering solution, most project owners selected the least expensive solution, even where this would imply greater environmental impacts on the affected area.

These facts outlined above explain the origin of legislation that has been created in the most developed countries and regions, to mandate that environmental impact assessments are conducted for all large engineering projects and that the results are considered in balance with the social and economic aspects of their implementation, to determine the best design solution Oliveira (1997).

For example, European Commission Directive 85/337/CEE was agreed in June 1985 and soon transposed to many countries even outside the EU. In Portugal, the first decree was Law 186/90, which was first updated in 2000 (Law 69/2000), and its most recent version is from 2013 (Law 151-B/2013). This last version decrees that the environmental impact study of a given project must be terminated at the same time as the completion of its basic design and that the two studies include, wherever possible, the analysis of alternative solutions. The authorities take those studies into consideration and they only approve the design when a favourable Declaration of Environmental Impact (DEI) is issued, based upon them.

For a large number of such projects, the engineering geological and geotechnical engineering roles are fundamental, depending on the degree of interference of the project with the ground as well as, often, with the groundwater. These roles are especially relevant when the studies are conducted by experienced professionals who make decisions based on the economics of the project as well as on its social, environmental and operational consequences Oliveira (2008).

Table 1.1 List of Engineering Works

Construction materials
Hydraulic undertakings
Linear surface works (highways, canals, pipes)
Underground works (tunnels, caverns)
Marine works (earth fills, breakwaters)
Bridges and viaducts
Natural and excavation slopes

1.2 Examples of Geotechnical Project Optimization that Takes Environmental Issues into Consideration

In order to give some examples of how it is possible to optimize civil engineering projects, taking environmental issues into consideration, while still preserving their economic and social value, a list of topics and works is shown in Table 1.1.

Most of the examples were constructed years ago and they show how it is possible for both sides of the problem to be compatible.

The first example relates to the use of geological construction materials, which is a subject transversal to most engineering works. In general their extraction is only possible through the excavation of soils (borrow areas) or the blasting of rock masses (quarries). These procedures always interfere, although to different degrees, with the environment and the landscape, and at the end of extraction the rehabilitation of the degraded areas should be mandatory. Unfortunately, in many countries the legislation does not yet mandate the rehabilitation of those areas, but that will certainly change in the near future.

In order to avoid using large volumes of geological materials for construction, research has been conducted recently to find ways to replace the natural geological materials with alternative products, such as recycled materials, quarry debris and geosynthetics.

Other important topics related to construction materials and the optimization of their use, both from economic and environmental points of view, are the management of the materials during construction and the compensation of volumes (excavations and fills) that have to be considered at the design stage. Good volume compensation reduces the amount of materials used and avoids unnecessary deposit of non-used materials, which requires free areas and generally interferes with the landscape.

Table 1.2 Construction materials

Dams	Extraction of soils and rocks in the reservoir area
Linear works	Excavations and embankments
	Compensation of volumes and materials management
Bridges versus embankments	Foundation conditions and the availability of embankment materials
Land reclamation	Dredging, hydraulic fills, borrow areas

In order to support the concept of optimization of the use of geological construction materials, Table 1.2 presents examples of dams, linear works (motorways and canals), bridges and coastal land reclamation.

With respect to hydraulic undertakings (i.e. dams and hydraulic structures), environmental impacts may be experienced during the construction stage (e.g. the excavation of large spillways) and/or during operations, upstream and downstream (e.g. landslides of reservoir slopes and waves, sedimentation of the reservoir, erosion of slopes).

An interesting example of an environmental impact caused during construction was the excavation of the 90 m high spillway of the Gargar dam in Algeria (Fig. 1.1). In this case, the impact could be minimized as a result of the sub-horizontal structure of the karstic limestone, that permitted a very steep slope, and also by the use of controlled blasting of the rock mass.

A very well-known example of environmental impact during operations relates to the large sedimentation of the reservoir in the Three Gorges dam, in China. To deal with the erosion of soils from the hydraulic basin, which are mainly transported to the reservoir by the tributaries, additional dams have been constructed in several of the water lines to retain sediments and thus to avoid them reaching the reservoir. A program of forestation upstream has also been implemented. Furthermore the design of the main dam takes into consideration the detrimental aspect of sedimentation by incorporating 22 vanes and 23 bottom outlets in order to allow the sediments to be expelled downstream when floods occur. Large discharges or floods, if uncontrolled, may erode the base of the natural slopes of valleys, downstream of dams, provoking instability of the ground.

With regards to surface linear works, examples of environmental optimization include the layout of alternative routes, crossing karstic zones, open excavations versus tunnels and landscaping, all of which must be considered while also taking into account the social and economic aspects of the land expropriation for each solution.

Good examples that show how the optimization of engineering solutions can improve the environmental conditions of the area include: the motorway A1, in the region of Fatima, the Lisbon Regional Outer Circular (CREL) and the motorway 24 to the Douro valley, all in Portugal and the Anilio tunnel part of the Egnatia highway in Greece. Figure 1.2 shows two alternatives which were studied 25 years ago, for the crossing of motorway A1, near Fátima, through a karstic limestone rock mass. The decision taken, based on the relative costs, was for the surface excavation of the rock mass with slopes of approximately 30 m high (Fig. 1.3). Nowadays this solution would not have been approved by the environmental authorities.

Fig. 1.1 Gargar dam (Algeria). Spillway

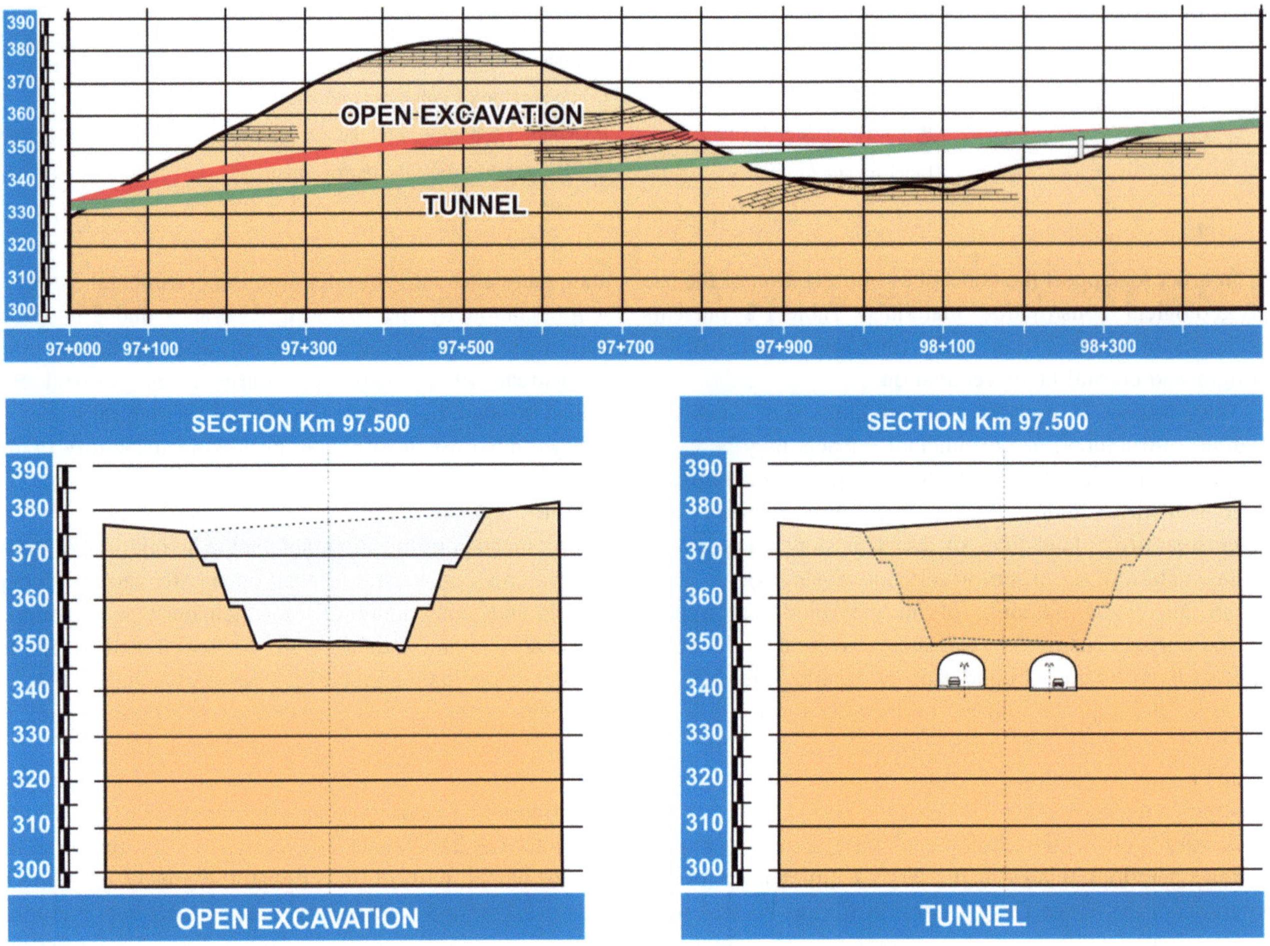

Fig. 1.2 Motorway A1. Alternative solutions

Fig. 1.3 Motorway A1. Solution open excavation

The excavation of tunnels, as part of engineering works, can create significant environmental problems that have to be anticipated at the design stage and solutions have to be found to mitigate their impact. Examples include their interference with the hydrogeological conditions of the geologic formations they cross and the destination site for any materials resulting from the excavations.

Those concerns also apply to the construction of underground caverns, which are opened for several engineering purposes, such as hydroelectric power plants, metro stations, fluid storage and dangerous waste disposal. Moreover, reference should also be made to the problems of ground pollution, subsidence and collapse that reflect at the surface of the ground.

One important issue mentioned previously in this paper is the final treatment given to excavation debris that is not used as a construction material, and is therefore a waste product; effective modelling during the design phase can enable proper landscaping of the storage area that will be affected. An interesting example is the study conducted by Eletricidade de Portugal (EDP) for the power plant Venda Nova II, located in the north of Portugal. The construction of the tunnels and of the deep power plant required the extraction of more than half a million cubic meters of rock that were not appropriate for construction. The deposit materials were modelled in such a way as to give the area a pleasant landscape (Fig. 1.4).

Finally, the topic of natural and excavated slopes is discussed. Firstly, some comments on the geological risks, resultant from the instability of natural slopes, which often affect people and property. In such cases, studies have to be conducted for the design of support works which assure the safety of the ground, while ensuring that the necessary works have minimum interference with the environment

Fig. 1.4 EDP Venda Nova II power plant. Modelling of deposit materials

Fig. 1.5 Serra da Arrábida. Retaining works and false tunnel

and the landscape. An interesting example is shown in Fig. 1.5, which presents stabilization works being conducted on a natural slope, more than 100 m high, of the Serra da Arrábida, south of Lisbon, as a result of large limestone blocks having fallen on the road below, from time to time, which imposed the closure of the road to all traffic for 2 years.

As this area belongs to a natural reserve, the solution chosen had to take into consideration that no significant excavation works could be performed to improve the stability

Fig. 1.6 Motorway CREL. Retaining curtain

of the rock mass. For excavated slopes, the stability analysis has to consider the geological, hydrogeological and geotechnical conditions of the ground for the feasibility study of possible solutions. The optimal solution will assure the stability of the slope, while requiring the least possible interference with the environment. Figure 1.6 shows an example of a high and long retaining structure constructed on an excavation slope of the Lisbon motorway CREL, supporting a very fragmented limestone rock mass, which was selected after taking into consideration the concern to choose a solution that would be well-framed in the landscape.

1.3 Conclusions

The rapid growth in the world's population (which has increased four times in the 20th century) and its concentration in urban areas (more than 50 % of the actual total world population) with many people living in megacities, continues to necessitate the construction of new infrastructure of all types in order to create adequate conditions for the economic and social development aspired to. However, such construction must be done with a mind to environmental preservation Oliveira (2000).

The purpose of the examples presented in this paper is to show that it is possible to reconcile development with environmental preservation, if and when qualified engineering geologists and geotechnical engineers use their knowledge to find the best possible solutions for the required engineering works.

References

Oliveira R (1997) Teaching environmental subjects in engineering geological education. In: Proceedings international symposium on engineering geology and the environment, Athens

Oliveira R (2000) Geotechnics, development and environment. Libro Homenaje a José Antonio Jiménez Salas. Ministerio de Fomento. Madrid

Oliveira R (2008) Geo-engineering education and training. The past and the future. In: Proceedings I international conference on education and training in geo-engineering sciences, Constanta

Part I

Engineering Geological Models

Convener: Dr. Jan Novotny—*Co-conveners*: Steve Parry

The session will discuss the use of engineering geological models within the framework of the total geological approach (Fookes et. al. 2000; Baynes et al. 2005; IAEG Commission 25). Such models allow the understanding and prediction of engineering geological conditions and processes, and allow uncertainties to be defined. Examples of the use of engineering geological models for different engineering projects and for different geological and geomorphological environments are expected.

Engineering Geological Models: Some Examples of Use for Landslide Assessments

2

Jan Novotný

Abstract

The Commission C25 of the International Association for Engineering Geology and the Environment is currently working on "The use of engineering geological models". This article presents examples of engineering geological models for landslides using both a conceptual and an observational approach. Generally speaking, the conceptual model forms the basis for the development of the observational model. However, there are cases where the relationship between the conceptual model and the observational model is not so unidirectional. The experience gained in developing the observational model in these cases can facilitate considerably the development of future conceptual models in the same type of engineering geological conditions.

Keywords

Landslides • Engineering geological model • Conceptual • Observational

2.1 Introduction

This paper constitutes a contribution to the discussion currently taking place within the Commission C25 of the International Association for Engineering Geology and the Environment, which is currently working on a paper entitled "The use of Engineering Geological Models" (Parry et al. 2014). The C25 considers two different methodologies for developing the models:

The *conceptual approach*, according to Parry et al. (in press), is based on understanding the relationships between engineering geological units, their likely geometry and anticipated distribution. This approach, and the models formed, are based on concepts formulated from knowledge and experience and are not related to real three-dimensional (3D) space or time. Importantly, the model is largely based on consideration of *geological concepts* such as age, stratigraphy, rock type, unconformity and weathering.

The *observational approach* is based on the observed and measured distribution of engineering geological units and processes. These data are related to actual space or time and are constrained by surface or sub-surface observations.

To illustrate these concepts, the article will present several examples from Cretaceous sedimentary regions of the Czech Republic.

2.2 Conceptual Models of a Slope Structure Comprising a Rigid Layer Above a Plastic Layer

In the Czech Republic, slope movements relatively often occur in a rock structure characterized by an upper layer (No. 1 on Fig. 2.1) consisting of rigid (competent) rock broken into blocks by the Tertiary tectonics, and by a thick

J. Novotný (✉)
ARCADIS CZ a.s., Prague, Czech Republic
e-mail: Jan.Novotny@arcadis.cz

J. Novotný
Faculty of Natural Science, Charles University in Prague, Prague, Czech Republic

G. Lollino et al. (eds.), *Engineering Geology for Society and Territory – Volume 7*,
DOI: 10.1007/978-3-319-09303-1_2, © Springer International Publishing Switzerland 2014

lower layer (No. 2 on Fig. 2.1) consisting of plastic (incompetent) rock. Typically, the upper layer is composed of massive sandstone and the lower layer of plastic claystone.

Figure 2.1 represents a conceptual model of a long-term evolution of a slope with this structure, as seen in various stages of development (Rybář and Nemčok 1968). In Fig. 2.1a erosion processes start to cut through the rigid layer. Figure 2.1b demonstrates that a narrow valley is prone to bulging (Varnes 1978). Further deepening and widening of the valley (Fig. 2.1c) leads to cambering—block-type movement on plastic underlying rock (Varnes 1978; Nemčok et al. 1972) in the upper part of the valley and to landslides of plastic rocks and derived soils in the lower part of the valley. Figure 2.1d represents a denudated slope prone to landslides triggered by river erosion at its base.

In Czech Cretaceous sediments, the most common slope state corresponds to the model stage "c" (Fig. 2.1c), which can be encountered also in Prague. Using archival research data, the general conceptual model can be further developed into a site-specific conceptual model for a particular location. Common features of site-specific models for the stage "c" comprise: (1) upper slope consisting of sandstone, often affected by block movements; (2) groundwater horizon developed in the sandstone above the impermeable clay, often drained ahead of the sandstone blocks on slopes consisting of fine grained soils; (3) ahead of the sandstone blocks, potential occurrence of landslides in the slope composed of fine grained soils.

An example of a site specific conceptual model is given in Fig. 2.2. In the village of Hrubá Skála, located in the NE of the Czech Republic, family houses were built on a seemingly favourable flat terrain which in reality consisted of unrecognized old landslides. A correct use of the principle of engineering geological models would have easily prevented damage to the houses (Novotný 2009).

A similar site specific conceptual model in Fig. 2.3 characterizes the Prosek district in the north of Prague (Pašek 2000 in Novotný 2009), affected not only by slope movements but also by historical undermining (Cílek 1999 in Novotný 2009), which should be taken into account when determining the scope of site investigation works needed for the correct development of the observational model. Houses constructed near the slope edge with disregard of the model were damaged by fissures and one of them had to be demolished (Lešner 2004 in Novotný 2009).

Generally speaking, all site investigation works should aim to answer questions raised by the conceptual model, and thus to elaborate the observational model. At the same time, the conceptual model itself can be used to determine efficiently the type and scope of site investigation works needed (when compared to a "grid-like" site investigation planned without knowledge of the site geology and processes).

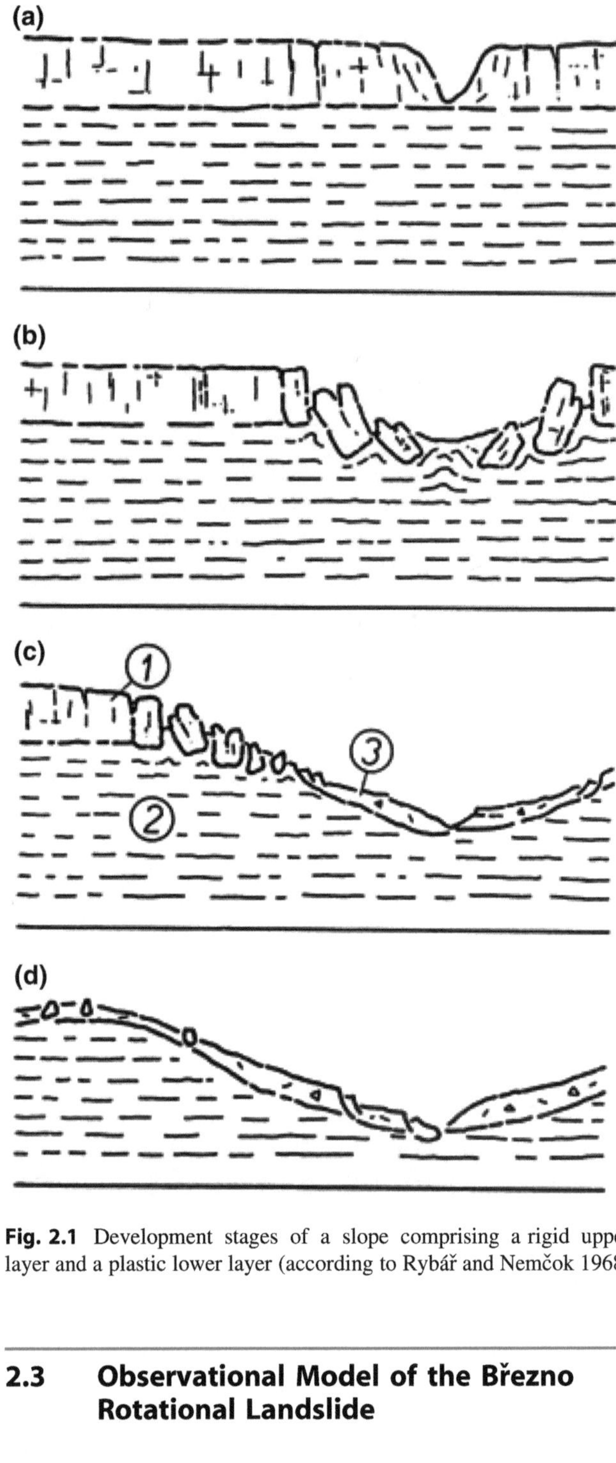

Fig. 2.1 Development stages of a slope comprising a rigid upper layer and a plastic lower layer (according to Rybář and Nemčok 1968)

2.3 Observational Model of the Březno Rotational Landslide

Figure 2.4 presents an example of an observational model of the central part of the Březno u Postoloprt landslide. The model was constructed by the author using a cross section from a site investigation report (in Pašek 1974), aerial view by Google and field mapping carried out by the author.

By sliding along the rotational surface of rupture ("rock slump" according to Varnes 1978, rotational landslide according to Nemčok et al. 1972), the Cretaceous marlstone block was pushed into the river bed, substantially narrowing

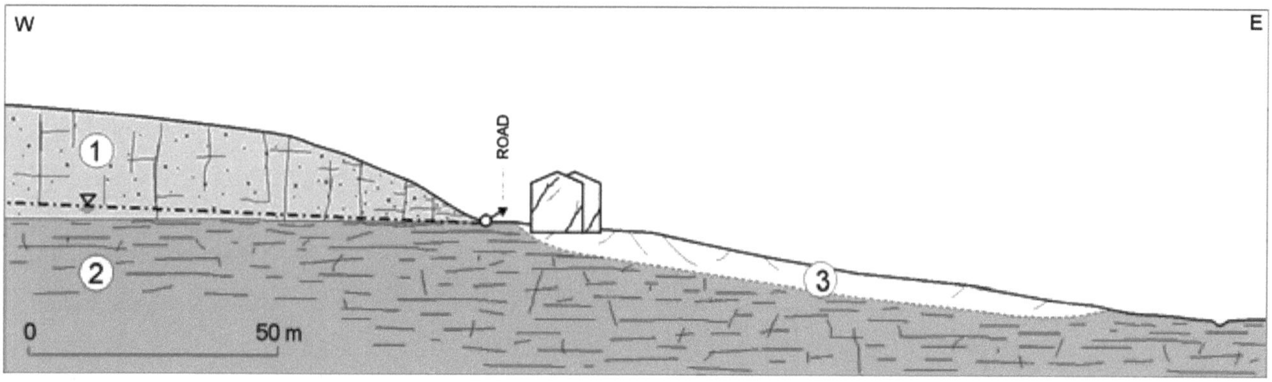

Fig. 2.2 Conceptual model of a slope in Hrubá Skála. *1* Cretaceous sandstones, *2* Cretaceous claystones, *3* landslide

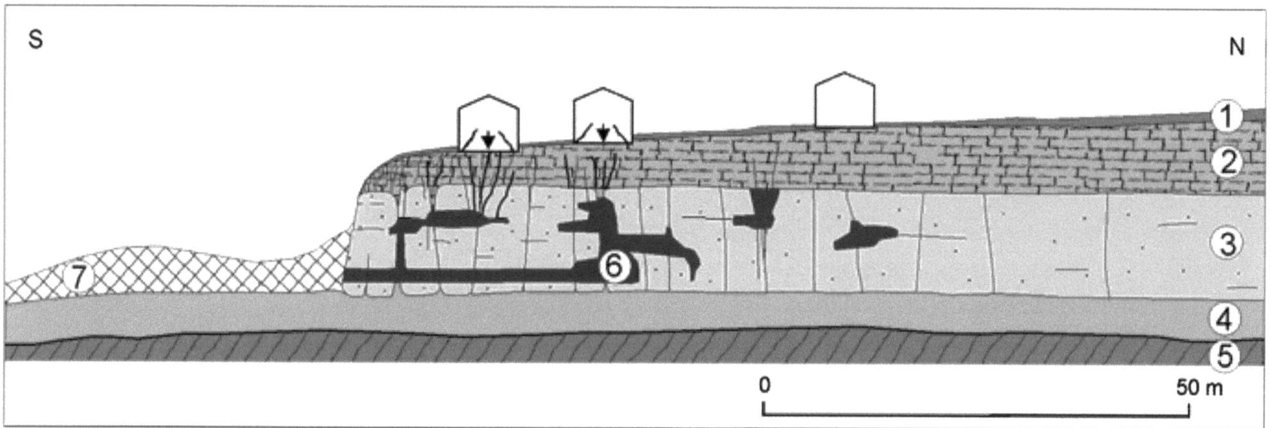

Fig. 2.3 Conceptual model of the edge of the Czech Cretaceous Formation in Prosek (after Pašek 2000 and Lešner 2004 in Novotný 2009). *1* Quaternary loess soils, *2* Turonian marls and marlstones, *3* Cenomanian sandstones, *4* Cenomanian claystones, *5* Ordovician shales, *6* mining cavities, *7* mounds

it in this area. The main cause of the landslide, besides lithology prone to sliding, is primarily the erosive action of the Ohře river which to this day maintains the whole landslide in an unstabilized state. A layer of baked clays located in shallow depth below surface also plays an important role, preventing the area from being denuded into a gentle slope less prone to sliding.

Unlike the simple structure of the rotational landslide, the morphology of the main scarp, resulting from various slope processes, is considerably complex. The main scarp above the rotated block is divided into a series of ridges separated by areas with periodical occurrence of minor landslides and notably earth flows. In long-term conditions, the ridges themselves are also unstable, prone to rock fall and opening of vertical tension cracks which can lead to rock topples of large blocks. The material from minor landslides, earth flows, rock falls and rock topples accumulates in the head area, adding weight here and destabilizing the entire landslide. In the upper part of the accumulations, minor scarps are formed; above them, the terrain locally dips towards the slope, creating undrained basins that further destabilize the slope by the process of water infiltration into the unstable masses.

2.4 Conceptual Model of the Head Area of the Březno Landslide

During the development of the observational model of the Březno landslide, a conceptual evolutionary model of processes in the main scarp area was also established. Without knowledge of the slope's history and evolution, the model would be much simpler in comparison with the following concept (Fig. 2.5).

Figure 2.5(1): State after the development of the rotational landslide. Figure 2.5(2): Irregularly, minor landslides and rock falls occur in the main scarp area, creating partial ridges in the scarp area and accumulations at the base of the scarp. Fig. 2.5(3a): In long-term conditions, local rock falls repeatedly occur in the ridges, heavily fractured by

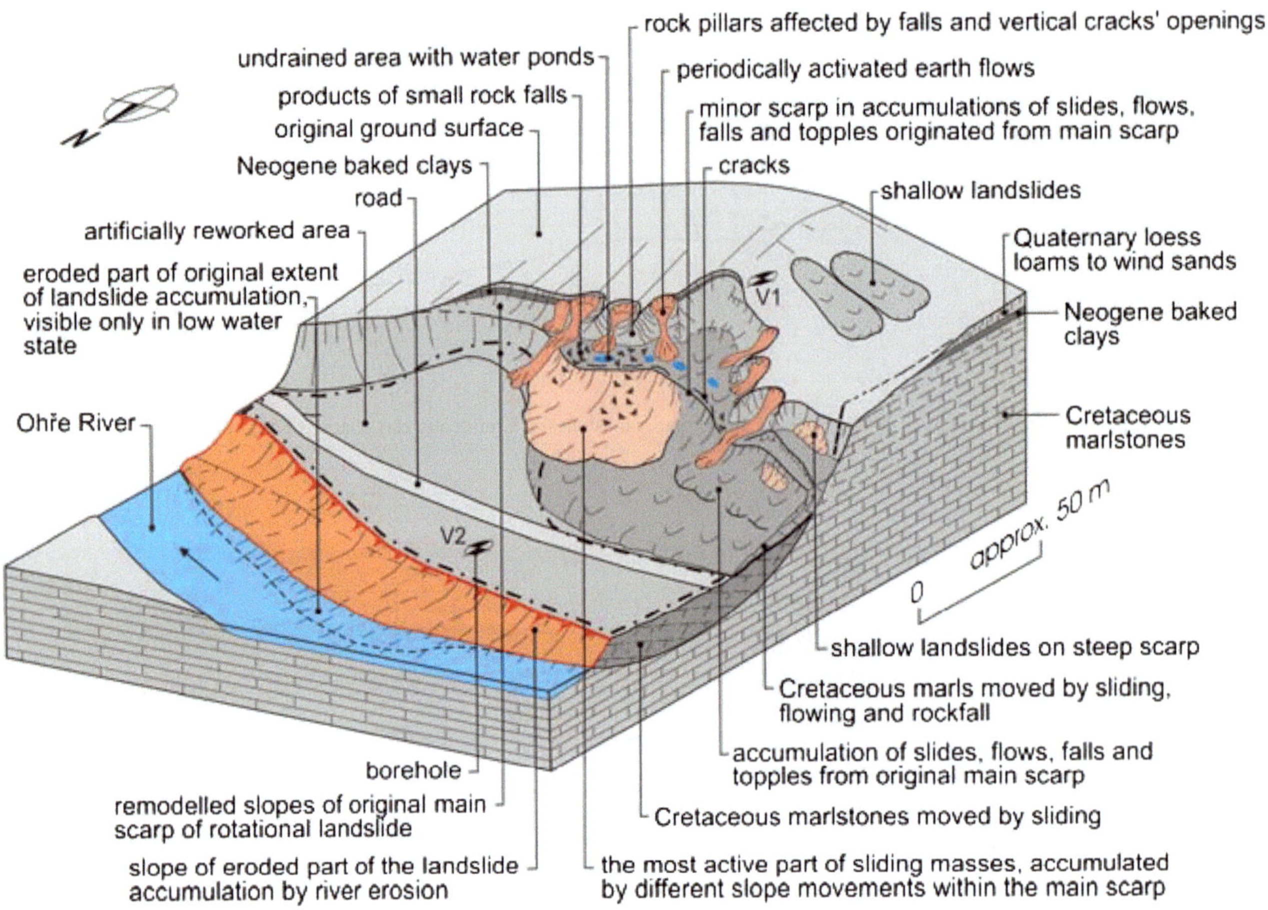

Fig. 2.4 Observational model of the Březno landslide in Cretaceous marlstones

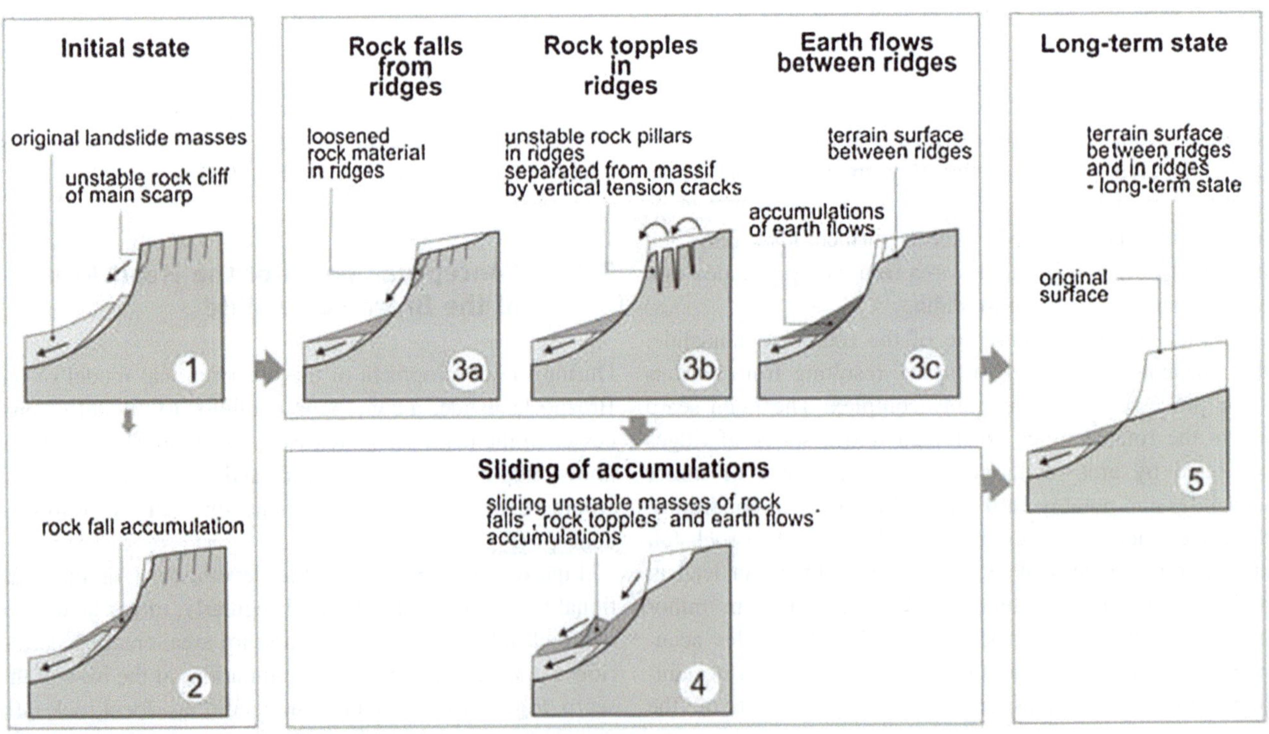

Fig. 2.5 Conceptual model of processes in the head area of the Březno landslide

discontinuities in the marlstone rock mass. Figure 2.5(3b): In long-term conditions, local topples repeatedly occur in the ridges. Figure 2.5(3c): Earth flows repeatedly occur between the ridges. Figure 2.5(4): Rock fall, rock topples and earth flow accumulations are unstable and prone to sliding. Figure 2.5(5): In the long term, the processes depicted in Fig. 2.5(3a–3c)–(4) result in a considerable retreat of the whole main scarp very far beyond its original position (an important factor influencing any potential construction projects in the area) and in the gradual increase of slope stability. In reality, the situation is even more complex than described by the conceptual model presented above.

2.5 Conclusions

A knowledge of the conceptual model of a slope consisting of a rigid upper layer and a plastic lower layer, including its future development, can be used not only to predict engineering geological conditions in all parts of the slope, but also to recognize all issues that have to be resolved by site investigation when compiling the observational model. Sufficient experience is needed to elaborate an appropriate conceptual model, and as shown in the example of the Březno landslide, the experience gained when developing observational models can also be useful.

References

Nemčok A, Pašek J, Rybář J (1972) Classification of landslides and other mass movements. Rock Mech 4:71–78

Novotný J (2009) Case studies of ground stability problems in family houses construction. In: 8th International congress on advances in civil engineering, 15–17 September 2008. Famagusta, North Cyprus, pp 319–326

Parry S, Baynes FJ, Culshaw MG, Eggers M, Keaton JF, Lentfer K, Novotný J, Paul D (2014) Engineering geological models—an introduction: results from IAEG Commission 25. In: Bulletin of engineering geology and the environment

Pašek J (1974) Sesuvy středního Poohří (Landslides in the middle course of the Ohře river valley). J Geol Sci Hydrogeol Eng Geol 11:53–79 (in Czech)

Rybář J, Nemčok A (1968) Landslide investigation in Czechoslovakia. In: Proceedings of the 1st session of the international association of engineering geology. Prague, pp 183–198

Varnes JD (1978) Slope movement types and processes. In: Schuster RL, Krizek RJ (eds) Landslides, analysis and control, Special Report 176. National Academy of Sciences, Washington, DC, pp 11–33

Engineering Geological Models IAEG Commission 25

3

Steve Parry, Fred J. Baynes, Martin G. Culshaw, Mark Eggers, Jeffrey R. Keaton, Kori Lentfer, Jan Novotný and Daren Paul

Abstract

IAEG Commission C25 on the 'Use of Engineering Geological Models' has recently completed a report entitled "Engineering Geological Models—An Introduction" to be published in the Bulletin of Engineering Geology and the Environment. This paper summarizes the key findings of this report. The report notes that the generation and use of engineering geological models should be a fundamental activity for any geotechnical project. Such models are an essential tool for engineering quality control and provide a transparent way of identifying project-specific, critical engineering geological issues and parameters. Models should also form the basis for designing the scope, the method and assessing the effectiveness of site investigations. However, whilst the idea of models in engineering geology has existed for several decades, there has been little published to date that systematically distinguishes the different model types and how and when they might be used. The reader is referred to the full report for additional background, examples and details.

Keywords

Engineering geological model • Conceptual • Observational

3.1 Introduction

The aim of C25 is not to provide a 'cook book' for generating engineering geological models. Rather, it is intended to present the philosophy behind the development and use of these models, suggest appropriate terminology to describe them and provide general guidance for their construction, primarily through the use of examples. The reader is referred to the full report for additional background, examples and details (Parry et al. in press).

C25's working definition of the term model as used in engineering geology is simply: *A model is an approximation of reality created for the purpose of solving a problem.*

S. Parry (✉)
Parry Engineering Geological Services Ltd, Derbyshire, UK
e-mail: parrysteve@gmail.com

F.J. Baynes
Baynes Geologic Pty Ltd, Malmsbury, Australia

M.G. Culshaw
School of Civil Engineering, University of Birmingham, Birmingham, UK

M.G. Culshaw
British Geological Survey, Nottingham, UK

M. Eggers
Pell Sullivan Meynink, Macquarie Park, Australia

J.R. Keaton
AMEC Americas, Los Angeles, USA

K. Lentfer
Coffey, Hamilton, New Zealand

J. Novotný
ARCADIS CZ a.s., Praha, Czech Republic

J. Novotný
Faculty of Natural Science, Charles University in Prague, Praha, Czech Republic

D. Paul
Golder Associates, Richmond, Australia

G. Lollino et al. (eds.), *Engineering Geology for Society and Territory – Volume 7*,
DOI: 10.1007/978-3-319-09303-1_3,

3.2 Approaches and Model Types

Typically, engineering projects develop in stages from pre-feasibility to feasibility, various stages of design, construction and through to operation. With each stage of the project more data become available. Consequently, a range of engineering geological models are required during the life of a project.

C25 considers that there are two fundamentally different methodologies for developing engineering geological models that are independent of the project stage—conceptual and observational. The different methodologies used for the generation of these model types are:

(a) *The conceptual approach*: which is based on understanding the relationships between engineering geological units, their likely geometry, and anticipated distribution. This approach, and the models formed, are based on concepts formulated from knowledge and experience and are not related to real three-dimensional (3D) space or time.

(b) *The observational approach*: which is based on the observed and measured distribution of engineering geological units and processes. These data are related to actual space or time and are constrained by surface or sub-surface observations.

The accuracy and completeness of observational models depends on the accuracy and completeness of the associated conceptual models; similarly, any analytical modelling will depend on the observational models. If the conceptual model is wrong, then any subsequent observational models and the analytical models are likely to contain errors or even be incorrect. Importantly, especially for those responsible for engineering design, it is most unlikely that any analytical modelling will be correct if the geology is not understood.

Whilst the engineering specifications and performance of the project must be known to the engineering geologist for the development of a model, C25 also believes that, regardless of the model type, it is absolutely essential that geological concepts must be the starting point for building models.

A fundamental objective of any engineering geological model should be to evaluate and, where necessary, investigate the potential 'unknowns', that is, ground conditions that consideration of the model suggests could be present and which could potentially affect the project, but which have not been specifically observed. By identifying potentially critical conditions, these can be factored into the site investigation and design, for example, through additional targeted ground investigation or by contingency planning. Ultimately the understanding of the project embodied in the engineering geological model becomes an understanding of the *site conditions that an experienced contractor could reasonably have foreseen,* with all of the contractual overtones associated with these words.

3.3 The Conceptual Engineering Geological Approach

The conceptual engineering geological approach and the resulting models typically provide input to the earliest stages of a project. The conceptual model is critical in assessing the potential engineering geological variability that may be present at a geographical location and, when combined with the specific engineering requirements of the project, has the potential to identify elements that can result in hazard to that project, i.e. it is site and project specific. Consequently, a fundamental purpose of the conceptual model is to identify what credible engineering geological unknowns may be present, so that these unknowns may be targeted for investigation and, if found to be present, to assess their potential for hazard to the project.

The resulting conceptual models can be broadly subdivided into two types:

1. *Conceptual models that deal with relationships in space*: these are extrapolated from existing knowledge of geological environments and processes. An important advantage of these models is the ease with which they can be used to communicate the geological conditions to engineers who may have little or no knowledge or understanding of geology but have to make critical decisions that are driven by geological factors.
2. *Conceptual models that deal with relationships in time*: these illustrate the geological evolution of a site or particular geological conditions or processes which are relevant to the project.

By its very definition, the conceptual approach and resulting models are associated with considerable uncertainty. The uncertainty is rather abstract in that it relates to whether or not the set of concepts that have been identified as being relevant are the most reasonable set of concepts, which is inherently difficult to judge. However, the power of the approach is that when a good conceptual engineering geological model is developed, it should be capable of anticipating most of the engineering geological issues that could potentially affect the project.

3.4 The Observational Engineering Geological Approach

The observational engineering geological approach and the resulting models are usually based on observations and data from project-specific ground investigations. These ground investigations should be designed using conceptual models and, in particular, should seek to verify the basic components of the conceptual models and target the uncertainties identified by them. Observational models may be developed directly from conceptual models or they may be developed following the acquisition of new, site specific, observations. An observational model is usually constrained by observations and/or measurements, even though some observations and measurements themselves are interpretations of incomplete information or remotely sensed data, such as geophysical measurements. These observations usually can be constrained in space by actual position (x, y, z) data; occasionally the model is constrained in time.

This approach is applicable to engineering geological tasks that range from core logging to regional mapping. Consequently, the resulting observational engineering geological models can take a wide variety of forms: graphical borehole logs (one dimensional), engineering geological cross sections and maps (two dimensional) and spatial engineering geological models (three dimensional).

It is important to note that the geological interpretation required to construct an observational model should be based upon the knowledge encapsulated in the conceptual model. Whilst observational data, such as boundaries in boreholes, are constrained in x, y, z space, the conceptual model is used to establish the relationships that support interpretation of geological surfaces between such points. Furthermore, the interpretation of the observation data themselves is based on a conceptual approach to differentiate the significance of each specific piece of observational data.

3.5 Models as They Relate to Construction

The most useful engineering geological models define uncertainties and unknowns so that they may be incorporated into the project analyses or so that the project cost estimate can include a contingency to cover the risks associated with them. This allows the potential sources of risks to the project from ground-related hazards to be identified, as far as possible, and investigated and evaluated, thereby reducing ground-related risk to the extent that is practicable. Commonly within the geotechnical industry consideration of engineering geological models with their implicit variability is rarely discussed. For example, in many cases owners choose to issue only "factual" information (i.e. borehole and test pit logs) in the belief that providing any "interpretations" will somehow increase their exposure to geotechnical risk. The members of C25 accept that this is an industry wide practice but are of the opinion that withholding interpretations from contractors can only reduce their ability to reasonably foresee the ground conditions that they might encounter.

3.6 Conclusions

An engineering geological model is any approximation of the geological conditions created for the purpose of solving an engineering problem and includes models which are based mainly on geological characteristics as well as models which are based mainly on engineering characteristics. In a general sense, the geotechnical risk faced by an engineering project is inversely proportional to the level of detail and accuracy embodied in the engineering geological model. The better the model reflects actual conditions, the lower the remaining risk. However, it is not possible to define every finite detail of the ground. So, ultimately, the objective of an engineering geological model, throughout the project, should be to provide sufficient detail and understanding of the ground, based on the data available at the time, to carry out the engineering to an acceptable degree of reliability.

Engineering geological models should form a fundamental component of any geotechnical project as they provide a systematic methodology to support all of the engineering geological thought processes that must be worked through for successful project completion. The use of models as an approach to solving engineering geological problems, with the inherent requirement for prediction and verification, is also ideally suited to training and education.

The authors are of the opinion that distinguishing between the conceptual and observational components of a model will create a better understanding of the type and range of uncertainties that are present within the model. Clearly, the uncertainty associated with the choice of geological details on which to base a conceptual model is very different from the uncertainty associated with the location of a geological boundary within 3D space for an observational model. By acknowledging these different approaches, the different types of uncertainty within the model can be appreciated and hopefully understood.

Finally, the knowledge encompassed within each type of model must be transferable between project stages, in particular from the site investigation, to engineering design, to project construction, and into facility operation, so all of the different types of models must seamlessly relate to each other. Engineering geological models are an ideal way to *communicate* what is known about the project as it progresses through different project stages.

Reference

Parry S, Baynes FJ, Culshaw MG, Eggers M, Keaton JR, Lentfer K, Novotný J, Paul D Engineering geological models—an introduction: results from IAEG commission 25. In: Bulletin of engineering geology and the environment (in press)

The Use of Ground Models for Probabilistic Assessment of Multiple Geohazards Along Lifeline and Transportation Corridors

4

Clark Fenton, Fergus Cheng, Andrew Davie, Andria Loppas, Scott Loudon, Rebekah Marsh, Matthew Pendlebury, Yeldos Sultanbek, Leonidas Tatarakis, Michael Teo, Charlene Ting and Xue Yao

Abstract

Lifelines and transnational transportation corridors often traverse diverse geological terrains and are exposed to a variety of geohazards. The evaluation of such hazards can be made difficult by a lack of data (epistemic uncertainty) and an overwhelming and often confusing juxtaposition of differing geological environments. Analysis of such long, linear routes is best achieved using terrain analysis, considering both the seismotectonic characteristics and geomorphological settings. The development of detailed conceptual ground models for each terrain unit can allow the rational and logical synthesis of prior data allowing preliminary (deterministic) assessment of geohazard susceptibility. Once the hazards have been identified, available prior information can then be used to assign magnitude and recurrence parameters to each hazard. This allows a first-order assumption of hazard recurrence. Coupled with a basic understanding of system fragility, this can be used as a means to ensure that robust front end engineering design (FEED) can progress. This allows the identification of significant hazards requiring either project rerouting, redesign or further study. The levels of hazard knowledge can be clearly demonstrated and significant epistemic uncertainty identified early in the hazard evaluation process, allowing timely modification to subsequent site investigation activities in order that the relevant knowledge gaps are addressed.

Keywords

Hazard • Uncertainty • Ground models

4.1 Introduction

Experience in both industry and in academia highlights the difficulty that many earth scientists and engineers have in incorporating uncertainty and variability into their understanding of natural systems. This becomes particularly important for successful engineering design incorporating geohazards considerations. Although we usually consider geohazards to include earthquakes, volcanoes, slope failure and their attendant manifestations, we should be considering any naturally occurring (and even man-made) ground conditions that have a potentially adverse effect on the engineered environment. The concept of geohazards is often introduced to students in a deterministic framework, with an emphasis on the location and maximum magnitude of each hazard. Often geohazards are treated on an individual basis, and the cumulative effect of hazards can be missed, leading to an erroneous understanding of project vulnerability. Contemporary risk-based engineering requires an integrated approach to hazard analysis, considering system fragility and a full probabilistic hazard evaluation of all geohazard

C. Fenton (✉) · F. Cheng · A. Davie · A. Loppas · S. Loudon · R. Marsh · M. Pendlebury · Y. Sultanbek · L. Tatarakis · M. Teo · C. Ting · X. Yao
Department of Civil & Environmental Engineering, Imperial College London, London, UK
e-mail: c.fenton@imperial.ac.uk

G. Lollino et al. (eds.), *Engineering Geology for Society and Territory – Volume 7*,
DOI: 10.1007/978-3-319-09303-1_4, © Springer International Publishing Switzerland 2014

Table 4.1 Geohazards evaluation projects

Project location	Infrastructure	Terrain units	Geohazards
Irkutsk (Russia) to Ulaanbaatar (Mongolia)	Oil and gas pipeline	1. Craton/intraplate plateau	Ground shaking
		2. Active rift margin	Fault rupture
		3. Rift basin	Liquefaction
		4. Foothills	Slope stability
		5. Permafrost plateau	Permafrost
		6. Loess plains	Scour
		7. Alluvial basin	Avulsion
			Inundation Loess
Kampala (Uganda) to Djibouti	Railway corridor	1. Craton	Ground shaking
		2. Active rift margin	Fault rupture
		3. Rift basin	
		4. Subaerial spreading centre	Liquefaction
		5. Wadi	Slope stability
			Scour
			Volcanism
			Expansive soils
			Collapsible soils
Kabul (Afghanistan) to Dushanbe (Tajikistan)	Water distribution system (pipe, canal and tunnel)	1. Strike-slip basin	Ground shaking
		2. Alpine bajada	
		3. Range-front foothills	Fault rupture
		4. Alpine valley	Liquefaction
		5. Himalayan mountains	Slope stability
		6. Loess plains	Loess
			Scour
			Avulsion

occurrences. Developing an integrated approach to hazard analysis requires an optimum understanding of the spatial and temporal distribution of geohazards; this is best accomplished using ground models. The development of detailed models, considering both the seismotectonic characteristics and geomorphological setting, allows the rational and logical synthesis of prior data thereby assisting preliminary (deterministic) assessment of geohazard susceptibility. Such an assessment is useful for initial route determination and/or corridor selection, but generally does not allow equal consideration of what can be very disparate hazards. Once the hazards have been identified, available prior information can then be used to assign magnitude and recurrence parameters to each hazard. When data are not available locally, or are of such a limited temporal extent that recurrence parameters cannot be reliably determined, then the principle of ergodic assumption can be used allowing data from geologically similar terrains from other locations can be used as a preliminary input into the model.

4.2 Geohazard Assessment

As part of the Engineering Geology MSc course at Imperial College London students are introduced to aspects of unknowns and uncertainty in the ground through modules covering ground model development, site investigation planning and geohazards evaluation. The latter is carried out in two stages: an introductory module covering the framework of probabilistic hazard assessment (concentrating on ground motion evaluation) and a subsequent module on practical geohazards assessment (e.g. Fenton et al. 2010). The latter module is divided into two elements: geohazards assessment for an urban setting and geohazards evaluation for a transnational lifeline corridor. These activities are carried out as individual and group exercises, respectively.

The urban geohazards exercise is used as a 'warm up', allowing the students to begin to integrate aspects of their site investigation course (recognising the value of and integrating

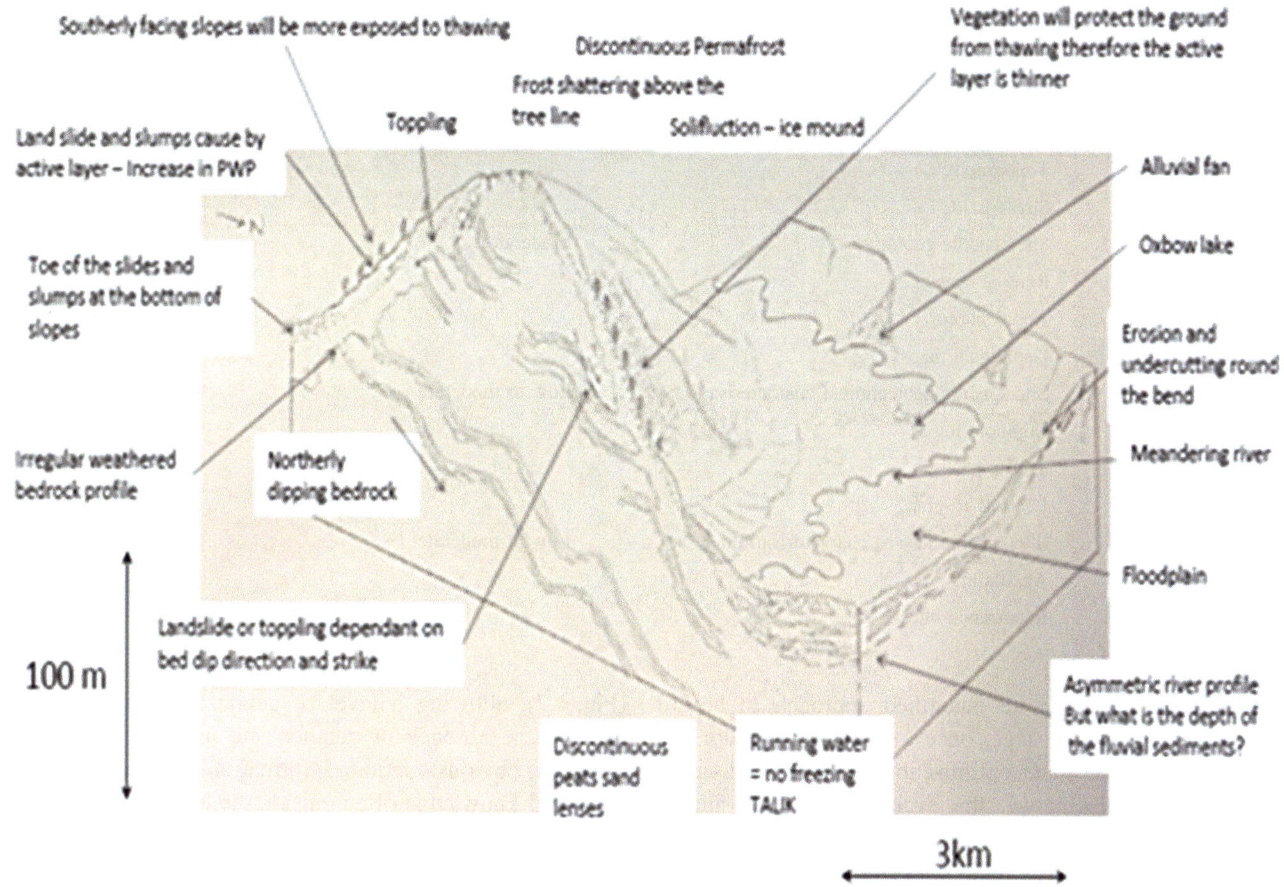

Fig. 4.1 Example ground model: developed for upland areas affected by permafrost along the Irkutsk to Ulaanbaatar pipeline project (Table 4.1)

prior information at a desk study phase), ground model construction, and preliminary (deterministic) hazard assessment. The product of this investigative stage is a data sheet covering the various geohazards for the area under consideration and a preliminary (deterministic or relative) assessment of hazards potentially affecting the site. This introduces the students to the value of prior information, how to evaluate the quality of this information, and begin to assess the uncertainties (both explicit and implicit) contained therein.

Evaluation of this work is twofold: weekly informal round table meetings allow the students to air their current level of understanding and respond to 'value of information' (VOI) questioning (e.g. What data are required and what improvements in geohazards understanding/value engineering will be accomplished as a consequence of acquiring these data?). These sessions are designed to focus the students' efforts by highlighting data gaps and focusing efforts on tackling epistemic uncertainty. This is the equivalent of a project progress meeting with the role of the client played by the course instructor. Although aleatory variability is discussed, time constraints usually dictate that at this stage this is a secondary concern. The second and final assessment is a series of individual oral presentations, usually accompanied by a final summary data sheet, outlining the hazards and their relative magnitudes.

The second part of the course is based on a series of hypothetical trans-national lifeline corridors (Table 4.1). This exercise is performed as a group based exercise, involving a team-based structure, not unlike that involved in a preliminary hazard assessment at FEED stage in industry. The projects are designed to cover a series of differing engineering demands, and cover a broad range of geological and geohazards environments. This provides the challenge of dealing with multiple geohazards, a wealth of differing ground conditions, and varying topographic and climatic conditions. The project locations are deliberately chosen in order that the quality and detail of data are also highly variable. The projects are loosely defined, often with a start and end point, but no corridor designation. Part of the initial evaluation is based on route selection, often allowing the evaluation of multiple corridor options. This is usually carried out in a manner that initial selection is based on the identification of 'showstopper' hazards. After this initial evaluation several corridor options may be evaluated in

Table 4.2 Ground model development

Ground model	Data requirements	Level of epistemic uncertainty	Hazard quantification
Conceptual	Regional geology:	Moderate to high	Relative
	Tectonic setting		
	Topography		
	Stress field		
Geological	Site-specific geology:	Moderate	Deterministic
	Remote sensing		
	Aerial photos		
	Geological maps		
Geotechnical	Site specific geotechnical characterisation:	Low to moderate	Deterministic
	Borehole logs		
	Trial pit logs		
	Lab test results		
Geohazard	Site-specific hazard characterisation:	Low to moderate	Probabilistic
	Age-dating		
	Recurrence rates		

parallel, allowing a more quantified approach to hazard evaluation. The third stage, once a preferred corridor has been identified, is route optimisation. Like the individual projects described above, the evaluation of the student performance is based on weekly project progress meetings, when the discussion may focus on VOI, fatal flaws and data gaps. Aleatory variability, and the means by which it can be quantified, is explored in an increasingly detailed manner as the course progresses. The concept of ergodic assumption is introduced, widening the areas of geohazards research beyond the geographic constraints of the chosen corridors. As well as evaluation from the course instructor, the students benefit from peer interaction in the form of questions, data exchange and the sharing of experience (gained from beyond the scope of these individual exercises); this follows the SSHAC methodology (Budnitz et al. 1997) of expert elicitation with clearly defined evaluation and integration activities.

There is a strong emphasis placed on the presentation of ideas and analysis to a mixed technical group. Student direction is that they should be able to present their conclusions (and their justification) to a technical, but non-expert audience. To assist with this goal, the preferred means of presentation is a series of geological models (Fig. 4.1), ranging from preliminary conceptual models, through to detailed, quantified geotechnical and geohazards ground models (Table 4.2). As well as being a means to convey what is often very complicated spatially and temporally variable information, the models also act as a framework within which the students can explore the shortcoming and gaps in the data sets that they have compiled. These models are tied to hazard and/or risk matrices (Fig. 4.2), allowing a level of geohazards quantification, even in the absence of detailed site-specific information. The latter obviously require information on system fragility, a level of knowledge often outside the level of expertise of the majority of students. Despite these shortcomings, the students are made aware of this important aspect, and that such data would be provided by the structural engineers involved in the design process. The important lesson to be learned at this stage is that geohazards assessment is not merely an academic exercise carried out in isolation from the remainder of the engineering design process. This further emphasises the importance of the VOI process, highlighting whether certain geohazards even need to be considered in light of system fragility. The importance of accuracy and precision in the hazard quantification process is also stressed, e.g. Do we really need to calculate fault slip rate to a certain level of precision, or indeed at all, or is it sufficient to understand that the fault is active and poses a potential seismic hazard? In understanding system fragility, the students are also introduced to the demands of differing design codes for different structures and their implementation (or not) across international boundaries.

4.3 Conclusions

Geological ground models, from preliminary conceptual constructions to detailed quantified geotechnical models provide a powerful tool for discussing uncertainty in geohazards determination. By allowing a three dimensional view of the ground, this allows a true realisation of the consequences of multiple geohazards. The models

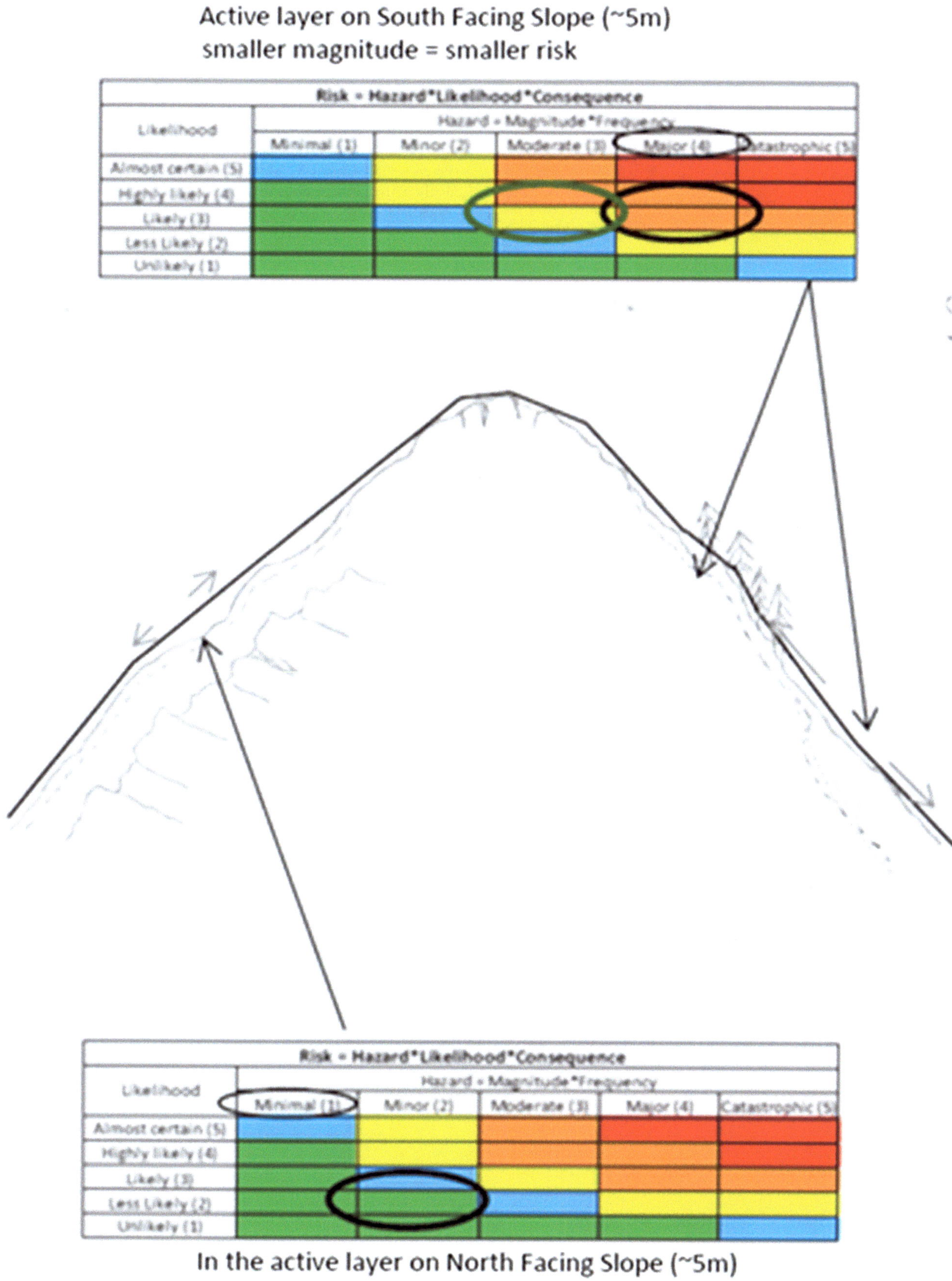

Fig. 4.2 Integration of ground models and geohazard risk matrices for upland areas along the Irkutsk to Ulaanbaatar pipeline corridor (Table 4.1)

themselves become a record of the evolution of understanding during the project, allowing the identification of data gaps, highlighting assumptions, and potentially identifying avenues for future research and investigation.

References

Budnitz RJ, Apostolakis G, Boore DM, Cluff LS, Coppersmith KJ, Cornell CA, Morris PA (1997) Recommendations for probabilistic seismic hazard analysis: Guidance on uncertainty and the use of experts. NUREG/CR-6372. Two volumes, US Nuclear Regulatory Commission, Washington, DC

Fenton C, Abassi S, Bardsley E, Cox J, Gkrozos C, Hickson T, Hoggard S, Jarvis L, Martins P, McCann P, Nee T, Richards M, Scales C, Shaw R, Shiathas A, Walsh M, Springer J (2010) Geomorphic evidence for recent movement and segmentation of the Chaman fault, Afghanistan and Pakistan. In: Williams AL, Pinches GM, Chin CY, McMorran TJ, Massey CI (eds) Geologically active, Proceedings of the 11th IAEG congress. CRC Press, Auckland, New Zealand, pp 163–170

Educational GIS-Project "World Experience in Site Investigation and Construction Under Different Engineering-Geological Conditions"

5

Tatyana Averkina, Natalia Pravikova and Natalia Kosevich

Abstract

GIS-project "World experience in site investigation and construction under different engineering-geological conditions" includes numerous data on engineering objects (engineering structures), and contains more than 20 layers divided into three blocks. Base layers include geographical and geological maps of the World and Russia, forming the first block of information. Map of engineering-geological structures of the Earth is the second block, it consists of 8 layers. The intersection of these layers gives taxonomic units of engineering-geological zoning of the four hierarchical levels. Engineering objects (buildings, dams, reservoirs, bridges, tunnels, etc.) form the third block. Attributive table contains data on every object and on its engineering-geological conditions: relief, composition and properties of foundation rocks, underground water, geological processes, the difficulties that arise upon construction, as well as the source data. This GIS-project can be used as a reference database and in educational activity.

Keywords

GIS-project • Layer • Map • Engineering-geological structure • Construction

5.1 Introduction

GIS-project "World experience in site investigation and construction under different engineering-geological conditions" is developed at the Department of Engineering and Ecological Geology (Geological Faculty of Lomonosov Moscow State University). This GIS-project can be used as a reference database, and for educational tasks for students.

Scientific and technical publications describing the engineering-geological conditions of construction of each engineering structure (or complex of structures) are the primary source data for the project. These are articles and few monographs. For example, the monograph by Anderson and Trigg (1976), Legget (1973), 4 parts of monograph "Engineering geology of the USSR" (1990–1992) and others. Archive data of the Department of Engineering Geology MSU is another source of information. Besides, a set of different maps for the territory of Russia and the World was selected for GIS project.

It is important to note that nowadays almost all engineering-geological agencies have their own GISes that contain fuller and more detailed information on engineering objects than our GIS does. The presented GIS is based mainly on short publications about engineering objects, but these are the objects, where problems and difficulties occurred during construction. This knowledge appears to be very important for students. In addition, we try to encompass the maximal variety of conditions in our GIS, to show the experience of different countries obtained in different years. This information is also very important for students.

T. Averkina (✉) · N. Pravikova · N. Kosevich
Geological Faculty, Lomonosov Moscow State University, Leninskie Gory, 1, Moscow, Russia 119992,
e-mail: averkinati@yandex.ru

N. Pravikova
e-mail: npravikova@mail.ru

G. Lollino et al. (eds.), *Engineering Geology for Society and Territory – Volume 7*,
DOI: 10.1007/978-3-319-09303-1_5,

Project is created in ArcGIS 9.3. Base scale is 1:25,000,000, but in the key regions data are more detailed—up to the scale of 1:2,500,000. Data are displayed in World Robinson projection.

5.2 Characteristics of the GIS "World Experience in Site Investigation and Construction Under Different Engineering Geological Conditions"

GIS database contains more than 20 layers divided into three blocks. In the first block, there are base layers including topographical, geological and tectonic maps. Map of engineering-geological structures of the Earth is the second block. The data obtained on engineering objects (engineering structures) form the third block.

5.2.1 Base Layers

Several base layers are templates from ArcView 3.2 and ArcMap 9.3. They are World Continents, Countries, Rivers, World Lakes and World Cities. Tectonic and geological maps of Russia are layers with base scale of 1:2,500,000. Tectonic and geological maps of the World are added in the raster format.

5.2.2 Map of Engineering-Geological Structures of the Earth

Engineering-geological structures are the regularly arranged parts of the lithosphere, formed under the influence of certain regional and zonal factors, and which are homogeneous with respect to some engineering geological parameters (Trofimov and Averkina 1996). By varying the combinations of regional and zonal factors (and, consequently, corresponding engineering geological parameters) and proceeding from the general to the particular, we may distinguish the engineering-geological structures with different content of various hierarchic levels. In this case, complex (two-member) foundation for division is the sign of the classification. We propose to name the largest structures of the first level as the engineering geological superstructures. For example, continental subaerial or continental subaquatic engineering-geological superstructures. Structures of the second level are named engineering-geological megastructures. For example, a platform with continuous permafrost, an orogen with unfrozen rocks. Structures of the third level are named engineering-geological macrostructures. For example, an old orogen with slightly moistened rocks. Structures of the fourth level are named engineering geological mezostructures. For example, a plate with seasonally frozen rocks or a shield with seasonally unfrozen rocks.

Information block "Map of engineering-geological structures of the Earth" consists of 8 layers. The intersection of those layers gives 4 layers of engineering-geological structures of the four hierarchical levels. It is made by means of Union operator from Geoprocessing Wizard. In fact, we get taxonomic units of engineering-geological zoning of the four hierarchical levels.

5.2.3 Data on the Engineering Objects

The data are displayed in two layers: point and linear engineering objects (engineering structures).

Layer "Engineering-points" is a layer of point objects, which contains engineering structures displayed as points in the scale map. Position of objects was defined using space imagery data of Google Earth and MSU Geoportal (Zimin and Turbalina 2012). High-resolution space images were useful for recognizing these objects.

Layer "Engineering-lines" is the layer of long linear objects—pipelines, canals, railways and highways. Position of lines was also determined using high-resolution space images.

Each object is described by attribute information. Attributive table of layers "Engineering-points" and "Engineering-lines" includes 12 fields: object number; L-code (by classifier); engineering structure class (by classifier); engineering structure type (by classifier); proper name of the object (if any), for example, St. Isaac's Cathedral; formation of pre-Quaternary rocks of the foundation (by classifier); the newest deposits of the foundation (by classifier); soil state, consistency of the foundation (by classifier); geological processes (if any) (by classifier); factor that caused the greatest problems (if any) (by classifier); engineering protection (by classifier), source data.

Every object is linked by Dynamic Hyperlinks to additional text and image files or web pages.

We collected data on several hundred engineering objects (engineering structures), constructed in different countries and continents. This database is constantly updated.

5.3 Use in Educational Activities

By means of the presented GIS, students can study experience in site investigation and construction obtained in various countries and under different engineering-geological conditions for particular cases.

Moreover, on the basis of this GIS students will solve engineering-geological educational problems. For example, they can make some generalizations. Information layers

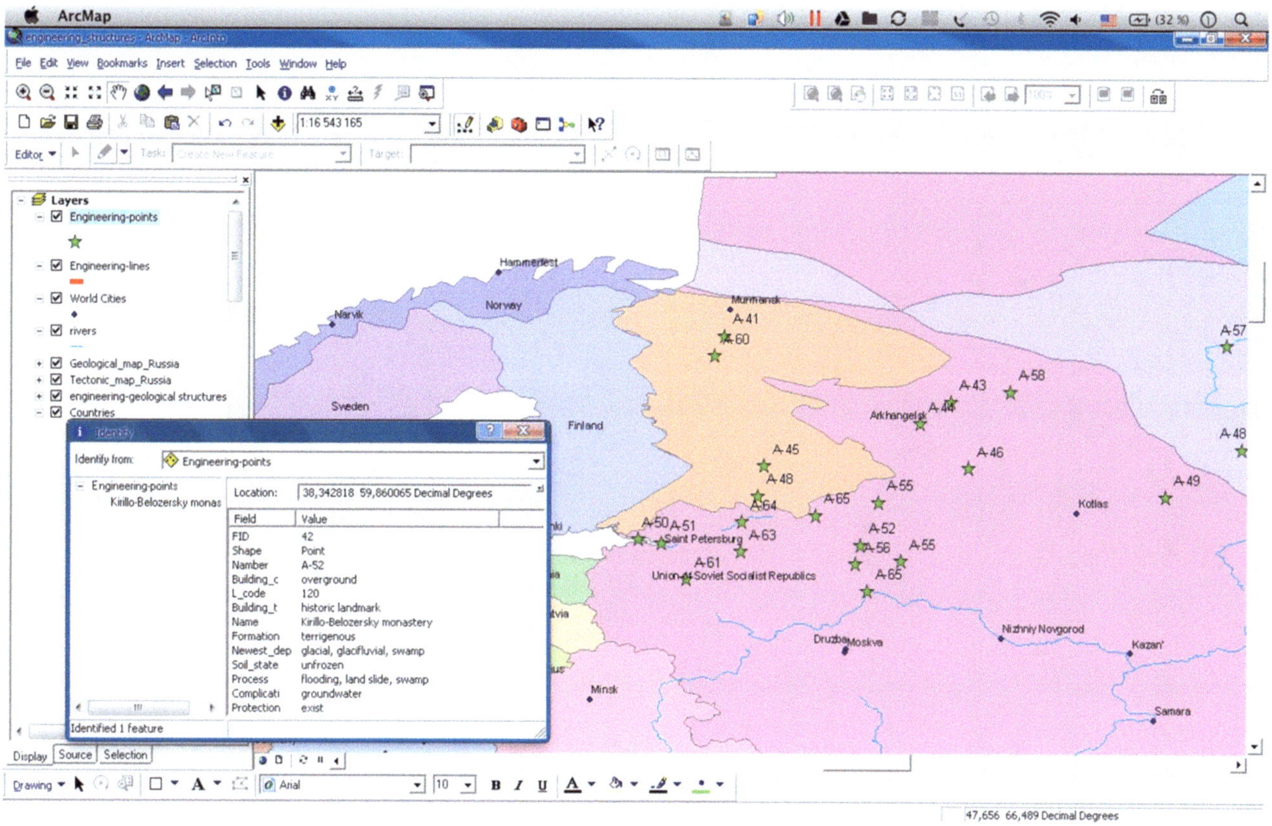

Fig. 5.1 Engineering-points layer in GIS-project

"Engineering-points" and "Engineering-lines" are the local data. If these data are considered in the context of point's location within region or different types of engineering-geological structures of the Earth, we can get some regional generalizations and regularities. Here are some examples of educational tasks for students.

(1) What engineering-geological factors usually complicate the construction of railways within orogens with continuous permafrost?
To perform this task, student must make query in GIS-project and create layer "Railways", superimpose it on a layer of "Engineering-geological megastructures of the Earth", and choose railways that are located within orogens with continuous permafrost. Most likely, these will be certain railways in the United States (Alaska), North-East of Canada, some regions of Russia, etc. Complicating engineering-geological factors can be found in the attributive table and in additional files linked to each object. After reviewing this information, a student should make the final synthesis.

(2) What methods of engineering protection of historic building from underflooding (waterlogging) are used within young platforms with slightly moistened rocks? This task is executed by the similar scheme.

(3) What engineering-geological factors complicate the construction of channels in Africa?

(4) Compare engineering protection of objects from landslides in the coastal areas in Russia and China.

(5) Compare engineering-geological conditions and engineering protection of hydropower complexes in UK and in the European part of Russia.

(6) Compare experience in construction of tunnels within carbonate formations in different countries?

A special complex of educational tasks deals with the selecting objects-analogues (Fig. 5.1).

References

Anderson JGC, Trigg CF (1976) Case-histories in engineering geology. Elek Science, London

Engineering Geology of the USSR (1990–1992) vol 1–4. Nedra, Moscow

Legget RF (1973) Cities and geology. McGraw-Hill Book Company, New York

Trofimov VT, Averkina TI (1996) Engineering-geological structure: hierarchy, types, and paragenetic series. Vest Mosc Univ Ser.4: Geol 4:5–26. (in Russian)

Zimin M, Tutubalina O (2012) MSU geoportal: history, results, future development. In: Earth from space—the most effective solutions, vol 14, pp. 54–61. (in Russian)

Part II

Geoethics and Natural Hazards: Communication, Education and the Science-Policy-Practice Interface

Convener: Dr. Silvia Peppoloni—*Co-conveners*: Janusz Wasowski, Giuseppe Di Capua, Tom Dijkstra, Peter Bobrowsky, Meng Wang

Among the critical ethical issues faced by Geoscientists and Engineers concerned with natural hazards (e.g., earthquake, volcano, landslide, and flood events) is the increasing death toll and cost owing to rapidly growing population, occupation of marginal/unsafe land, and misuse of land. These problems are often aggravated in developing countries. Many natural disasters can be prevented and/or their impact reduced. We must share and communicate our knowledge more effectively with all private and public stakeholders involved, paying attention to providing balanced information about risks and addressing inevitable uncertainties in natural hazard mapping, assessment, warning, and forecasting. The search for balance between short-term economic issues and wider social impacts from natural hazards is an increasingly urgent need. The scope of this part is to raise awareness on the above issues and foster greater interdisciplinary collaboration that is needed to fulfill our ethical obligations.

Geoethics: Perspective Bangladesh

6

Mohammad Omer Faruk Khan

Abstract

Geologists are becoming more socially involved professionals as their activities are increasingly focusing on social safety, ensuring that societal interactions with the planet are done in an environmentally friendly, safe and sustainable way. As a consequence, geologists need to be aware of the ethical perspectives of their profession and this awareness is most successfully enhanced through professional collaborations and a strong community. Unfortunately, these are lacking in Bangladesh, largely as a result of only very few geologists employed in a small subject sector that is much undervalued, partly as most people perceive the local geology to be rather monotonous and the potential valuable contributions of geologists not being appreciated. Still, there is much scope for geological sciences to perform vital services in Bangladesh, from both a societal and an environmental perspective. Geologists can serve other professionals such as agriculturists, economists, environmentalists, engineers, doctors, policy-makers as well as general people in a scientific way. Ensuring that these services are developed appropriately requires a leading role of geoethics to provide a relevant philosophical context and guidance.

Keywords

Bangladesh • Development • Geoethics • Geoscientists • Society

6.1 Introduction

Geological sciences originate in an environment where it was mainly seen as an outdoor science, focusing on remote locations and the search for valuable materials. Recently, however, geology has evolved more towards the realms of social sciences, with activities directly impacting on society making the work of geologists much more conspicuous (Peppoloni 2012). Geologists are now dealing with the safeguarding and well-being of societies in the face of natural and anthropogenic anomalies, alongside their more traditional roles of discovering geo-resources for the material comfort and development of humankind. This orientation towards social sciences has recently led to recognition that there is a need to put their motivation and dedication in the context of an ethical code of conduct (Lucchesi and Giardino 2012). Compared to the rest of the world, Bangladesh has a limited breadth of geo-scientific work. It occupies one of the largest deltas (Fig. 6.1) in the world and supports a large population (more than 160 million) at one of the greatest population densities, globally. It does so with very limited natural resources and has to deal with frequent, severe hazards resulting in national-scale disasters—a situation which appears to get worse as global climate and environmental change progresses. These hazards retard the economic growth and development of the country in every possible way.

M.O.F. Khan (✉)
Geological Survey of Bangladesh (GSB), 153, Pioneer Road, Shegunbagicha, Dhaka, 1000, Bangladesh
e-mail: faruk_geology@yahoo.com

G. Lollino et al. (eds.), *Engineering Geology for Society and Territory – Volume 7*,
DOI: 10.1007/978-3-319-09303-1_6, © Springer International Publishing Switzerland 2014

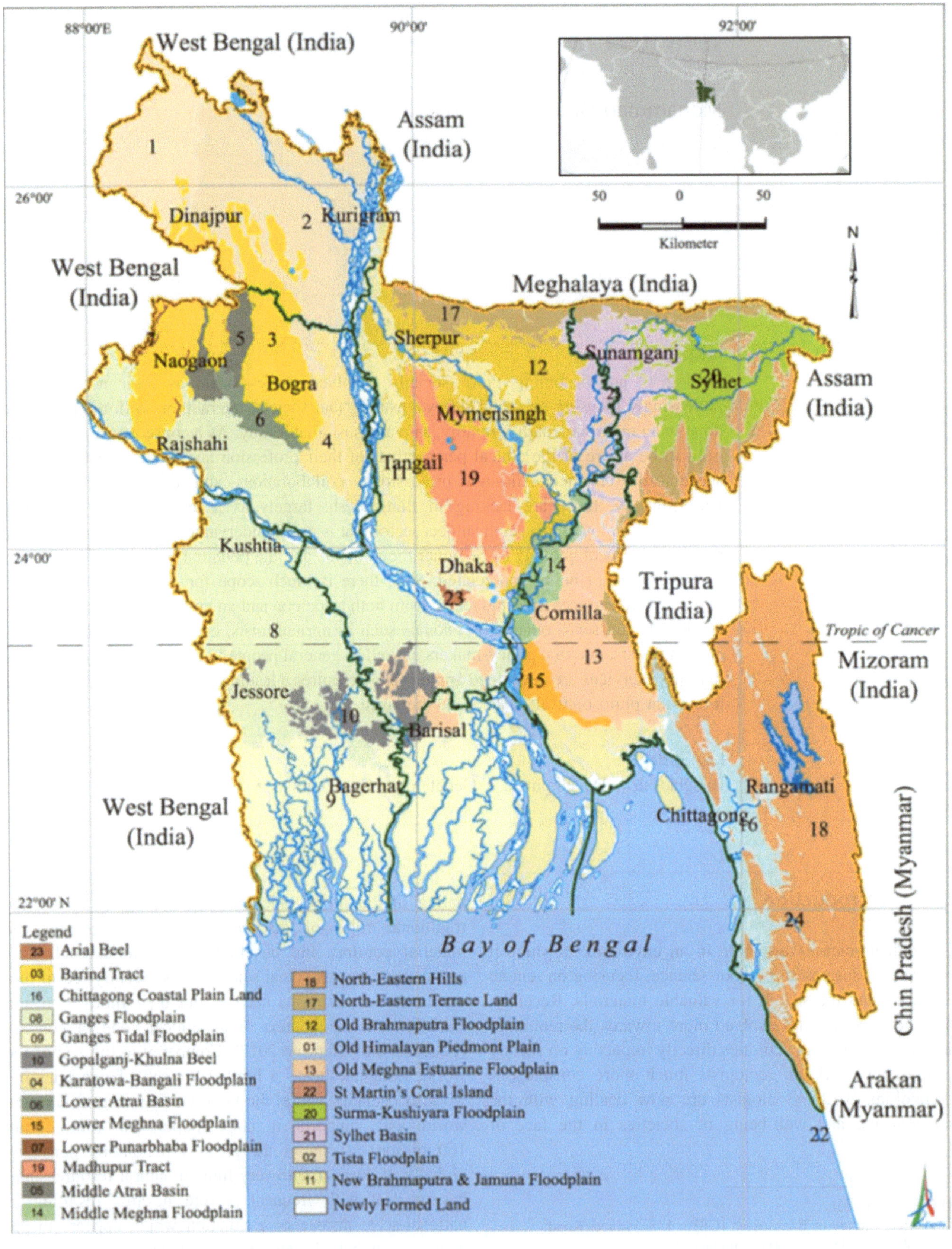

Fig. 6.1 Physiographic features of Bangladesh (*Source* Banglapedia and Wikipedia)

In addition, a lack of proper geological and geotechnical practice leads to widespread damage to infrastructure and construction taking a heavy toll on both lives and livelihoods. Therefore, sustainable development is a term that makes much sense in the context of Bangladesh. To achieve this, now more than ever, is the time for the Bangladeshi geo-scientists to really contribute to the well-being of their fellow countrymen and step forward with their service offers. This can be best achieved through applying in a meaningful way the vision of geoethics (Peppoloni and Di Capua 2012).

6.2 Some Notable Features of Bangladesh

Bangladesh is a land of rivers situated in the tropics, in the northeastern corner of the Indian Subcontinent in southeast Asia. It occupies the Bengal Basin, which is one of the largest sedimentary basins of the world. The basin accommodates a thick (±22 km) Early Cretaceous-Holocene sedimentary succession (Curray 1991; Curray and Munasinghe 1991). The country lies in the close proximity of two major thrust bounded tectonic subduction zones; the Himalayan subduction zone in the north and the Burmese subduction zone (Maurin and Rangin 2009) in the east which form the Himalayan orogeny (Alam et al. 2003) and Indo-Burman orogeny (Rahman and Faupl 2003) respectively. The country is bounded by low lying Tertiary hills in the north and east. The rest of the country is mostly low lying plain land drained extensively by the Ganges–Bhramaputra–Meghna river system, which forms the world's largest fluvio-deltaic system, the Bengal Delta which is one of the largest deltas of the world, ending up in the Bay of Bengal to the south (Alam et al.). The Bay of Bengal hosts the Bengal Fan, which is the largest submarine fan in the world (Curray et al. 2003). The southern part of Bangladesh is fringed by the Sunderbans, which is the largest mangrove forest and only tiger mangrove land in the world (Manna et al. 2010). This low lying subsiding plain of Bengal may thus aptly called a land of water where the hazardous natural phenomena as landslide, flooding, inundation, cyclone, drought, frequent earthquakes, sea-level rise, saline-water/sea-water encroachment, coastal erosion, river-bank erosion (Fig. 6.2) etc. actively shape the livelihood and economy of the country. Although Bangladesh is gifted with plenty of natural resources as fresh water, fish stock, livestock, grazing fields, agricultural resources, largest sandy coastline, largest mangrove forest, the progresses of the country is consistently hindered by the disasters caused by the natural hazards mentioned above. The situation is further complicated by the high density of population of Bangladesh. Thus, an understanding of the low-lying deltaic system is very important for the sustainable development of the country taking the soft, subsiding landscape into consideration in a background of changing climate and land submergence due to sea-level rise.

6.3 Problems Regarding the Professional Reputation of Geoscientists in Bangladesh

The influence of a particular subject area on the performance of a society is very much dependent on the renown of this field of study within a country, and how well it is supported by an appropriate number of skilled professionals of high repute and standing. Unfortunately, in Bangladesh the field of geology as a scientific discipline is not that well recognized, both among the general public and, more importantly, also among the policy-makers. There are only few professional geologists in the country and they are not provided with sufficient power or influence to make significant contributions to policy making. Although, individually, these geology professionals are very talented, these talents are not recognized in their own country and many have greater success when they go abroad for work or higher studies.

Empirical observations and personal experiences have revealed a rather unpleasant picture. The general public does not have much knowledge of the science of geology; often it is confused with 'zoology' as this sounds similar and is one of the major scientific disciplines of the national

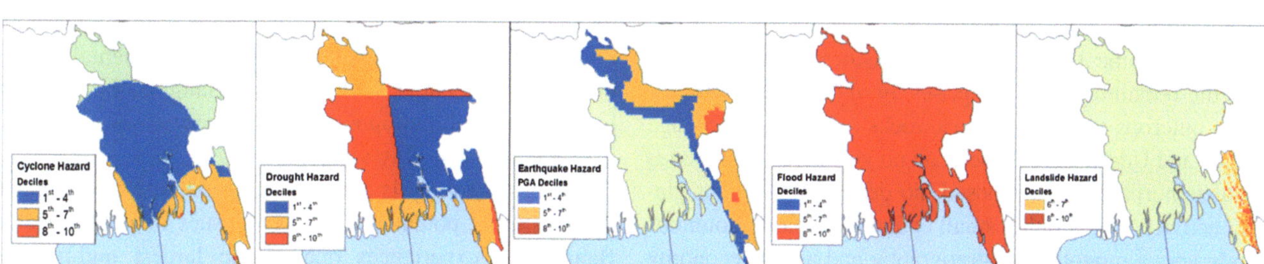

Fig. 6.2 Different hazard maps of Bangladesh (*Source* http://www.ldeo.columbia.edu/chrr/research/profiles/bangladesh.html)

curriculum. When it is explained to them the general interpretation is still one aligning it with soil science or geography, rather than appreciating it as a scientific discipline on its own. As a consequence, it is difficult to approach the public from a geological perspective and inversely, the people thus remain largely unaware of the beneficial services that geological knowledge can provide. Most frustratingly, it is not only the general public that has this mindset. Also other professionals and policy makers who may directly benefit from the services on offer by geologists have demonstrated little interest in this regard. This includes professionals such as civil engineers, public health officials, agriculturalists, environmentalists, etc.

This situation may be partly explained by the lack of opportunities to study geology in a national context. Geology as a full-time course curriculum starts at the undergraduate level and only three public universities (University of Dhaka, Jahangirnagar University and University of Rajshahi), out of almost fifty, offer geology degree courses. Although there are many specialized universities and institutions for medical, engineering, agricultural and other technical disciplines, there are none for geology. There is not even one geological institute—a very alarming situation. Even more alarming is the situation that for some 50 years now private universities are developing an increasingly strong position and none of these offer geology as a full-time course. At best, some of these universities offer highly specialized and commercially viable topics such as disaster management.

The number of geology students is very small, only some two hundred students graduate each year in a country of some 160 million people. The limited appreciation of the subject also limits the job opportunities on offer for our wonderful topic of geology as the job sector is not that flourished. This remains a self-reinforcing situation where too few graduates are produced by a restrained academic environment to make an impact on a job situation where their skills are under-appreciated and where only the very few can progress to attain positions of influence to change the status quo.

Interestingly, the above situation could possibly be related to the overall geology of the country. Bangladesh is a low-lying deltaic plain that does not offer much in terms of bedrock geology. Almost the entire country is covered by recent alluvial deposits, amounting to some 82 % of the land area. Some 10 % comprises Pleistocene terrace levels and the remaining 8 % comprises an undulating landscape with outcrops of sedimentary rocks of Quaternary age (Islam and Uddin 2002). There are no volcanoes, no igneous or metamorphic rocks exposed, no deserts, no ice nor glacial activity. It is a rather small country with a monotonous landscape, a monotonous stratigraphy and a poor economy and this situation does not help in the development of a flourishing geological science environment in a holistic way. As a result, where it is being practiced, geological sciences have developed in rather idiosyncratic way, rather than covering the full breadth of topics on offer. Thus, the already small community is shrinking further and many graduates and professionals prefer to study abroad and take up employment there, rather than serve their nation.

6.4 Opportunities and Scopes: Not All Hopes Are Lost

Recent issues, including environmental degradation, climate change and natural hazards have opened up new avenues of research and study for geologists. This deltaic lowland of large rivers and wetlands provides for extensive river and aquifer systems that play critical roles in the nation's socio-economic, cultural and environmental conditions. As a result, there is tremendous scope for the application of hydro-geo-sciences such as hydrology, hydrogeology and hydro-geochemistry. These major areas of study still offer plenty of room for future development.

As a developing country, Bangladesh needs further investment in infrastructure development. However, there have been several recent incidents of buildings being damaged or even collapsing (Fig. 6.3) due to limited or even absence of appropriate geological understanding and application of engineering geological practice (GSB). There is thus a need to integrate these disciplines into general development of the country.

Bangladesh is situated in the tropics and occupies an active tectonic hotspot of the Indo-Burmese-Eurasian plate junction. As a result it is frequently plagued by a number of geo-hazards including earthquakes, landslides, floods, cyclones, storm surges, river bank erosions, saline water encroachments, coastal landscape changes, tsunamis etc. and Bangladesh has long been known as a land of natural hazards. Risk management (e.g. Fontana et al. 2012) and disaster management (Parkash 2012) form the modern developments of applied geological disciplines that need appreciation in the context of Bangladesh. There is also a lot of scope for further development of water resource management and land resource management.

The country is suffering from river water withdrawal by its neighbor and there are disputes concerning the maritime borders (ITLOS). The resources and environmental quality of the Bay of Bengal, to the south of Bangladesh, are yet to be fully studied and appreciated. In these cases, geo-political practices can play vital roles.

Medical geology is another important new development that has great potential to address issues of public health. Bangladesh experiences one of the largest outbreak of arsenic contamination of the world (Kinniburgh and Smedley 2001). Other health issues related to geological

Fig. 6.3 Example of building collapse (*Source* The Economist)

causes can be well served by professional geologists trained in medical geological applications.

A further area where geologists should find employment is alongside social development professionals developing proper resource management practices and disaster management/mitigation strategies. Proper resource management is essential for a developing economy such as Bangladesh that has a rapidly growing population and a relatively small land mass. Furthermore, as an agriculture based economy, there are plenty of opportunities for agricultural geological work in Bangladesh.

That these opportunities exist is clear, but to make these work in practice requires the coordinated efforts of geoscientists and professional in Bangladesh to convince policy-makers and communicate with the general public. Moreover, as many of the above mentioned fields are multidisciplinary, there is an urgent need to communicate with scholars and professionals from other fields. Geoethics can play a pivotal role in this regard as it provides formal and substantial value to the commitment of science for the benefit of citizens and institutions (Peppoloni and Di Capua 2012).

6.5 The Possible Roles of Geoethical Values in the Duties of Bangladeshi Geoscientists

Bangladeshi geologists can benefit themselves from the philosophies offered by geoethics (Graziano 2012) and the Hippocratic oath for the geologist (Matteucci et al. 2012). These highlight that there are other vital roles to fulfill well beyond the professional arena. First of all, their learning and appreciation of the vitality and practical application of their profession in a social context will be highly beneficial. Getting acquainted with and applying the ideas and values of professional ethics and the duties and responsibilities these carry for the well-being of their fellows will help enhance their reputation. Communicating these with society, including other professionals and policy-makers, in a proactive way will help further strengthen the geologists' professional dignity and honor. Geologists must endeavor to create demands on their professional services among the general public. A key point could be the issue of proper land management and resource use (Badiali and Piacente 2012) as the population of Bangladesh continues to grow rapidly against an increasingly constrained land resource. A balance between these two conflicting drivers must be achieved to sustain social, cultural and environmental harmony and peace. Another demand of geoethics is that there should be greater awareness among as many people as possible and this demand can be addressed by involving the general public through grass-roots education and the generation of more geo-professionals.

The fact of taking care of the Earth is largely a duty of geologists as they are the ones who pierce her for resources and know the consequences of such activities along with the effects of other natural and anthropogenic disturbances and are capable of providing possible mitigation strategies. A full appreciation of this duty is a must for them. So, they should play the role to guide the society toward a safer and sustainable future where geoethics comes to play with its visions of responsibilities and ideals. It is up to the geologists to step forward and take the leading role where there is a call of duty that stems from the teachings of geoethics.

6.6 Conclusion

Geology is no more a routine work and geologists are no more just resource explorers. They need to serve the society in every possible way and thus their duty spread beyond traditional roles. This is a matter of mass awareness where geoethics come into play and this is particularly true in the context of Bangladesh.

References

Alam M, Alam MM, Curray JR, Chowdhury MLR, Gani MR (2003) An overview of the sedimentary geology of the Bengal Basin in relation to the regional tectonic framework and basin-fill history. Sediment Geol 155:179–208

Badiali F, Piacente S (2012) The study of landscape: from a holistic approach to a social concept of knowledge. Ann Geophys 55(3). doi:10.4401/ag-5539

Curray JR (1991) Geological history of the Bengal geosyncline. J Assoc Explor Geophys 12:209–219

Curray JR, Emmel FJ, Moore DG (2003) The Bengal Fan: morphology, geometry, stratigraphy, history and processes. Mar Petrol Geol 19:1191–1223

Curray JR, Munasinghe T (1991) Origin of the Rajmahal Traps and the 85 E Ridge preliminary reconstructions of the trace of the Crozet hotspot. Geology 19:1237–1240

Fontana M, De Amicis M, Rossetti M, Garcia C (2012) Risk management: a proposal for communication strategies. Ann Geophys 55:3. doi:10.4401/ag-5521

Graziano GV (2012) Professional ethics: know-how, deontological caode and legality. Ann Geophys 55(3). doi:10.4401/ag5560

GSB. http://www.gsb.gov.bd/english/index.php?options=com_contents&overview=category&layout=blog&id=7&Itemid=27

Banglapedia. http://www.banglapedia.org/HT/P_0184.htm

ITLOS. http://www.itlos.org/index.php?id=108

Islam MN, Uddin MN (2002) Country paper on hydrogeology section in international workshop on arsenic issue in Bangladesh, 14–16 January 2002. http://phys4.harvard.edu/~wilson/arsenic/remediation/Reports/Countrypaper.doc

Kinniburgh DG, Smedley PL (2001) Arsenic contamination in groundwater of groundwater in Bangladesh. BGS technical report WC/00/19, vol 1

Lucchesi S, Giardino M (2012) The role of geoscientists in human progress. Ann Geophys 55(3). doi:10.4401/ag-5535

Manna S, Chaudhuri K, Bhattacharyya S, Bhattacharyya M (2010) Saline Systems 2010, 6:8. http://www.saliesystems.org/content/6/1/8

Matteucci R, Gosso G, Peppoloni S, Piacente S, Wasowski J (2012) A Hippocratic oath for geologist? Ann Geophys 55(3). doi:10.4401/ag-5650

Maurin T, Rangin C (2009) Impact of the 90°E ridge at the Indo-Burmese subduction zone imaged from deep seismic reflection data. Mar Geol 266(1–4):143–155

Parkash S (2012) Ethics in disaster management. Ann Geophys 55(3). doi:10.4401/ag-5633

Peppoloni S (2012) Ethical and cultural value of the Earth sciences. Interview with Prof. Giulio Giorello. Ann Geophys 55(3):343–346. doi:10.4401/ag-5755

Peppoloni S, Di Capua G (2012) Geoethics and geologic culture: awareness, responsibility and challenges. Ann Geophys 55(3). doi:10.4401/ag-6099

Rahman MJJ, Faupl P (2003) $^{40}Ar/^{39}Ar$ multigrain dating of detrital white mica of sandstones of the Surma Group in the Sylhet Trough, Bengal Basin, Bangladesh. Sediment Geol 155(2003):383–392

The Economist. http://www.economist.com/blogs/banyan/2013/04/disaster-bangladesh

Geoethics and Communication 2: Ethics, the Notably Absent from the Internet

7

Stefano Solarino

Abstract

The statistics show that the number of people having access to the internet is growing quickly, even in lower income countries. They also show that internet is becoming a source for information, sometimes even the unique one, for many people, especially the students. Blogs, discussions on social media, chats, virtual forums are to be considered public interactions among people. They represent the newest and powerful way to "speak" in public. Conversely to real life, where the selection of experts is made on the basis of education and experience, in virtual "round tables" any surfer can express ideas or opinions. In principle this is a tremendous piece of democracy but is it not trouble free. The identity of speakers, their culture and skills are not pre-requisites and can anyway be mystified; the bloggers often express personal opinions, manipulate (on purpose or in ignorance) the information; they often propose for natural phenomena an alternative explanations that comply with their political or social ideas; they often disregard the experience of other people. In other words, the internet-era has made establishing trust an increasingly complicated issue. All this is a clear offence to the ethics that denies voluntary or involuntary manipulation of information, requires recognition of the professional competence, claims the assumption of responsibilities, avoids conflict of interest, places emphasis on self integrity, honesty and objectivity. In one sentence, ethics is the notably absent. In this study a few case histories are presented to introduce a discussion on the possible solutions to gain more geoethics on the internet. The conclusions lead to the assumption of responsibilities of the scientists and a call to a more unbroken presence on the web.

Keywords

Geoethics • Communication • Internet • Science blogs

7.1 Introduction: Information, Education and the Web

The number of internet surfers rose from 16 millions to about 3 billions in the last two decades. Figure 7.1 (Internet World Stats 2013) shows the increase per year and displays that, although there have been peaks, the growth is steady during all period. The amount of people accessing to the internet is now about 40 % of the world population (blue bars in Fig. 7.1). However the distribution of web users in the world is rather proportional to the technological features of the relative continent more than to the population, so we may expect a different "geography" of the internet in the next future, when the technology will offer the chance also to less developed countries.

The reasons for this tremendous increment are beyond the scope of this article, but it is clear that many different variables other than technical evolution concurred to it. For example the availability of web browsers on mobile

S. Solarino (✉)
Istituto Nazionale di Geofisica e Vulcanologia, Genoa, Italy
e-mail: stefano.solarino@ingv.it

G. Lollino et al. (eds.), *Engineering Geology for Society and Territory – Volume 7*,
DOI: 10.1007/978-3-319-09303-1_7,

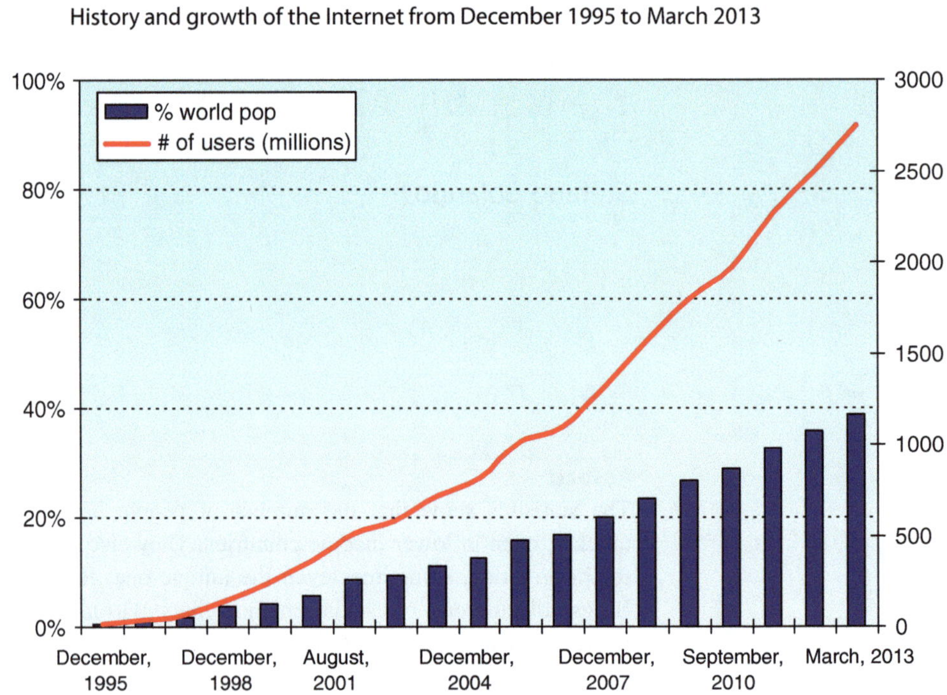

Fig. 7.1 History and growth of the Internet. Data extracted from Internet world stats, 2013

telephones and tablets made the internet more friendly and easily available, but the decrease of costs for the connections and the establishment of wi-fi networks were also of paramount importance.

It is also well acknowledged that in the last years there has been a change of role of the internet users, that switched from "spectators" to protagonists. In fact the birth of several social networks has given the chance to the internet surfers to interact with other users and especially to manage their own blogs and thus share their ideas and knowledge.

A statistics (We are social, http://wearesocial.sg 2012) shows that in Europe and North America almost 50 % of the internet surfers have joined and actively take part to social networks. Through the social media people can share ideas, opinions, documents, knowledge in a very (too?) easy way. In principle, the participation of "everyday" people to discussions and forums is a great proof of democracy on the one side and a good chance to enlarge perspectives, to overpass physical boundaries (borders, language, distances, cultures) and acquire knowledge on the other side. However it must be remarked that this evolution (1) applies to a virtual world (2) happened too fast, giving no time to establish clear rules (3) is governed by technical aspects that render it very different from previous forms of communication and cultural exchanges (4) does not require any background; at present few, if any, educational institutions teach the art of proper digital communication.

The internet is slowly substituting other classical means of information. One key question is that the internet, conversely to a conference, a speech, a printed issue or a TV broadcast, is active 24 h a day; thus the information contained within the web pages (reliable or not) remain there forever, can be reproduced or reintroduced unlimitedly, in other words cannot be "physically" discarded.

At present there is no way to a priori discriminate between good and reliable from bad and unreliable sources of information on the internet. When searching for information, the top sites are those that are most visited or the most appealing, not necessarily the most reliable. Moreover, conversely to what happens in real life where the selection of speakers is made on the basis of skills and experience, in virtual "round tables" on the internet the identity of surfers is often unknown (nicknames are fantasy names and do not give insights about the real identity). The culture and skills of web surfers are not known or in some cases mystified; the bloggers often express personal opinions, manipulate (on purpose or in ignorance) the information. They often searches for an alternative explanations to questions or open problems that comply with their political or social ideas; they often take the chance to disregard the experience of other people.

All this is a clear offense to ethics, that denies voluntary or involuntary manipulation of information, requires recognition of the professional competence, claims the assumption of responsibilities, avoids conflict of interest, places emphasis on self integrity, honesty and objectivity. Anyone substituting a scholar in explaining the science behind a natural phenomenon without having enough credit or giving advises about how to be prepared for a natural hazard without a consolidated experience commits an offence toward the principles of ethics.

In particular, it must be underlined that in Geoscience an incorrect or partial information may also have consequences on the safety of citizens. Adopting inappropriate behaviours during an emergency may compromise the health or have unwanted effects on the environment and on other people.

Unfortunately, on the web there are several evidences of miscommunication. They also apply to projects above any suspect, like Wikipedia in its original (but partly also in its current) form, and to popular social media when the moderators are not professionals.

In the following paragraphs a few case studies are presented to introduce a discussion on the possible solutions to gain more geoethics on the internet.

7.2 A Few Case Studies

From what stated above it is clear that the world of digital communication differs very much from standard forms of information and fetches new and unprecedented issues. In order to get the feeling and some evidences about the problems all this bears a couple out of the many examples of improper (from the point of view of ethics but not only) communications on the web are here described. They are selected on the sole criterion of being at the top of a search for topics on the web, and they are not better or worse than others. Finally it must be remarked that I am not against these sources of information, but I believe that a caveat on their use is necessary and should be firmly commended by the scientific community.

7.2.1 Wikipedia

Wikipedia is the free internet encyclopedia written by volunteers around the world. The history of the project is thoroughly described on the web page http://en.wikipedia.org/wiki/History_of_Wikipedia and references within. Here I simply recall that the original project (Nupedia) was to publish on the web an encyclopedia solely edited by experts and with articles subjected to peer review. This organization of the work proved to be too slow (in the first year only 12 articles were completed), so the project switched to voluntary contributions to be edited out with a specific software on a platform open to any internet surfer. As of September 2013, Wikipedia still allows anonymous editing: contributors are not required to provide any identification, or even an email address. This poses a major question about the reliability of the information provided by Wikipedia and especially it introduces the problem of vandalism (Solarino 2010). The discussion about the quality and reliability of the articles published on Wikipedia is out of the aims of this study; the topic has been reviewed by Wikipedia itself and can be found on the web page http://en.wikipedia.org/wiki/Reliability_of_Wikipedia.

It is noteworthy that people at Wikipedia know that the information can be distorted or not totally reliable, as it is shown by the long discussion about how to correct blunders and incorrect information which reads in the web page mentioned above. In my view, this is against the principle of ethics. Moreover the discussion about the reliability of the source reads the following sentence: *It is the responsibility of those who intend to use such a dynamically changing, multi-authored source to ascertain the quality and reliability of articles, and the degree of usefulness, misinformation or vandalism which might be expected, in order to decide what reliance to place upon them. A helpful safeguard is always to reference Wikipedia accurately when it is quoted to allow false or unreliable material to be identified and corrected.* It is my personal opinion that such a point of view is a pure offence to ethics for several reasons:

- it implies that the editor knows that there might be mistakes and biases.
- it implies that the reader (which is often a non-technical) is able to recognize and discard the errors. Why should then a reader search for information on any encyclopedia if his/her skills are higher level?
- it denies the charge of responsibility that a regular editor should assume by common sense and especially by law.

7.2.2 Earthquake-Prediction Sites

On the internet the surfers find quite a number of web sites that announce next earthquakes in the world. Some are very popular (www.quakeprediction.com; www.world-earthquakes.com/index.php) and generally well designed; conversely some are very naïve in appearance and based on very imaginative theories (http://earthquake.itgo.com, prediction of earthquakes using sunlight). It is out of the aims of this paper to describe these sites in details and especially the weak points of some of the methodologies their predictions are based on. I simply underline the many failures in the field of ethics they show (not all these points apply to all sites).

- they do not describe in details the methodology they use for prediction, nor are they published in scientific journals.
- although they capture the attention of surfers with the word "prediction" in both URL address and text, they do not list predictions (since a prediction should contain accurate information about exact time, position and magnitude which are not shown on the sites).
- they state that it is the responsibility of the surfers how they interpret the information published, while generally speaking the charge to users is how they use the information, that must be non interpretable.

- they contain significant errors and dangerous suggestions on how to behave. One of these sites, for example, suggests to leave from a building during the seismic shake to avoid aftershocks.
- they link to institutional scientific sites or contain information taken from them, giving the fake idea to surfers to deal with a scientific research agency.

7.3 Discussion and Conclusion

In the last decade the massive use of the internet has introduced an innovative way to communicate, but it also created unprecedented moral queries.

The revolution of digital communication cannot be denied nor can the advantages it carries be forgotten. Getting information from the web is a simple and fast issue, especially nowadays that being on-line is cheap and easy. An evidence is that digital encyclopaedia, blogs and thematic web sites are the most popular source of information for students. In principle there is nothing wrong with it. Digital communication is fast, always up to date and, thanks to the hyperlinks, complete, but not always accurate, ethical or trustable.

Earth scientists did not take active part in the recent evolution of the web potentialities and are still not completely aware of the power of such a mean of information There seems to be a big gap between scientific communication and communication on the web. Scientists used the web, especially in early times, as a platform to publish data and results rather than envisage a chance to interact with the public. The gap between science and society has meantime been fulfilled by non-scientists, mostly amateur people practising science, endorsed by the massive spread of the social media. They were able to conquer a big virtual audience by designing appealing sites and giving the surfers the chance to express their opinions by simplifying the interaction among them. Any interested surfer can join a virtual discussion: the attendees will simply focus on the comments and not on the skills of who is posting them.

Often also scientific topics are discussed in blogs and forums and this poses a big problem both in terms of geoethics and disaster risk reduction. The information shared by surfers may be not correct and biased by ignorance or false beliefs but are often more attractive then the boring and reliable science. In the human imaginary being able to predict earthquakes is much more fascinating than learning safety rules or be prepared on how to live with a potential natural disaster. The idea behind is that if we would be able to predict catastrophes we should not worry about the occurrence of an earthquake or a tsunami because we would apparently be able to avoid their consequences by evacuating.

Since scientists cannot make forecasts, they must be blamed and are considered untrustworthy. All this distract people from learning the main safety rules and lead them to believe that public money spent on preparedness are wasted: these money should be spent on prediction studies.

As a consequence of what stated above, incorrect-appealing information spread fast and are very difficult to eradicate; the intervention of an "expert" in a blog or in a discussion is often criticized more than necessary and it sometimes leads to a denial of the correct information because the audience does not trust in the bearer.

Murray (2010) in an article about new aspects of science communication points out that "the current phenomenon of bloggers should be of serious concern to scientists"; in fact bloggers and scientists are two separate entities. "Scientists who also blog apparently do not exist". With the megaphone of internet bloggers (but I extended this sentence to all internet active surfers) can reach a much larger audience and this magnifies the problem in assessing credibility. Writing can be done for any purpose (political, religious, business) without the constraint of truth. And of ethics, I add.

The analysis of Murray dates back to 4 years ago; things slightly changed meantime. Nowadays some science scholars do write on blogs (www.wired.com or www.science20.com, for example; the latter with a full section devoted to Earth Sciences) and many others cooperate with Wikipedia editors to improve the quality of scientific topics. However these improved versions of blogs and virtual encyclopedia are designed and have contents suited for other scholars than for the non-scientific population. In fact a part of the comments in the blogs are from other scientists sharing their personal experience and discussing the content of the article of the blog itself. Most of these blogs were born to host articles and discussion from experts and are organized like scientific journals. They have the worship to have opened scientific discussion to the world outside the scientific environment but are not yet ready to be considered of public domain. They make "everyday" people feel shy to interact, and although they are partly contributing to good communication and dissemination of scientific topics, they also partly contribute to reinforce the wall between science and society.

It seems that Murray's sentence about "the separate entities" is still valid since in principle blogs by "normal" people are not attended by scientists and viceversa.

The problem of assessing reliability on the web is not easy to solve. A more serious assumptions of responsibilities must be carried on by the scientific community. The task is not easy, due to the huge number of information available on the web. One possible solution to ensure trustable origin of information would be to mark reliable sites (like USGS, CSEM, IRIS, NOAA and so) with a special icon (like that used for SSL transactions) or better to establish a dedicated www domain. The agencies quoted above already use a specific domain (.gov, .org) that indicates the belonging to governmental institution or the nonprofits, but their use has a different meaning than presenting the relative web sites as reliable or scientific sources. Sometimes belonging to one of these domains may even appear negative to people in bad faith. Moreover, the .org domain has lost its original meaning, it is now used in several environments and does not even have any real restriction to registration.

The use of a devoted domain (e.g. .sci) would instead flag specific requirements and aims. Of course the right to use this domain would be released only upon accreditation, like it happens for the .edu domain, to entitled scientists. The public should of course be educated to utilize the sites flagged under these domains prior to any other source of information. The blogs should be moderated by the scientists or better by surfers under the supervision of scientists; the articles within the sites should be aimed at showing that science rules many of the aspects of everyday life; the topics treated in these sites should have a direct link to practical aspects; the main message should aim at convincing people that science has limits and that knowing how to behave in emergencies is the best life saving device.

Eventually, educating public to "virtual" ethics should start from educating it to find the right and reliable information; to teach to wonder about the reliability of sources, to persist in the effort of making bloggers and surfers to adopt the principle of ethics; to avoid that hidden wishes, like the chance to predict earthquakes, cancelled common sense. It also will require scientists to have more availability toward the educational projects on the internet; but most of all to establish a better, new relationship with the web.

References

Internet World Stats, Usage and Population Statistics (2013) Internet growth statistics. http://www.internetworldstats.com/emarketing.htm

Murray R (2010) Science blog and caveat emptor. Anal Chem 82(21):8755

Solarino S (2010) Seismologists and the web: is Wikipedia an ally or an enemy? XXXII General Assembly of the European Seismological Commission

A Road Map for a Deontological Code for Geoscientists Dealing with Natural Hazards

8

Eduardo Marone and Luis Marone

Abstract

Many professions have codes of conduct or deontological rules, but geoscientists dealing with natural hazards have been working on their own in the last decades not yet completing the task. A series of recent natural hazards that have hit the society in several parts of the world have made the call. Deontology is one of the main ethical decision-making approaches for driving actions leading to what is moral/ethically correct. To contribute with the construction of such a code, we present a road map to formulate this code based in the question-driven approach. Some considerations may be presented as guidelines: (a) it cannot be designed to the self-protection of the geoscientists but to safeguard society and the environment (even if it will contribute to); (b) it has to be constructed collectively; (c) it requires deep changes in educational systems, better preparing the citizens to deal with some scientific concepts. Without (c), (a) and (b) would not be enough. In addition, in constructing a geoethics code, we have to consider the three periods in the time-line of an extreme event: before, during, and after. Each period has particularities that will indicate different actions and conducts. Finally, the code, constructed collectively among practitioners, cannot be just a collection of "steps to be followed" but a real bridge between the moral consequences of being the privileged small part of the society who hold the scientific knowledge and the full society and the environment, for which we use to say we work for.

Keywords

Deontology • Natural hazards • Geoethics • Education

8.1 Introduction

While many professions have well established codes of conduct and/or deontological rules, geoscientists dealing with natural hazards have perceived the need of constructing one in recent times due to the consequences of their actions. This need has been widely exposed by a series of natural hazards that have hit the society in several parts of the world. Deontology is one of the main ethical decision-making approaches for driving actions leading to what is moral/ethically correct (Mitcham 2005). The relevance of deontological ethics to issues in science and technology is not obvious, mostly because proper scientific knowledge, in most practical cases, is constructed trying to following the epistemological "knowledge that" and when, and if, arriving to the "knowledge how", and because the communication is done mostly using academic language for targeting peers or colleagues. In such a sense, it seems that the (mis)concept under which the scientific knowledge is independent of moral considerations has helped obfuscating

E. Marone (✉)
Centro de Estudos do Mar da UFPR/IOI, Parana, Brazil
e-mail: edmarone@ufpr.br

L. Marone
Ecodes IADIZA—Consejo Nacional de Investigaciones Científicas y Técnicas—FCEN UNCuyo, Mendoza, Argentina
e-mail: lmarone@mendoza-conicet.gob.ar

G. Lollino et al. (eds.), *Engineering Geology for Society and Territory – Volume 7*,
DOI: 10.1007/978-3-319-09303-1_8, © Springer International Publishing Switzerland 2014

the intrinsic nature of why we do science and, mostly, how and to whom we communicate our discoveries.

The fact that our moral duties are not universal and unconditional constraints of universal practical reason does not forgive us for not following proper rules of duties (and being accounted for not doing that).

The point here is that we, geoscientists dealing with natural hazards, do not have such a deontological code and the need for it is painfully urgent. We do recognize that codes of ethics have been established for geosciences in different places by organizations, scientific societies or group of researchers (IUGS 2014; CONFEA 2014; CPRM 2014; CSPG 2014, etc.). However, the case of Natural Hazards is so specific that needs its own rules. It is about that we will present our case.

8.2 The Base-Lines

An ideal geoethics code regarding science of natural hazards has to define right actions in terms of duties and moral rules, focusing on the rightness of the acts and not just on what results from those acts. The crucial point is that right action is required, and that the goal of moral behaviour is that it is performed. The code cannot consist in just a list of duties and rules, but the list has to be put forth as instrumental for competent or even excellent conduct.

The main reason to construct the code has to be the rightness of the actions and, at any cost:

(a) *It cannot be designed to the self-protection of the geoscientists but to safeguard society and the environment.*

Obviously, the proper deontological code will help to protect the right actions of the geoscientists, but it is neither morally nor ethically acceptable to construct it "ad personam".

In spite of the discussions on duties or rules are often quite abstract and sometimes address personal morality, hence seeming ill suited to deal with broad and complicated questions in technical fields like those of geosciences, the deontological code for geoscientists in the case of natural hazards is a real need and cannot result in a sort of contradiction between words and deeds.

(b) *It has to be constructed collectively.*

The code has to be collectively constructed among practitioners and cannot be just a collection of "steps to be followed" but a real bridge between the moral consequences of being the privileged small part of the society who hold the scientific knowledge and the full society and the environment, for which we use to say we work for.

Natural hazards embrace many branches of natural and social sciences, not only those traditionally included under the wide umbrella of geosciences. If we consider the task objectively, any well-prepared individual scientist could embrace the challenge and emerge with a good code after some hard work. However, geosciences comprise so many areas, and the science behind natural hazards so many others, each one with their own peculiarities, that it will be wise to have a diverse group of scientists to enrich the range of the code and to avoid substantial bias. On the other hand, the bottom-up approach is always useful in any case where the rules being proposed will have to be followed by a great number of people. Additionally, as much wider the sense of parenthood of the rules one may have, the greater the acceptance of the rules will be among the community.

8.3 The Main Reasons

Many groups, at all levels, have already defined their deontological codes (being the Hippocratic Oath in medicine the best known). In geosciences, several countries, national or international associations, etc. have developed similar codes of conduct (which are available online as cited before: IUGS 2014; CONFEA 2014; CPRM 2014; CSPG 2014; etc.). These already existing codes of ethics are important for any geoscientist, even in the case of natural hazards, and they serve as a general umbrella. Such initiatives, however, do not deal usually with the specificities related with the natural hazards situations we would like to develop herein. Due to the particular characteristics of natural hazards, we would like to show that they could not be enough. It is time to come with a code for geoscientists dealing with natural hazards, as being proposed by scientists and associations like the IAPG-International Association for Promoting Geoethics (Matteucci et al. 2012).

The connection between duties and our practices in geosciences seems to be hard to grasp, and deontology can and should play an important role (Martínez-Frías et al. 2011), particularly if we consider that the consequences of not following best practices are often impossible to anticipate and very difficult to repair at the real world.

In addition, we have the "imperative of responsibility" (Jonas 1984), mandatory for creating new formulas of duty because traditional ethical theories could not be enough in the task of protecting the human species in light of the power of modern technology, capable of producing irreversible damage to the geosphere and hence to the environment and the human species. Thus, regarding natural hazards, we have to consider widening the concept to the same threats (earthquakes, storm surges, hurricanes, landslides, floods, etc.) when man induced or potentiated, because we owe something to the next generations.

8.4 Communication Issues

At the same time we have to radically change the way we communicate with the external (and real) world out the academy: The media, the stakeholders, the governments, NGOs and the most important part of society, the citizens. It is real that the pressure for publishing in the academy has been increasing in recent years and that an uneven weight is given to those works presented in scientific terms compared to those in plain words. One of the consequences of that is that we developed a new language (we could name it "scientifish"), which is particularly appropriate to communicate among peers, but mostly useless outside the academy.

Nevertheless, not all is a fault of the scientists. In spite of Albert Einstein worry about the "miracle that curiosity survives formal education" at the first half of the 20th Century, it seems that the second half served to deactivate the curiosity at any cost, dispreparing the youths to deal with classical concepts as uncertainty, critical thinking, reflexive behaviour or the intrinsic limitations of scientific knowledge, and so on.

The construction of a deontological code for geoscientists dealing with natural hazards will not be a fully operational tool if we do not alert and prepare society into a new educational model.

We scientists have to learn to better express our discoveries in plain words, but it will be necessary to review the education and social models today in place too. We have to offer the youths with intellectual tools needed to understand not only the "knowledge that" (as in the advertisement of a smartphone: it could help you to find the nearest restaurant), but also the "knowledge how" (why and how it will help you). It requires deep changes in educational systems, better preparing the citizens to deal with concepts of scientific prediction and forecasts, the errors associated with any scientific procedure and concepts like recurrence times, resilience, preparedness, scenarios, precision versus accuracy, precautionary principles, moral and ethics, although not with a patronizing approach.

(c) *It requires deep changes in educational systems.*

These changes are needed at all levels, training the youth at schools to understand some important scientific and still classical concepts and, at the university level, training the future geoscientists to follow some rules of conducts and to interact properly with the non-scientists when communicating their discoveries.

All practitioners should realize that the base lines (a) and (b) above are the piers of the arch that will support the deontological code; however, (c) is as important as the keystone of that arch.

8.5 Time-Lines in Natural Hazards

In addition, in construing a geoethics code of conduct, we have to consider the three periods in the time-line of an extreme event: before, during, and after. This peculiarity is the main reason that differentiate a general code of ethic for geoscientists and the one needed for their actions regarding natural hazards. Firstly, we have to ponder how to better delivery the scientific knowledge to the society, making readable for non-scientist people results of studies on the topics of early warning, the assessments of risk and vulnerability, scientific forecasts and their meaning, particularly regarding accuracy and precision. Only on that way we will help the society to have access to the best available (and readable) information to abet them to be organized against natural hazards, building preparedness and social and environmental resilience.

During the second and third moments, the most critical ones, geoscientists have to put their knowledge at the service of society and the environment, and not at that of a particular group, avoiding to be caught in the spiral of chaos that use to prevail during the course of and following extreme events. In these cases, many agencies (UNDRO, UNOCHA, Red Cross and Red Crescent, governments and NGOs, etc.) have well established rules of engagement and procedures which could be used by geoscientists on the way to their own deontological code.

8.6 The Question-Driven Approach

In Table 8.1 we include several questions that need to be answered in the way of building the geoscientists' deontological code related with natural hazards. As mentioned above, the proper answer could be different if the questions are formulated before, during or after a natural hazard, and therefore they have to be considered separately for every stage.

Although we have some tentative answers to these questions, we believe that responding them has to be a collective exercise, and we left it as a way of inciting the geoscientist community to go beyond.

Some general questions for preparing a deontological code have been already formulated and can be used to initiate the process (Mitcham 2005):

- What is the content of duty?
- Which rules direct us to morally right actions?
- Why must we follow exactly those duties and rules, and not others?
- What grounds them or validates them as moral requirements?

Table 8.1 Questions to be answered in order to produce a code of duties for geoscientists dealing with natural hazards

Questions to be answered for a code of geoethics on natural hazards	Before	During	After
How much knowledge we have on a particular natural hazards?	?	?	?
How certain we can forecast a given natural hazards?	?	?	?
How highest is our degree of precision and accuracy?	?	?	?
What has to be the approach in relation with authorities?	?	?	?
What has to be the approach in relation with the media?	?	?	?
What has to be the approach with agencies/NGOs/Media?	?	?	?
What has to be the approach in relation with the stakeholders?	?	?	?
What has to be the approach in relation with the citizens?	?	?	?
What new education approach needs to be constructed?	?	?	?
What is the best way to communicate or scientific knowledge?	?	?	?
Who have to be the target of this communication?	?	?	?
What are the core duties of the researchers?	?	?	?
Other questions	?	?	?

- What is the logic of these duties or rules?
- Can their claims on us be delayed or defeated?
- Can they make conflicting claims on us?

8.7 Conclusions to Move to the Code

The foundation of the code for geoscientists dealing with natural hazards has to be linked with the higher ethical concept that the Earth is a common heritage of humankind.

We propose a basis and a methodology to construct a deontological code, through a question-driven exercise. We suggest the code has to consider three periods (before, during and after a natural hazard) to properly address all the necessary issues.

The success of this construction, first, and during its use and application, after, will depend on following some guidelines:

(a) *It cannot be designed to the self-protection of the geoscientists but to safeguard society and the environment.*

(b) *It has to be constructed collectively.*

(c) *It requires deep changes in educational systems.*

Even if (a) will safeguard the right actions of geoscientists, it cannot be constructed properly just as a protection umbrella. This shelter will be a natural consequence, but it should not be the reason for doing a code.

It could be wise if some well-respected international organization is tempted to form an ad hoc team with the highest diversity of members to ensure (b), as a condition to produce a deontological code with the widest foundation and consensus.

Finally, the right actions of geoscientists dealing with natural hazards, even after the construction of the best of the codes, will still have barriers that cannot be overcome if the society does not realize (c). The educational systems should prepare the citizens to better understand certain concepts, recovering some of the classical philosophical ones and promoting the curiosity that leads to the "knowledge how and why", because the utilitarian behaviour we see nowadays is a result of the conformism that moves the society to consider the "knowledge that" as the end of the road to happiness.

References

CONFEA (2014) http://normativos.confea.org.br/downloads/anexo/1002-02.pdf. Accessed April 2014

CPRM (2014) http://www.cprm.gov.br/etica/Codeticaset2009.pdf. Accessed April 2014

CSPG (2014) http://www.consejogeologia.org.ar/index.php/legal/codigo-de-etica. Accessed April 2014

IUGS (2014) http://tg-ggp.org/professionalism-in-geoscience/codes-of-ethics/. Accessed April 2014

Jonas H (1984) The imperative of responsibility. In search of an ethics for the technological age. University of Chicago Press, Chicago

Martínez-Frías J, González JL, Rull Pérez F (2011) Geoethics and deontology: from fundamentals to applications in planetary protection. Episodes 34(4):257–262

Matteucci G, Gosso G, Peppoloni S, Piacente S, Wasowski J (2012) A Hippocratic Oath for geologists? Ann Geophys 55(3):365–369. doi:10.4401/ag-5650

Mitcham C (ed) (2005) The encyclopedia of science, technology, and ethics, vol 4, 1st edn. Gale Group Publishing, pp 1800. ISBN 9780028658315

Reducing the Gap Between Science, Policy and Practice: The Role of Civil Protection

9

Appiotti Federica, Eleonora Gioia, Gabriella Speranza, Maurizio Ferretti and Fausto Marincioni

Abstract

Climate change confronts human society with a variety of new challenges. Among the various hazards in prospective scenarios, there is an increasing pressure on soil due to climatic variations, effects and impacts. This paper offers a perspective in the debate about science-policy-practice interface, about climate change and emerging hazards, analyzing a 50 years' time-series of the Marche Region's (central Italy) meteorological data and perception data. The analysis of meteorological data highlights a significant increasing trend in temperature both annually and seasonally, and a significant decreasing trend in daily precipitation in all seasons except autumn, which are causing a reduction in soil water availability. Moreover an historical analysis on the number of the regional landslides shows an increasing trend in the number of shallow landslides in the last 30 years. The perception data where obtained through the analysis of approximately 800 questionnaires carried out in the Marche Region in May–June 2011 to residents, policy makers and emergency managers. The analysis of questionnaires revealed a high awareness of climate change and apprehension about it causes and possible effects. Moreover the analysis highlights a differentiated concern about the increase in the number of landslides and their socio-economic impacts among the three groups. This could be due to an ineffective information exchange among scientific community, public administration, emergency managers and citizens. Therefore, we suggest that the "brokerage" role of the Civil Protection Agency in the information exchange process should be increased.

Keywords

Climate change • Risk perception • Risk communication

A. Federica
Department of Design and Planning in Complex Environment, Università IUAV, S.Croce 1957, 30135 Venice, Italy
e-mail: federica.appiotti@gmail.com

E. Gioia · F. Marincioni (✉)
Department of Life and Environmental Sciences, Università Politecnica delle Marche, Ancona, Italy
e-mail: fausto.marincioni@gmail.com

E. Gioia
e-mail: e.gioia@univpm.it

G. Speranza · M. Ferretti
Dipartimento per le politiche integrate di Sicurezza e per la Protezione Civile, Regione Marche, Ancona, Italy
e-mail: maurizio.ferretti@regione.marche.it

9.1 Introduction

The climate is changing and the current climatological variation does not appear to have precedents in the planet's climatological record (IPCC 2007). The effects of these changes are actually visible and very often destructive. The amount and type of consequences related to the ongoing climate change are closely dependent on individual eco-systems and differ strongly on spatial scale, in relation to the geographic area considered and its physical and social characteristics. Thus, the identification of local climatic trends and its potential future climate impacts is becoming a key factor to better prepare Society for emerging hazards

G. Lollino et al. (eds.), *Engineering Geology for Society and Territory – Volume 7*,
DOI: 10.1007/978-3-319-09303-1_9, © Springer International Publishing Switzerland 2014

and risks. People's predisposition to adopt personal mitigation and adaptive strategies to face with climate change, and to support the government to do likewise, is not influenced by the real emerging risk but by people's perception of it (O'Connor 1999; McDaniel et al. 1997). The perception of current changes and emerging risks are strongly related with the transfer of information about the risk, for example direct and indirect communication, personal knowledge of cause and consequences, past experiences, and to the response mechanism of society (Kasperson et al. 1988).

Thus, in order to deal with climate change effects is essential to increase people knowledge about the problem, enhance communication and knowledge transfer, reduce the existing gap between scientists and practitioners and, above all, strengthen the information network among all the stakeholders involved (Adger 2003; EEA 2008; Kasperson 2011).

The aim of this study is to highlight the current climatological trends in the Marche region (central Italy), resident's perception of the phenomena and the in- formation exchange. This is made in order to identify if and where are the main gaps that slow down the risk reduction and climate adaptation processes.

Table 9.1 Climatic characteristic, in terms of mean temperature and precipitation, of the different regional geomorphologic zones

	Mean annual temperature (°C)	Mean annual precipitation (mm/year)
Coastal	15	600–800
Valley/low hill	14	850/1000
High hill/mountain	13	Over 1000

Murri and Fusari (1987)

9.2 Study Area and Methodology

The area selected for this study is the Marche Region, located on the East coast of central-east part of Italy, with a total surface area of 9,366 km^2 and 1,565,000 residents. The region is bordered on the east by the Adriatic Sea and on the west by the Apennines mountains.

Due to its geologic composition and climate variability the Marche Region is characterized by a high flood and landslide hydro-geologic hazards; 14.8 % of the territory is exposed to landslides (Grassi et al. 2009). Proceeding from East to the West is possible to subdivide the region in three different climatic and geomorphological zones: coastal, valley and low hill, mainly composed of clayey sand-stone sediments, and high hill/mountain, with a large part composed of massive limestone. From a climatic perspective each of these zones shows different temperature and precipitation trends (Table 9.1). Autumn is the most rainy season with the exception of the third zone (high hill and mountains) where the rainy season is winter (Grassi et al. 2009).

Local climate trends over the last 50-year period (1960–2009) were studied collecting daily temperature and precipitation data from 21 and 61 regional stations respectively distributed across the study area. The data were

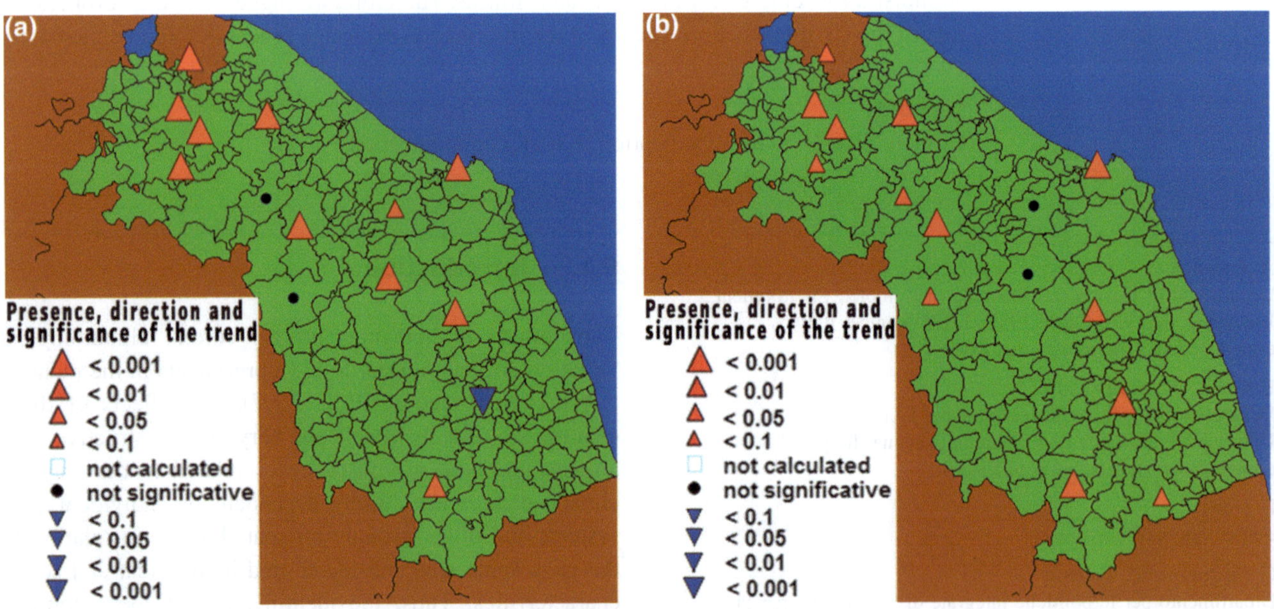

Fig. 9.1 Tendency of annual minimum temperature (**a**) and maximum temperature (**b**) over the period 1960–2009 calculated using Mann Kendall statistical test and mapped using GIS. *Triangle direction* and size are related to trend direction and significance respectively

selected on the base of general criteria used by the European Climate Assessment and Dataset (ECAD). Time series were investigated using the Mann-Kendall statistical test (Mann 1945; Kendall 1975) considering four different significance levels. The results of trend analysis were then mapped using GIS software.

Conversely, the perception of climate-change risk among the local residents, policy makers and emergency managers, was studied through interviews and questionnaire distributed using a purposive stratified sampling method (Stalling 2002). All collected data were analyzed using qualitative and quantitative methods.

9.3 Results and Discussion

The statistical analysis conducted on annual maximum and minimum temperature shows a statistically significant increasing trend in almost all the available stations (Fig. 9.1). The increasing trend of temperatures is statistically significant also at seasonal level. Seasonally, spring and summer show the higher level of confidence in positive trend definition.

The statistical analysis of precipitation data highlights a significant decreasing trend both annually and seasonally, with the exception of autumn (Fig. 9.2). As a matter of fact

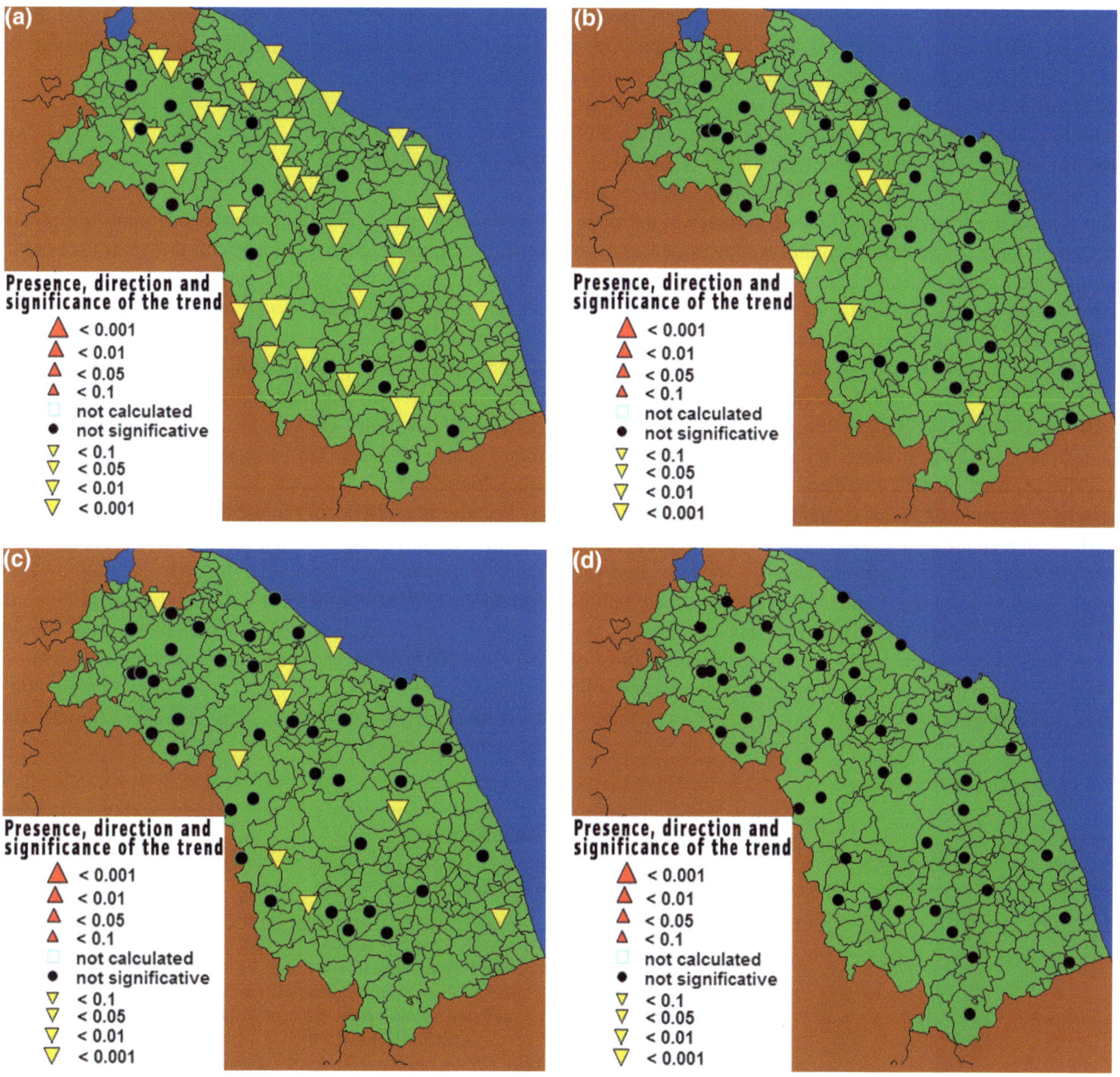

Fig. 9.2 Tendency of seasonal precipitation over the period 1960–2009. **a** Winter; **b** Spring; **c** Summer; **d** Autumn calculated using Mann-Kendall statistical test. Seasons have been divided considering J-F-M as winter season. *Triangle direction* and size are related to trend direction and significance respectively

the results show that no station in autumn has a significant trend either increasing or decreasing, while, winter seems to contribute mostly to the decrease in annual rainfall. Combined, the temperature and precipitation data analysis shows that the region is affected by significant temperature rise, especially in spring and summer, jointed with a significant reduction in rainfall amount except for autumn season.

The landslides statistical analysis is referred to the years 1980–2010 that represent the longest consistent period of landslides' reports available at the Regional Civil Protection. In the 1980–1990 decade almost all the landslides were triggered in winter, whereas in the next decades we noticed a progressive homogenization in the period of initiation (Fig. 9.3). Landslide distribution over the three studied decades appears to be correlated with rainfall data (Gioia et al. 2013). As a matter of fact, we log increasing number of landslides while recording decreasing amount of rainfall; possibly this is related to the increased number of the high intense precipitation phenomena (Appiotti et al. 2013).

Regarding the perception of climate change risks, causes and potential effects, the analysis of interviews and questionnaires shows little differences among the three identified groups and a widespread awareness about the topic. 91 % of interviewed citizens affirm to currently observe the local effects and consequences of climate change such as an increasing number of extreme weather and hydro-geologic events. 85 % of respondents believe that the climate changes will significantly affect their life, and the 12 % of them stated that this will be in terms of landslides hazard increase.

The analysis of policy makers' answers reveals a consistent evaluation of socio- economic consequences of climate change in the local territory related to extreme weather events, floods and landslides increase. 15 % stated that will likely be an increasing frequency of heavy precipitations and 12 % of landslides. However this concern is not translated into concrete activities, as the identification of potential future risk scenarios, long term planning or public awareness programs.

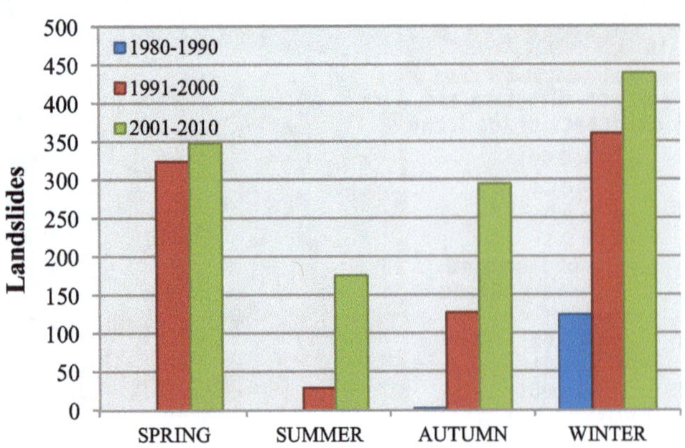

Fig. 9.3 Number of landslides occurred in the Marche region since 1980 up to 2010 during the different seasons

Emergency managers on their own revealed a good perception of what will be the future local impacts of climate change, especially in terms of increasing pressure on the response emergency system. This group feared landslides more than the other two groups; 14 % of responding emergency managers identified land- slides as one of the main local climate change effects. This is consistent with physical studies of exposure and vulnerability. In spite of this relatively accurate perception of climate change risks, confusion emerge about who should do what; who is responsible to manage and mitigate climate risks? Finally, the analysis of questionnaire data highlighted also a good exchange of information and knowledge between the population and the Regional Civil Protection system, and a rather weak exchange between the population and the scientific community.

9.4 Conclusions

Considering the multidimensional and multidisciplinary aspects of climate change, along with the great uncertainty connected with the global and local effects of this phenomenon, it is fundamental to develop a key figure (individuals or organization) capable to interface and bridge science, policy and practice. An effective communication channel between institutions and citizens could enhance a participative network capable to identify and pursue common goals in terms of climate risk prevention and adaptation.

In Italy such interfacing figure is partly performed by the Civil Protection System, which connects policy, science and practice in complex situations. Indeed, creating or reinforcing the authority of a local Civil Protection figure could enhance communication and climate-change risk knowledge exchange. Moreover, assuming such an information-brokerage role, the local Civil Protection System could increase citizens' trust and sway sounder climate risk reduction strategies.

References

Adger WN (2003) Social capital, collective action, and adaptation to climate change. Econ Geogr 79(4):237–404

Appiotti F, Krželj M, Russo A, Ferretti M, Bastianini M, Marincioni F (2012) A multidisciplinary study on the effects of climate change in the northern Adriatic Sea and the Marche region (central Italy). Regional environmental change. Published online on March 2013. doi:10.1007/s10113-013-0451-5

EEA (2008) Impacts of Europe's changing climate, 2008 indicator-based assessment EEA Report No 4/2008. European Environment Agency, Copenhagen, Denmark

Gioia E, Speranza G, Ferretti M, Marincioni F, Godt JW, Baum RL (2013) Rainfall induced shallow landslide forecasting in large areas: application of the trigrs model over a broad area of post-orogenic quaternary sediments. Geological Society of America, Abstracts with Programs, 45(7):775. https://gsa.confex.com/gsa/2013AM/webprogram/Paper226319.html

Grassi K, Zepponi A (2009) Da Kyoto a Copenhagen. Giunta Regione Marche, Servizio Ambiente e Paesaggio. URL:http://www.regione.marche.it/Portals/0/Ambiente/EnergiaKyoto/2009_abstract_relazione_assessore_kyoto_copenhagen.pdf Last accessed Oct 2011

IPCC Climate Change (2007) Impacts, adaptation and vulnerability. Contribution of working group II to the fourth assessment report of the intergovernmental panel on climate change. In: Parry KL, Canziani OF, Palutikof JP, van der Linden PJ, Hanson CE (eds) Cambridge University Press, Cambridge, UK

Kasperson RE, Renn O, Slovic P, Brown HS, Emel J, Goble R, Kasperson JS, Ratick S (1988) Social amplification of risk: a conceptual framework. Risk Analysis 8:177–187

Kasperson RE (ed) Charaterizing the science/practice gap. In: Kasperson RE, Berberian M (eds) Integrating science and policy: vulnerability and resilience in global environmental change. Washington DC, USA, pp 4–20

Kendall MG (1975) Rank correlation methods, 4th edn. Charles Griffin, London

Mann HB (1945) Non-parametric test against trend. Econometrica 13:245–259

McDaniels TL, Mark S, Kamlet G, Fischer W (1992) Risk perception and the value of safety. Risk Anal 12:495–503

Murri A, Fusari R (1987) Tipi di tempo in grande e condizioni meteorologiche sulle Marche. Centro di Ecologia e climatologia di Macerata

O'Connor RE, Bord RJ, Fisher A (1999) Risk perceptions, general environmental beliefs and willingness to address climate change. Risk Anal 19:461–471

Stallings RA (ed) (2002) Methods of disaster research. Xlibris Corporation, USA

http://www.regione.marche.it/Portals/0/Ambiente/EnergiaKyoto/2009_abstract_relazione_assessore_kyoto_copenhagen.pdf. Accessed September 2013

Geoethics and Communication 1. Scientists, Politicians, Journalists, Media Entertainers, Sociologists: Can Ethics Help to Establish Their Exact Role in Communicating Geoscience?

10

Stefano Solarino

Abstract

The purposes of communication range from simple transmission of information to motivation of people to act, to giving advises, suggesting actions or, in particular circumstances, even giving orders and directions. Often any of these targets needs intermediate steps, with the distinct or integrated intervention of more "experts" or communicators, but generally speaking these aims are independent and achieved separately. In either cases for each aspect there is an "expert" who has the knowledge, the experience and possibly the power to carry out the relative duty. Boundaries must be then designated to make sure that any speaker deals with his/her own expertise and does not interfere with the others'. Failures in doing this turn in a bad-incomplete-ineffective-inaccurate or even wrong communication. The recent grow of interest of the public towards Geoscience increased the need for information and, as a consequence, the episodes of bad communication. In this presentation I discuss whether ethics may represent a tool to fix boundaries between spokesmen and be used to avoid interferences. The subject is complicated by the nature of ethics, that applies to the moral and not to the rational behaviour; but there are other aspects that render the discussion intricate, including the different principles of ethics for every kind of communicator and the fact that the concept of ethics, and its principles, must be flexible (since it somewhat depends on the circumstances). The results of the analysis described here can be briefly summarized as follows. The fundamental principles of ethics would probably be enough for distinguishing an appropriate conduct and achieve a fair communication; unfortunately the personal expectations, the fear to appear unprepared, the desire for greater visibility and the unconscious will to be more helpful than the situation would require make communicators to fail some of their ethical obligations. Moreover, the emergency conditions, the difficulty of having available many speakers from different fields, the strict time for communication, the pressing demands of the media make much more difficult to strictly follow the ethical principles. In my view, the main deviation from ethics is when the speaker expresses personal opinions; this is quite normal for a scientist presenting his/her study within a workshop or a scientific meeting, but becomes dangerous when presented in public. It is even unfair for journalists or politicians. It becomes very harmful if uttered from media entertainers. And, above all, this failure affects one of the main obligations common to all ethics, a sort of root of moral behaviour, that is to ensure impartiality.

S. Solarino (✉)
Istituto Nazionale di Geofisica e Vulcanologia, Rome, Italy
e-mail: stefano.solarino@ingv.it

G. Lollino et al. (eds.), *Engineering Geology for Society and Territory – Volume 7*,
DOI: 10.1007/978-3-319-09303-1_10, © Springer International Publishing Switzerland 2014

Keywords
Geoethics • Communication • Media

10.1 Introduction

Geoscience is probably the branch of Science where communication has the most numerous actors. This is a direct consequence of the fact that the results in this field may have a significant impact on everyday life. A strong earthquake, a major volcanic eruption or a destructive tsunami are at the same time scientific, political, economic issues, and then of great interest for the media too.

As in other disciplines, the purposes of communication range from simple transmission of information to motivation of people to act, to giving advises, suggesting actions or, in particular circumstances, even giving orders and directions. Often any of these targets needs intermediate steps, with the distinct or integrated intervention of more "experts" or communicators, but generally speaking these aims are independent and achieved separately. In either cases for each aspect there is an "expert" who has the knowledge, the experience and possibly the power to carry out the relative duty. For example, scientists can make formation and contribute to the knowledge of citizens about natural phenomena; media entertainers or sociologists can increase motivations for particular actions or behaviours, politicians can set up rules or laws and describe their consequences on the society or simply argue about the reasoning behind their establishment. Boundaries must be then designated to make sure that any speaker deals with his/her own expertise and does not interfere with the others'. Failures in doing this turns in a bad-incomplete-ineffective-inaccurate or even wrong communication.

The recent grow of interest of the public towards Geoscience, caused by the great amount of natural phenomena of the last decade, increased the need for information and, as a consequence, the episodes of bad communication. Journalists, for example, often manipulate information mainly because they do not have the culture to understand the science behind the occurrences but also to increase the impact of the news; media entertainers select and insist on aspects that are more familiar to them or offer more chances to use their oratory experience, and the consequences are to place emphasis on secondary aspects of the natural phenomenon or to distract the audience from the main safety cunnings.

The worst situation comes up in emergency conditions, when most of the aims of communication should concur to the final and hopefully life-saving action/decision. In fact in such a situation there is the need to inform, motivate, suggest and sometime order to the public, all in a very short time and to a very broad audience. Recent episodes showed that in many cases there has been great interference among all actors of communication, with severe consequences for the society.

In this presentation I discuss whether ethics may represent a tool to fix boundaries between spokesmen and be used to avoid interferences. The subject is complicated by the nature of ethics, that applies to the moral and not to the rational behaviour; but there are other aspects that render the discussion intricate, including the different principles of ethics for every kind of communicator and the fact that the concept of ethics, and its principles, must be flexible (since it somewhat depends on the circumstances).

10.2 Are There Rules for a "Good" Communication ?

The "normal" flow of communication for natural hazard is intuitive, even from the point of view of ethics: scientists study the problem, find results and describe the phenomenon. They report to other colleagues in a scientific and technical language (in meetings, workshops or personal communications) and get comments to improve. When they believe to have achieved an interesting and sound result, they publish it in a scientific journal. The peer review ensures objectivity, quality and performance. After the publication scientists are allowed to present their findings to the public (intended as citizens, media, politicians). This way of conducting research and communication is ethical. By doing this the scholar ensures himself not to have any overlap with studies already existing or in progress and gets hints on possible errors and future developments of the research.

Some slight differences can apply. For example a scientist can describe to the public the aims of a research before or during its development, the methodology applied, the innovation of the approach; he or she even can, under certain conditions, dare to describe preliminary findings provided that the audience does not use them as they were definitive. In the last case the border between "ethical" and "non-ethical" is very thin.

In the last years, there have been numerous cases of scholars that presented their results to the public without any previous scientific acknowledgement. They were sometimes facilitated by the internet, the role of which is discussed in another paper of this volume (Solarino 2014). In most cases they are amateur scientists; their experiments were not conducted in a rigorous way and are sometimes non repeatable.

In the next lines I will describe a few examples without any will to judge the reliability of the scientific methods used to obtain the results. My aim is only to underline miscommunication and the possible problems that may arise from an incorrect dissemination and use of the information.

Earthquake prediction is one of the topic where incorrect or inappropriate communication is present most. In early 2011, rumours sparked that the town of Rome, Italy, was going to be hit by a strong earthquake the following May. This news had no scientific fundaments, although it was supposed to be a prophecy by an amateur seismologist, Raffaele Bendandi who was quite well known for having apparently predicted other earthquakes in Italy (Georgiadis. and Pescerelli Lagorio 2012). A description of the studies conducted by Bendandi is out of the aim of this paper, however it is acknowledged that his past prophecies were too generic to be considered real predictions and that the quake in Rome was not even mentioned in his papers. However the announcement was quickly disseminated, with the cooperation of newspapers and tv but especially through the internet. Although the scientific community did its best to convince people that the prophecy was fake, and that no prediction is feasible, there has been a lot of confusing information that lead some 18 % of employees not to go to work on the "disaster day" and many citizens to leave the town for that day. This is clearly a case of miscommunication where media played an important role and contributed to the false alarm by giving a similar credibility to the scientific community and to the prophecy of unknown origin (Nostro et al. 2012). It is also evident that media were reporting the news with the aim to attract a certain attention of the readers without really caring about the consequences of their actions.

Everybody probably remembers the date 21/12/2012. That day was supposed to be last the last day of our planet, according to a Mayan prediction. The news was present on all media worldwide since the end of summer 2012; many internet sites were born to explain how to survive to the disaster or how to build a properly resistant structure. There was no real agreement about what would have caused the end of the world: a meteorite impact according to some, a catastrophic earthquake or a giant tsunami according to some others. It is surprising to realize that there was no real agreement even on the source of the prophecy, but for the time being nobody seemed to realize that. Again, media played a fundamental role in spreading out the news, and the scientific community was not able to defeat this fake prophecy. This happened also because the "enemies" were much stronger, like the movie "2012" that, released worldwide a month before the catastrophe, contributed to the idea that we were close to the end. In principle, disaster movies are shown in advance or filmed with the assistance of scientists to avoid miscommunication and scientific blunders. But it didn't probably happen this time.

Finally, the internet is the perfect place for miscommunication. It is full of sites where amateurs present their predictions or innovative theories on the dynamic of the earth. For example, the site http://www.world-earthquakes.com/ has a prediction section that includes a list of the earthquakes that will possibly strike the world in the next 48 h and a more comprehensive index of the seismic events that will hit in the next years. The predictions are completely automated and vary frequently within a few minutes; they are based on a secret and very powerful algorithm winner of seven awards. Unfortunately, these awards are not listed and it is very difficult to understand why they have been granted, but it makes the surfer more confident toward the information processed by this algorithm. Again, I consider it a case of inappropriate (or at least incomplete) communication. A disclaimer on the top of the page warns users about the exact meaning of the data published on the web page and the declination of responsibilities about the misinterpretation of the predictions. One statement is especially noteworthy: "Our mathematical algorithms cannot predict exact time, location or magnitude of an earthquake. There is no knowledge of such thing to do that". The site is thus, from the point of view of ethics and in relation to the predictions, very adherent to moral principles. However since the predictions change frequently, and I personally witnessed that it happens, the users may get a very different forecast depending on when they browse the site. For example in less than an hour the prediction for Greece changed several times from "ATTENTION: Pay attention in this zone for the next 48 h" to "DANGEROUS RISK: High seismic activity may occur for the next 48 h and back". Again, this must be listed as a case of bad communication.

Browsing the internet will take the interested reader to find predictions for likely earthquakes based on elusive oarfish, toads and other animals. In most cases these intriguing theories are also described in other sites, probably as curiosity items or description of funny things. But giving them room in another site automatically give them more credibility, and causes miscommunication.

Although it was not difficult for the scientific community to recognize the real (and low) scientific impact of the studies described above, nevertheless the findings (or better the hypotheses) described by these pseudo-scientists were of great interest for the public and especially for the media, more inclined to get an exclusive news than the scientific, boring truth (Solarino 2009). In those cases journalists and media entertainers clearly ignored the standards of journalism. In fact in the situations above described we must of

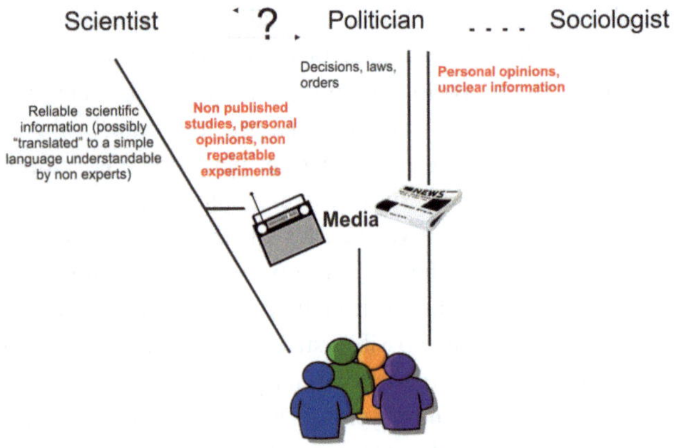

Fig. 10.1 A raw sketch of the relationships between scientists, stakeholders, sociologists, media and society. In black, normal characters the correct flow as based on the principle of ethics that should rule the interactions. In red, bold those that have sometimes been applied in cases of bad, non-ethical communications. Cliparts are public domain images from http://www.wpclipart.com/

course blame the scientists, but most of the charge is on media actors since, conversely to what happens in science, the category should comply with the "canon of journalism" which contain statements drafted by ad hoc organizations.

Scientists are responsible for producing results that shape public policy. For example, mapping areas of natural hazards has the consequence of rendering actions to mitigate the risk necessary. These kind of decisions, and their economical and social consequences, should be taken by the politicians and the local governors. However in some cases the scientists have been accused to be too optimistic or pessimistic in their view and to have lead the politicians to wrong or rushed decisions. When science and politics intersect, roles and boundaries are often muddied. But the competences cannot be confused.

Finally, the contribution of sociologists to the communication is an aside matter; their role is to study the social processes that influence the way humans think, feel, and behave and to help individuals to shape the social forces they face. However, there is no real interaction between scientists, politicians and sociologists except when the latter are asked to make analysis on what has failed in a communication process. To say it all, when it is too late.

In Fig. 10.1 I try to grossly summarize the relationships and interactions between the kind of actors (scientists, media people, politicians, sociologists and psychologists) in a "perfect" observance of ethics. Next chapter will be discussing the failures, which are also reported in the figure with bold-red characters.

10.3 Conclusions

A thorough discussion on effective and correct communication of Geoscience would require a very deep analysis of the society. In this article I simply summarized what every actor is expected to do and pointed out some weak points.

The results of this very naïve analysis can be briefly summarized as follows. To answer to the key question of this paper, the fundamental principles of ethics would probably be enough for distinguishing an appropriate conduct of each participant to the tough task of communication and, thus, to ensure a fair and reliable information; unfortunately the personal expectations, the fear to appear unprepared, the desire for greater visibility and the unconscious will to be more helpful than the situation would require make communicators to fail some of their ethical obligations. Moreover, the emergency conditions, the difficulty of having available many speakers from different fields, the strict time for communication, the pressing demands of the media make much more difficult to strictly follow the ethical principles.

In my view, the main deviation from ethics is when the speaker expresses personal opinions; this is quite normal for a scientist presenting his/her study within a workshop or a scientific meeting, but becomes dangerous when presented in public. It is even unfair for journalists or politicians. It becomes very harmful if uttered from media entertainers. And, above all, this failure affects one of the main obligations common to all ethics, a sort of root of moral behaviour, that is to ensure impartiality.

References

Georgiadis T, Pescerelli Lagorio P (2012) The communication of science as an ethical issue: the case of Raffaele Bendandi. Ann Geophys 55:3. doi: 10.4401/ag-5561

Nostro C, Amato A, Cultrera G, Margheriti L, Selvaggi G, Arcoraci L, Casarotti E, Di Stefano R, Cerrato S, and the May 11 Team (2012) Turning the rumor of the May 11, 2011, earthquake prediction in Rome, Italy, into an information day on earthquake hazard. Ann Geophys 55:3. doi: 10.4401/ag-5559

Solarino S (2009) Research and press: an (im)possible relationship? Geoitalia 2009, VII Forum italiano di Scienze della Terra, Rimini 9–11 September. Epitome 3:354 Geoitalia

Solarino S (2014) Geoethics and communication 2: Ethics, the notably absent from the internet. In Lollino G, Arattano M, Giardino M, Oliveira R, Peppoloni S (eds) Engineering Geology for Society and Territory, vol 7 Education, Professional Ethics and Public Recognition of Engineering Geology. doi: 10.1007/978-3-319-09303-1_7

11 Geoethical Aspects in the Natural Hazards Management

Giuseppe Di Capua and Silvia Peppoloni

Abstract

The defence against natural hazards involves many actors with different roles: geoscientists, decision makers, local authorities, mass media, citizens. A proper management of georisks requires that each role is well-defined and governed by shared operational protocols, especially during the emergency phase, so that overlapping and misunderstanding don't jeopardize population safety and economic activities. To achieve good results in this direction, it is necessary to undertake a careful evaluation of the limits and expectations of each component of society and the respect of legitimate aspirations and prerogatives. An effective defence system against natural hazards should be planned rationally and based on scientific data, in order to avoid alarmism among citizens, misleading sensationalism by media, careless decisions by politicians, as well as approximation in managing different phases of the risk cycle. Taking into consideration geoethical aspects related to natural hazards can be helpful to make geoscientists aware of their responsibilities towards society and to clarify the role they can play in the interaction with other actors, aiming at more efficacious actions for georisk mitigation.

Keywords

Geoethics • Natural hazards • Risks • Society • Responsibility

11.1 Introduction

With the increasing impact of human activities on the environment, it has become urgent to follow a respectful and pragmatic behaviour towards the geosphere. Humans are both an active part within Nature, as a factor which conditions Nature itself, and a passive element towards Nature, since they are exposed and forced to coexist with natural phenomena. This involves a risk for their life, productive activities and artistic and historical heritage.

Geological activity has evident repercussions on society (Wysession and Rowan 2013). So the adoption of ethical principles and standards is essential. Geoscientists are expected to put society's needs first in their activity, since they possess appropriate knowledge and skills and this implies moral obligations, especially considering practical consequences, that are issues dealt with by geoethics. In particular, geoethics consists of research and reflection on those values upon which to base appropriate behaviour and practices where human activities intersect the geosphere.

G. Di Capua (✉) · S. Peppoloni
Istituto Nazionale di Geofisica e Vulcanologia (INGV), via di Vigna Murata 605, 00143 Rome, Italy
e-mail: giuseppe.dicapua@ingv.it

S. Peppoloni
e-mail: silvia.peppoloni@ingv.it

G. Di Capua · S. Peppoloni
International Association for Promoting Geoethics (IAPG), via di Vigna Murata 605, 00143 Rome, Italy

G. Lollino et al. (eds.), *Engineering Geology for Society and Territory – Volume 7*,
DOI: 10.1007/978-3-319-09303-1_11,

Geoethics deals with the ethical, social and cultural implications of using Earth sciences for societal benefits. But above all, geoethics represents an opportunity for geoscientists to consider their activities in an ethical sense and also as a tool for increasing the awareness of society regarding problems related to geo-resources, geo-environment and geo-hazards.

Among its issues, geoethics deals with problems related to the management and mitigation of georisks and to the information provided to the public; fosters the proper and correct dissemination of the results of scientific studies; aims to improve the relationships between the scientific community, mass media, policy-makers and citizens.

In this perspective, the socio-cultural role played by geoscientists is fundamental. Through their activities, they can persuade people that the geo-environment constitutes a common heritage, which should be considered for its scientific, cultural and educational value, as well as a social capital. Moreover, they can teach to society that a defence against many natural phenomena, based on a rational approach, is possible (Peppoloni and Di Capua 2012).

11.2 The Role of Geoscientists

Since the beginning, geosciences have shown that natural phenomena are not metaphysical events, but rather normal events that demonstrate the vitality of planet Earth. They can be analyzed from a scientific point of view and their effects can be partly foreseen (Zoback et al. 2013).

We all know that the risk is defined as the symbolic product of hazard, vulnerability and exposure. It is quantified such as the loss produced on an element or group of elements at risk as a consequence of the occurrence of a given phenomenon of a given intensity. The hazard is the probability that a phenomenon of a given intensity occurs in a certain area in a given time interval. The vulnerability is the capability of an element to resist to a given phenomenon. And finally the exposure is the value of the elements at risk (in terms of human lives, or economic value or historical-artistic value) in a certain area.

These concepts have been introduced to analyze the impact of natural phenomena on humans and their effects are quantified using mathematical tools, for example, the probability calculus and evaluation of errors and uncertainties (Marzocchi et al. 2010; Albarello 2013).

Disasters always scared populations, the fear cannot be eliminated, but the proper dissemination of scientific knowledge and an adequate preparedness can help to transform the fear into respect for the natural processes that govern the geosphere. The scientific approach, based on quantitative assessments of risks and probabilities of occurrence (Marzocchi et al. 2010; Jordan 2013), helps to find strategies for mitigating their effects (Albarello 2013). It is also an effective way to limit the scope of irrationality and uncertainty.

The damage due to geo-hazards is not entirely avoidable, but can be reduced through correct land use and respect for natural processes, through prevention and mitigation efforts, and through effective information to the population.

And geoscientists possess the appropriate knowledge for bringing science closer to society (Allington and Fernandez-Fuentes 2013). They have an ethical responsibility towards both citizens and the scientific community to which they belong (Peppoloni and Di Capua 2012). It includes:

- making data and results of their studies public, easily accessible and user friendly;
- transferring advanced knowledge to industry and authorities;
- participating in educational campaigns for the population, paying attention to simplify concepts, without making them banal;
- assuring their ongoing professional training;
- collaborating in the training of the skills of technicians and professionals;
- conducting their studies, verifying the sources of information, the adherence of results to observations and the related uncertainties and errors;
- accepting a fair debate with hypotheses and theories that disagree, without being overconfident in their own results.

Without an ethical approach, geosciences run the risk of becoming a body of conventional knowledge, not oriented towards the common good and the human progress (Peppoloni and Di Capua 2012).

11.3 Georisks: Actors Involved

Who are the main actors involved in a society exposed to geo-risks and what are the weaknesses of their roles?

The actors are the same all over the world, but problems in their relationship could be different considering countries, or due to cultural and economic differences.

Decision-makers are responsible for the prevention and mitigation of natural hazards, but often have completely different skills than those required by their role. So, they often ignore the limits of a scientific study regarding the prediction of the hazard and the level of seriousness with which a warning could be issued to the public. In many cases, this results in a discharge of responsibility on the scientific community, which is asked to provide "truth" scenarios, while only probabilistic ones are possible, each of them with its own probability of occurrence (Albarello 2013).

Media has a crucial role in our culture, but the language of media is quite different from the language of scientists. In particular, on the one hand journalists generally have a poor qualification in Earth sciences. In addition, the media often use sentences given by scientists out of the context in which they were originally stated, and thus can transform the meaning of their words in a sensationalist manner.

Citizens have the legitimate right to demand actions in defence of their safety and to be appropriately informed about risks. In fact, in general they are poorly prepared in science (especially regarding aspects and concepts of "probability"; Albarello 2013) and thus have a low resilience towards a natural disaster.

Often geoscientists do not pay sufficient attention to the communication of risks and they don't succeed in making the population able to understand the scientific and technical language. Moreover, in some cases they haven't an open-minded attitude towards the discussion and comparison with new hypotheses of study. In order to gain social credibility, they should base their new theories on well-grounded observational data and propose them to the scientific community. Referring to natural hazards, new models should have a predictive level higher than those already in use or at least comparable to those in use.

11.4 Consequences of a Society Unprepared

The 2009 earthquake in central Italy and the convictions of scientists for negligence in seismic risk assessment (known as "The L'Aquila earthquake-case"; Hall 2011; Amato et al. 2013), certainly have produced a negative effect on people: the feeling that it is possible to do something in the short term to reduce the seismic risk. This is false from a scientific point of view, but in the medium and long term many initiatives can be adopted for the defence of our society against geo-hazards.

Moreover, there are also other possible consequences for future crises and problems in risks management, due to the severe epilogue of the L'Aquila earthquake-case:

(a) prolonged alarmism could have as an extreme consequence that a threat is not perceived as such all the time;
(b) if the precautionary threshold becomes too high, the costs of prevention become excessive and therefore there arises an attitude of resistance to risk mitigation policies.

The extreme effect will be the development of a culture of emergency rather than a culture of prevention to face geo-hazards, with an increase of victims and economic repercussions of disasters on future generations.

If society is not sufficiently involved in the scientific knowledge, we could have two negative consequences:

- the cultural and social marginalization of scientists, with a loss of sense of the role they can play in protecting society from natural hazards (and the "L'Aquila earthquake-case" is emblematic: society did not feel helped and protected by the scientific community and thus the judge condemned scientists for negligence);
- the tendency of people to lose confidence in science, to embrace preconceived ideas in a non critical way, ideas sometimes provided by the media, often incorrect.

As a consequence, without a society scientifically prepared, it is not possible to develop risk reduction strategies that are really effective and widespread.

It must be remembered that the assessment of the level of acceptable risk and the consequent action to be taken is up to decision-makers and not up to the scientists (Albarello 2013; Jordan 2013). Geoscientists have only the expertise necessary to provide accurate data and risk assessments based on deterministic (Wyss 2013) or probabilistic models (Marzocchi et al. 2010), but they can never be expected to predict an event with an accuracy of few hours or a few days. The technical and operational decisions should regard the civil protection authorities and not the scientists (Jordan 2013).

11.5 Conclusions: What Could be Done by Geoscientists

Geoscientists have a great responsibility towards society in the natural hazards management and their role is crucial to reduce the impact of natural phenomena and to improve the resilience of communities to future disasters (Peppoloni and Di Capua 2012; Jordan 2013; Wyss 2013).

The scientific and professional communities are still not fully aware of their social role, in a world where natural disasters are increasing and there is an urgent demand for an ethics of prevention and communication. Moreover, there aren't clear institutional procedures that facilitate a good interaction between geoscientists and society. Nevertheless geoscientists' action is fundamental in the balance among citizens, politicians, media, since they have the proper ability and skills to transfer knowledge to society for a more effective defence against geo-hazards (Allington and Fernandez-Fuentes 2013).

Moreover, the general public tend to have little confidence in science and therefore to assume a fatalistic attitude towards risks.

What could be done by the scientific community?

- Organizing a communication strategy before, during and after the emergency phase, strengthening the communication by the scientific community, without copying the style of traditional media and opening out to new communication tools (like social networks).
- Making research outcomes public, with explanatory information targeted to the population, distinguishing clearly scientific observations from working hypothesis.
- Learning to communicate science without trivializing it, showing also different hypotheses and theories.
- Making citizens involved in the process of construction of knowledge with educational campaigns that aim to develop an analytical and critical attitude, rather than transfer absolute certainties on scientific theories. The population should be informed also about the limits of the scientific methods used, so that it can better understand and share the decisions taken to deal with a natural hazard. Informing the population on natural risks should be prioritized for geoscientists, which is their ethical commitment.
- Increasing the synergy with government agencies and local administrations, through the development of operational protocols and the definition of an encoded stream of information from the scientific community to the authorities (Jordan 2013).

The improvement of own professional skills (Allington and Fernandez-Fuentes 2013) and the respect of research integrity values (Mayer and Steneck 2011) could help in this direction. Geoethics may be the foundation on which to establish a new and profitable relationship between science and society.

References

Albarello D (2013) Communicating uncertainty: managing the inherent probabilistic character of hazard estimates. Geophys Res Abs 15(EGU2013-13269). EGU General Assembly 2013

Allington R, Fernandez-Fuentes I (2013) The importance of professional skills alongside scientific and technical excellence to underpin ethical geoscience practice. Geophys Res Abs 15(EGU2013-9986-1). EGU General Assembly 2013

Amato A, Cocco M, Cultrera G, Galadini F, Margheriti L, Nostro C, Pantosti D (2013) The L'Aquila trial. Geophys Res Abs 15(EGU2013-12140). EGU General Assembly 2013

Hall SS (2011) Scientists on trial: at fault? Nature 477:264–269

Jordan TH (2013) The value, protocols, and scientific ethics of earthquake forecasting. Geophys Res Abs 15(EGU2013-12789). EGU General Assembly 2013

Mayer T, Steneck N (eds) (2011) Promoting research integrity in a global environment. Imperial College Press / World Scientific Publishing, Singapore, pp 317–328. ISBN 978-981-4340-97-7

Marzocchi W, Schorlemmer D, Wiemer S (eds) (2010) An earthquake forecast experiment in Italy. In: Annals of geophysics, vol 53. p 3

Peppoloni S, Di Capua G (eds) (2012) Geoethics and geological culture. In: Reflections from the geoitalia conference 2011. Annals of geophysics, vol 55. p 3

Wysession ME, Linda Rowan R (2013) Geoscience serving public policy. Geological society of America, Special Papers, 501, pp 165–187. doi: 10.1130/2013.2501(07)

Wyss M (2013) Lessons from the conviction of the L'Aquila seven: The standard probabilistic earthquake hazard and risk assessment is ineffective. Geophys Res Abs 15(EGU2013-3402). EGU General Assembly 2013

Zoback ML, Geist E, Pallister J, Hill DP, Young S, McCausland W (2013) Advances in natural hazard science and assessment, 1963–2013. Geological society of America, special papers, 501, pp 81–154. doi: 10.1130/2013.2501(05)

Need to Harness Geodiversity to Combat Disaster from Natural Hazards

12

Partha Sarathi Datta

Abstract

In most of high seismic zones parts of the world, limited knowledge on the heterogeneity in landform changes and geological formation causes threat from natural hazards disasters to the population. In absence of hazard resilient coherent long-term management policies, disasters result due to human failure to introduce proper preparedness measures, and tendency to manage vulnerability and risks by ad hoc and tactical approaches. To manage, prevent or mitigate hazards, efforts should be reoriented and consolidated to educate all stakeholders (including politicians, policy-makers and public) that they have to consider 'Geo-diversity' in their choices, and how they can overcome impacts by critical analysis of geo-ethical dilemmas, guided by the choices from the Geo-techniques based best obtainable detailed scientific information, and socio-economical considerations. Efforts by reliable, competent, capable and transparent institutions, to characterize the geo-regime spatiotemporal variability by integrated disciplinary approach, can help to harness the 'Geo-diversity' and ensure proper search and selection of the location of sites for planning the construction and maintenance of infrastructures. For better preparedness strategies, management plans and policies, discouraging corrupt and unscrupulous practices for private gain, can help to enhance efficiency and promote harmony by individuals trust, knowledge, transparency, respect of interests, share of responsibility and resources. This critical subject matter on how to harness the geo-diversity to combat hazard has been discussed here, through broader implications of an example of hazards and disasters in North India.

Keywords

Geoethics • Geo-diversity • Natural hazards • Disaster

12.1 Introduction

Globally, in many parts, availability of abundant natural resources has been always associated with increasing population, urbanization, and industrialization. Preserving extensive landscapes with such plentiful geo- and bio-diversity is well-recognized as an important integral part of sustained long run human existence and development process. For example, in the mountains regions run-of-river projects, to harness large amount of water from melting snow and glacier ice, plans for construction of the project facilities, access roads, and transmission systems/lines are made hoping minimum risks from deforestation, disturbance to hill slopes, soil erosion, and disruption of flora and fauna. However, if such areas are in the high seismic precarious zones, the environment and population remain at risk from natural hazards (Storms, earthquakes, landslides, floods, etc.), with catastrophic consequences. Over the last decades, throughout the world, there has been increase in

P.S. Datta (✉)
Independent Consultant on Water and Environment, 87, Kadambari Apartments, Sec-9, Plot-19, Rohini, New Delhi 110085, India
e-mail: psdatta1950@gmail.com

G. Lollino et al. (eds.), *Engineering Geology for Society and Territory – Volume 7*,
DOI: 10.1007/978-3-319-09303-1_12, © Springer International Publishing Switzerland 2014

natural hazards and their impacts on society, leading to disasters causing destruction of geo- and bio- diversity, vast network of levees, dams, navigation structures, and hydroelectric and nuclear power facilities. Although, it is vital for the public including scientists, disaster managers and policy makers to be aware of this for survival and safety, yet, generally, they have inadequate knowledge, awareness, and different perception of natural hazards, vulnerability and risks. Combating disasters from hazards is a great concern for planners and managers.

As an example, in the Uttarakhand State, India, located in the high Seismic Zone-5 (Fig. 12.1) with abundant natural resources; for developmental activities, in the recent years there has been a boom in reckless building of roads, bridges, multi-storey hotels, apartments, religious centers and hydroelectric projects. Over the next decades, in the Himalayas, the Government of India aims to build 292 dams to contribute to the current national hydropower capacity and the energy needs (Grumbine and Pandit 2013). A majority of which are run-of-the-river projects, each requiring construction of roads, bridges, tunnels, through the mountain side roads, townships and deforestation. This could lead to disappearance of species of bio-diversity, and may also put the population at high risk from natural hazards (earthquakes, landslides, floods, etc.). The Kedarnath Area in the State recently faced a flash flood on June 16–17, 2013, from which an unprecedented deluge destroyed roads and bridges along the banks of the rivers, inundated many towns and villages, and submerged some buildings under several feet of mud, smothering life. In this context, the need to harness geo-diversity to combat hazards has been discussed here, through some broader implications of hazards and disasters.

12.1.1 Geo-ethics and Geo-diversity

Ethical perception may vary from person to person, among societies and countries. Nevertheless, in order to achieve sustainable development, resources has to be managed taking into account the environmental, economic, social, geographical and political aspects, and govern the social actions and practical decisions of individual/group ethically. In this perspective, 'Geo-ethics' (Peppoloni and Di Capua 2012) is

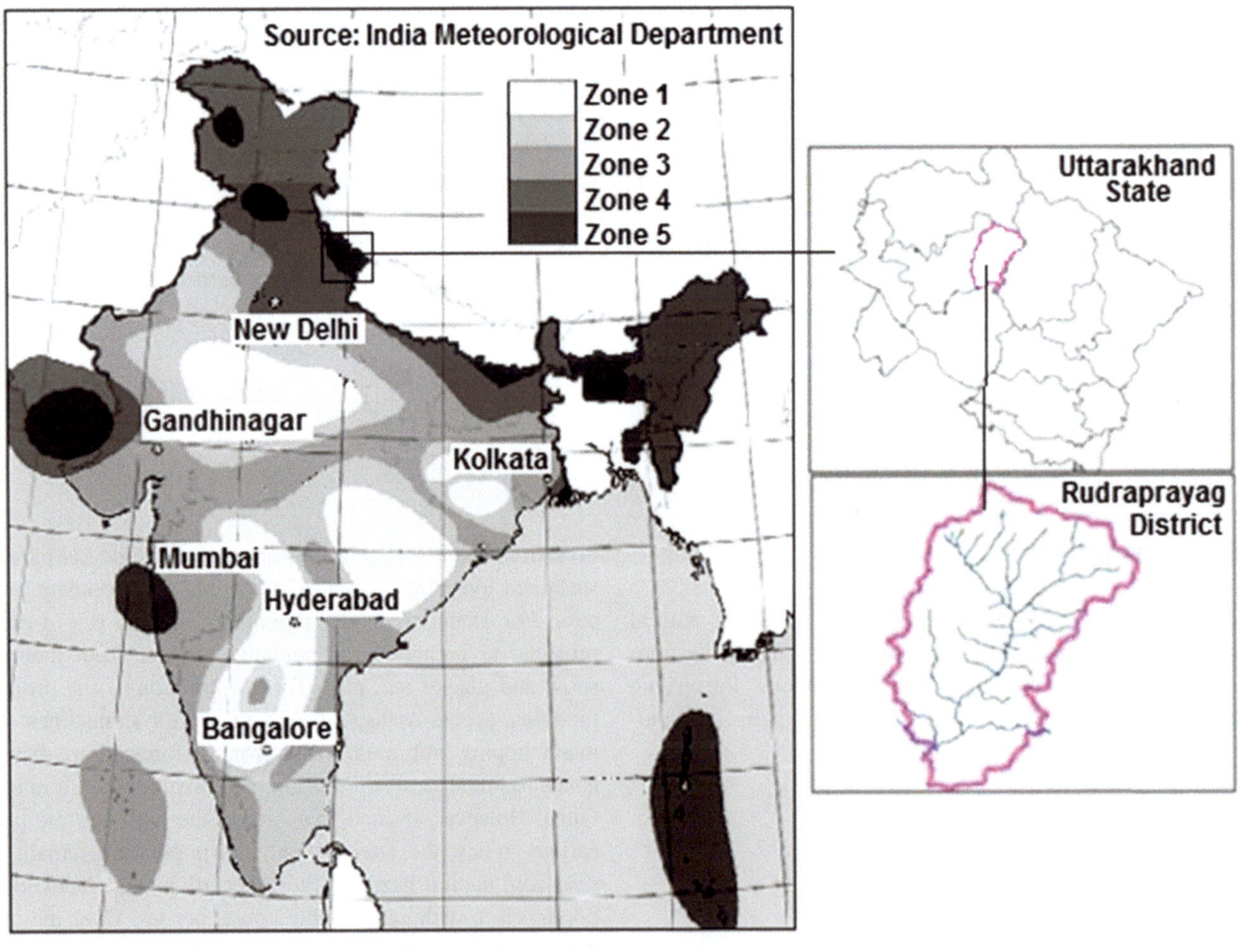

Fig. 12.1 Map of India showing seismic zones, and Uttarakhand state

essentially 'Geo-diversity' protection with moral values (honesty, trust, and treating others fairly and loyally). Geo-ethics (Peppoloni and Di Capua 2012) gives an opportunity for the geo-sciences to raise public awareness on the importance of abiotic parts of ecosystems. Moreover, geo-ethics can help to understand the great complementarity and interconnection between the two following systems:

(i) Natural System—Rocks, soils, sediments, minerals, water, landforms (folds, faults), fossils, biosphere (the flora and the fauna), and the processes (e.g., tectonics, sediment transport) that shape them throughout the geological time;
(ii) Human System—Link between people, their culture and the natural landscape.

12.2 Physiography and Rainfall Pattern in the Study Area

The Kedarnath town (3,546 m asl) is located in the R. Mandakini Valley (Fig. 12.2), Rudraprayag District, Uttarakhand, in a relatively young mountain range 'Central Himalayas' (30°44′6.7″N; 79°04′1″E) which has a fragile landslides prone geology (Uttarakhand Space Application Centre 2013). Several new buildings have sprung up on the flood plains of the Mandakini and Alakananda rivers. Two glaciers—Chaurabari and Companion occupy the upper reaches of the Valley. In the higher reaches, at the slopes of the Kedarnath Peak, from an altitude of ~6,000–3,800 m, The Chaurabari Glacier (30°44′50″N and 30°45′30″N, 79°1′16″E and 79°5′20″E; 3895 m asl, ~7 km length, ~11° slope facing south) covers ~38 km^2 area with ~5.9 km^2 ice cover. It is hypothesized that the original single glacier while receding split into two snouts—(i) the source of the Mandakini River at 3,865 m, and the (ii) at 3,835 m drains into the Chaurabari Lake—a moraine dammed Glacial Lake. Above 3,800 m asl altitudes glacial processes dominate, and between 3,800 and 2,800 m asl glaciofluvial processes are dominant (Dobhal et al. 2013).

Dobhal et al. (2013) reported that the rainfall data (2007–2012) from an automatic weather station near Chaurabari snout showed Indian Summer Monsoon rainfall of 1685, 1513, 734, 1662, 1348 and 1115 mm for the respective years. In the Kedarnath Area, 325 mm rainfall occurred on June 16–17 within 24 h. This isn't unprecedented (Comptroller and Auditor General of India 2013), and many times in the past, Uttarakhand recorded >400 mm single-day rainfall. In August and September 2012, Uttarkashi and Rudraprayag districts respectively experienced rainfall induced landslides, leading to devastation and other damages.

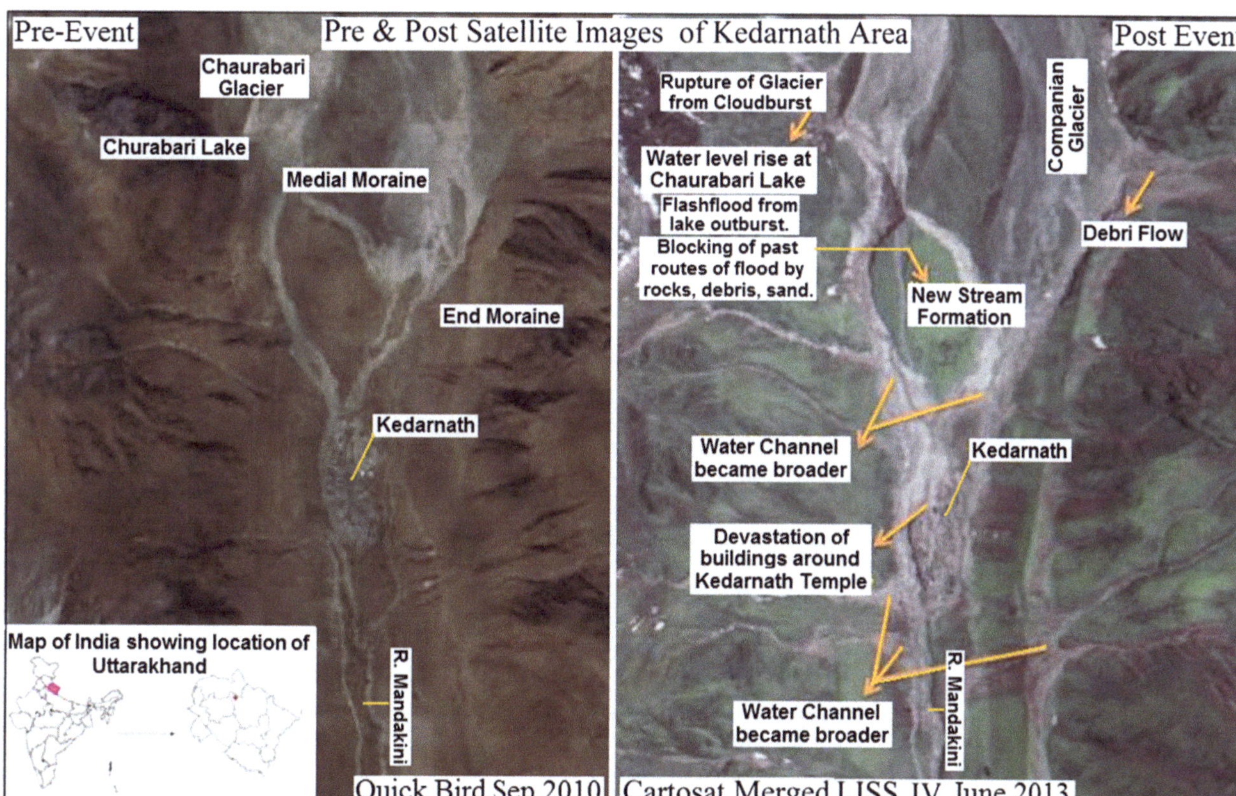

Fig. 12.2 Pre and post satellite images of Kedarnath, Uttarakhand, India

12.3 Causes of Disaster in the Study Area

Since, a natural disaster occurs as a result of multiple happenings together, it is a complex task to analyze the scientific sequences of a disaster and its impacts. The recent disaster is reportedly (Dobhal et al. 2013) due to a cloud-burst pressure of water induced rupture of a glacier, an entire mountainside collapse, sloughing off a side of a large section, entry of rocks and debris in the Chaurabari Lake increasing the lake water level, and its subsequent outburst causing flashflood laden with large amount of mud silt, huge boulders and debris (Fig. 12.2), creating several meters wide debris flow at many places. Since, the flow routes, which the flood waters used to take in the past, were blocked with sand and rocks, the debris and mud got deposited on the east bank of the R. Mandakini (Fig. 12.2), covered many houses under 10–12′ high rubbles, and the flood water washed away many houses. Evidently, despite many landslides in the past, the planners and policy makers paid little attention for early warning, and framed no proper plan to cope with disaster.

As per the Comptroller and Auditor General of India (2009, 2013) reports, the disaster resulted mainly due to indiscriminate, haphazard, unregulated, and unplanned infrastructure development along the rivers and on the flood plains; without proper geo-environmental study on the region's fragile geology and earthquake vulnerability; and social impact assessments at project or basin level; reckless sand-mining on the riverbeds; building of roads, bridges, and hydropower projects, each entailing underground tunnels that need blasting through the roads, townships and deforestation on landslide-prone ridges and steep slopes, and carelessly dumping of excavated muck and debris. Therefore, undoubtedly, hazards result in disasters due to human failure to introduce proper preparedness measures, and tendency to manage vulnerability and risks by ad hoc and tactical approaches with lack of ethical framework (Datta 2013). Absence of coherent long-term policies on control of communication by mass-media with half baked information, and generic opinions by all kinds of opinion-makers for a pervasive populism, also creates threat and panic in people for possible disaster (Datta 2013). This clearly suggests safety failures due to lack of proper harmony between the natural and human systems caused the disaster (Fig. 12.3).

12.4 Gaps in Disaster Management

Disaster management involves preparation activities prior to disaster; intervention activities following the hazard event; response activities during the disaster; recovery activities following a disaster; and prevention and mitigation activities to reduce the effects of disaster (Fig. 12.4). With reference to the Kedarnath Area recent disaster, the responsibility for the failures and gaps in disaster management rest on many institutions (Ghose 2013), including among others: the National and State Governance, Administration and Disaster management Authority and Unit, Indian Meteorological Department, Central Ministry of Water Resources, Ministry of Environment and Forests and its Expert Appraisal

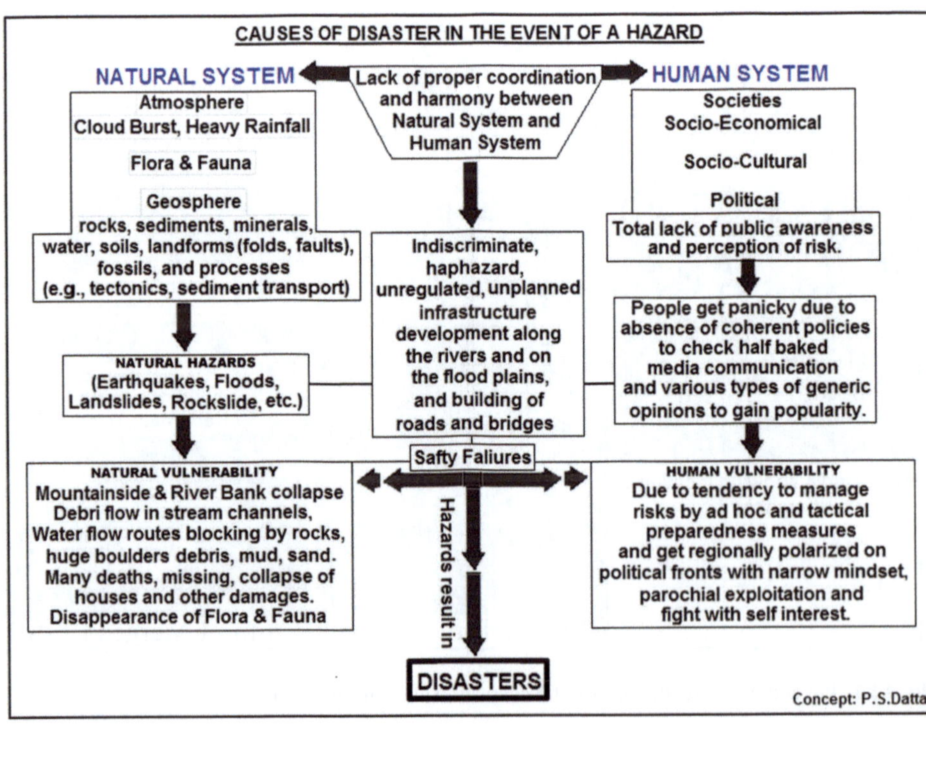

Fig. 12.3 Causes of disaster from natural hazard

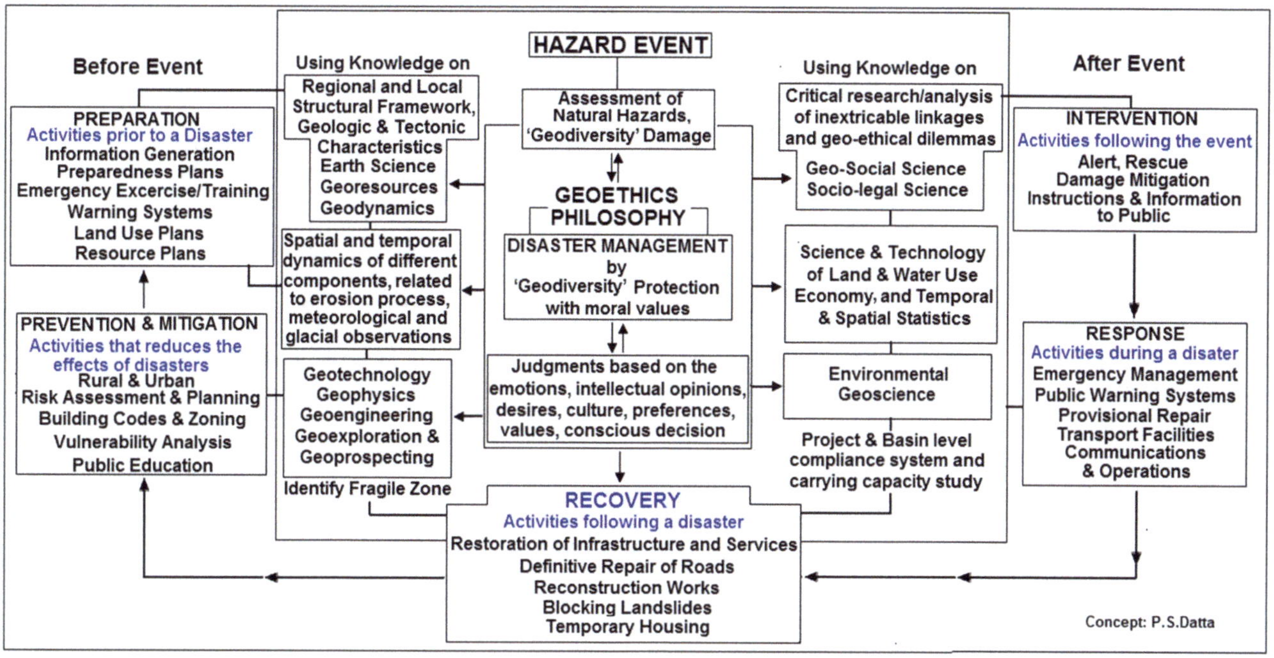

Fig. 12.4 Schematics of bridging the gaps in disaster management

Committee on River Valley Projects, Disaster Management, Central Water Commission, and National Ganga Basin Authority. Due to regional polarization from social and political angle with parochial interest, keeping 'only self interest' to the fore, development in the Himalayas has long been contentious, particularly from the point of declaring a stretch along the river as eco-sensitive zone, banning construction work along the river (Comptroller and Auditor General of India 2009, 2013).

12.4.1 Role of Geoscientists to Bridge the Management Gaps

Occurrence of sudden flood in the rivers near mountain sources suggests sudden morphing of the usual streams into torrents, totally altering the landscape, rocks, foothills, channels and the streams heights. The status of knowledge on aggregate situation of heterogeneity in the geological system and landform is limited at the level at which a hazard resilient management response is required (Datta 2013). In future, even normal rainfall in the area is likely to change the flooding, stream flow, and the bio-diversity. In order to manage, prevent or mitigate disasters from such hazards, the changes in the topography, geology, land contours, streams and rivers flows, flow paths, space of river beds and flood plains need to be remapped and analyzed for the clear land classifications, preliminary engineering designs, and hydrologic calculations. Monitoring programs and regulation zones need to be put in place by participatory decision-making. The geoscientists have significant role to create awareness in public (including the researchers, administrators, mining/industry operators, etc.) through the advancements in technical and scientific knowledge, for ensuring proper search and selection of the projects location, planning, construction, infrastructures maintenance, and preparedness strategies to manage activities. This would help in limiting development to safe places, by stopping deforestation, and improper construction of roads.

Through new communication strategies, and by interaction with politicians, media and citizens, the geoscientists can disseminate to them the geo-heritage available; and promote the ethical, cultural, social and educational values in study and practice of Earth Sciences. This would help in changing their mindsets; strengthen relationship among them, and framing policy guidelines on prevention, mitigation and recovery (Datta 2013). They should also educate the public about significance of proper use of the underground waters, hydrocarbons, natural gas, geothermal waters, etc.; and explain how to harness resources in integration with geo-information at different spatial and temporal scales, for risk assessment from hazards induced possible impacts on safe extraction and supply from groundwater collecting works, geo- and hydro- spheres contamination, etc. Both in mountainous areas and other areas; partnerships of geoscientists with states, communities, and private sector may be helpful to generate new resources and more efficient methods, especially in flood-risk management, hydropower generation, and port and harbor maintenance (Datta 2013).

12.5 Concluding Remarks

Since, most of the essential elements that form the building blocks of the humans' life are influenced by geological conditions; the geoscientists through wider experience and maturity should educate the public the necessity for the human beings to consider Geo -diversity when they prepare strategies to manage disaster from any hazard. An urgent need should be to monitor the possible causes of the disaster; to address issues emerged from previous event, to review ongoing and future projects, and to promptly disseminate the information on the carrying capacity of the region to the administration and the local people. Efforts should be reoriented and consolidated to educate all stakeholders how to promote harmony and enhance management efficiency by healthy relations based on individuals trust, knowledge, transparency, respect of interests and share of responsibility and resources. The politicians and policy-makers should be educated how to develop a problem-solving holistic integrated-disciplinary approach, planned in association with the public, by critical analysis of geo-ethical dilemmas, guided by the choices from the scientific information, and ethical and socio-economical considerations. Reliable efforts by transparent, competent, and capable institutions with non-involvement of corrupt and unscrupulous practices working hand-in-hand with greedy investors putting private gain before public welfare, can contribute to better preparedness strategies, management plans and policies to combat disasters.

References

Comptroller and Auditor General of India (2009) Report of performance audit of hydropower development through private sector participation. Chapter-5 [Environment Impact], pp 25–37. http://www.cag.gov.in/html/cag_reports/uttranchal/rep_2009/pa_chap5.pdf. Accessed 02 Sept 2013

Comptroller and Auditor General of India (2013) Report "Disaster Preparedness in India", April 23, 2013, PRS. http://saiindia.gov.in/english/home/Our_Products/Audit_Report/Government_Wise/union_audit/recent_reports/union_performance/2013/Civil/Report_5/Report_5.html. Accessed 02 Sept 2013. (

Dobhal DP, Gupta AK, Mehta M, Khandelwal DD (2013) Kedarnath disaster: facts and plausible causes. Curr Sci. 105(2):171–174

Datta Partha Sarathi (2013) Geoethics and geoscientists role to protect geodiversity for societal development. In: GEOITALIA, Session Geopolicy—L 2 Geoethics and Society: Geosciences Serving the Public. Accessed 16–18 Sep 2013

Ghose Joyita (2013) In: Disaster management in India PRS legislative research, institute for policy research studies, New Delhi , August 6, 2013. http://mla.prsindia.org/sites/default/files/policy_guide/Disaster%20Management%20in%20India.pdf. Accessed 02 Sept 2013

Grumbine Edward R, Pandit, MK (2013) Threats from India's Himalaya Dams. Science 339(6115):36–37. doi:10.1126/science.1227211 (Downloaded on 02.09. 2013)

Peppoloni S, Di Capua G (eds) (2012) Geoethics and geological culture. In: Reflections from the Geoitalia conference 2011. Annals of Geophysics, vol 55(3). ISSN 2037-416X

Uttarakhand Space Application Centre (2013) Dehradun, NRSA/ISRO, India Satellite based observation of devastation caused by torrential rainfall in Kedarnath Dham and Mandakini Valley of Rudraprayag District. Scientific Report USAC/SR/02/2013. Accessed 16 and 17th June 2013

What's the Seismic Risk Perception in Italy? 13

Massimo Crescimbene, Federica La Longa, Romano Camassi, Nicola Alessandro Pino and Laura Peruzza

Abstract

Risk perception is a fundamental element in the definition and the adoption of preventive counter-measures. In order to develop effective information and risk communication strategies, the perception of risks and the influencing factors should be known. This paper presents preliminary results of the first survey on seismic risk perception in Italy. The research design combines a psychometric and a cultural theoretical approach. More than 5,000 on-line questionnaires have been compiled from January 23rd till July 25th, 2013. The data collected show that seismic risk perception in Italy is strongly underestimated; 86 out of 100 Italian citizens, living in the most dangerous zone (namely Zone 1), do not have a correct perception of seismic hazards. From these observations we deem that extremely urgent measures are to be taken in Italy to find effective ways to communicate seismic risk.

Keywords

Risk perception • Seismic hazard • Hazard communication • Seismic risk

13.1 Risk Perception

Risk perception research in the domain of technical risks has shown that peoples' perception of risk is subject to many influencing cognitive, personal, situational and contextual factors (Sjöberg 2000a). Because of its complexity, it is very difficult to deduce general statements or a general theory of risk perception (Wachinger and Renn 2010). Nevertheless, knowledge about the risk perception of persons living in risk prone areas is relevant whenever risk management strategies have to be developed or applied.

A number of approaches and concepts have been applied in risk researches to study risk perception: the approach known as psychometric paradigm (Fischhoff et al. 1978a; Slovic 1987, 1992) and similar concepts (Lindell 1994), the cultural theory of risk perception (Douglas and Wildavsky 1983; Thompson et al. 1990; Dake 1991, 1992), trust-oriented concepts (Slovic 1993; Siegrist 2000a; Siegrist and Cvetkovich 2000b, Siegrist et al. 2000c), the mental models approach (Lave and Lave 1991), concepts to include associations and affect (Peters and Slovic 1996), demographic variables (Savage 1993; Barke et al. 1997), gender (Gustafson 1998; Greenberg and Schneider 1995) and others. Risk perception has been investigated with various methods on different risk levels (individual personal risk or risk for the general society), using various risk measures

This study has benefited from funding provided by the Italian Presidenza del Consiglio dei Ministri—Dipartimento di Protezione Civile (DPC). This paper does not necessarily represent DPC official opinion and policies.

M. Crescimbene (✉) · F. La Longa
INGV, Via Di Vigna Murata 605, 00143 Rome, Italy
e-mail: massimo.crescimbene@ingv.it

R. Camassi
INGV, Via Donato Creti 12, 40128 Bologna, Italy

N.A. Pino
INGV-Osservatorio Vesuviano, Via Diocleziano 238, 80124 Naples, Italy

L. Peruzza
OGS, Borgo Grotta Gigante 42/C, 34010 Sgonico, Trieste, Italy

G. Lollino et al. (eds.), *Engineering Geology for Society and Territory – Volume 7*,
DOI: 10.1007/978-3-319-09303-1_13,

(magnitude of risk, overall risk rating, probability of an event, estimated fatalities per year) and several risk dimensions (probability of damage, personal death/injury, property loss, interference with work, social disruption; see Rohrmann 1999; Sjöberg 2000b; Lindell and Perry 2000). For the investigation of risk perception from windstorm, flood and earthquake the psychometric approach (Fischhoff et al. 1978b; Slovic 1987; Slovic 1992) and theoretical concepts of cultural theory (Thompson et al. 1990; Dake 1991) were applied to reveal the underlying cognitive structure of risk and the influence of social values and worldviews.

Due to the complexity of risk perception, in order to build a seismic risk perception test we needed an agile and flexible method, able to investigate complex variables and to return measurable values. For the above-mentioned reasons, we opted for the semantic differential method (Osgood et al. 1957), which is conceptually similar to the *semantic images*[1] indicated as a very promising approach for risk perception studies by Wachinger and Renn (2010). Furthermore this method has already been used in similar research on risk perception, (e.g. Plapp and Werner 2006).

In conclusion our method can be framed as psychometric and cultural theory of risk approach, with in addition some components to obtain a better overview on possible influences on risk perception: causes attributed to disasters, images of and associations on nature and environment (Szalay and Deese 1978), several personal and demographic characteristics, and experience from past events.

13.2 The Questionnaire

Within the project S2-2012—*Constraining Observations into Seismic Hazard* financed by the Department of Civil Protection (http://sites.google.com/site/ingvdpc2012progettos2/home), an on-line questionnaire on the perception of seismic risk was prepared and tested. The questionnaire was constructed by the method of semantic differential, based on bipolar scales of opposing adjectives or terms (Osgood et al. 1957). The subjects had to indicate, on a scale of 7, "which of the two poles" the object of investigation was closer to in their opinion (Plapp and Werner 2006; Crescimbene 2008; Zacchi and Crescimbene 2010; Crescimbene et al. 2013).

The questionnaire was constructed on the factors that determine the seismic risk: hazard, exposure, and vulnerability. Other factors related to Institutions and People and to Earthquake perception in general were also considered.

The whole questionnaire consists of an informative part and seven sections respectively dedicated to:

1. Hazard
2. Vulnerability (home and workplace)
3. Exposure (with reference to territory perception)
4. Perception of institutions and people
5. Earthquake perception
6. Risk information and their sources
7. Comparison between earthquake and other natural hazard.

Assigned to each factor is a set of scales to which it is possible to assign a score from 1 to 7, Likert scale (Likert 1932).

The questionnaire makes it possible to obtain a perception score for each factor: Hazard, Exposure, Vulnerability, Institutions and People Perception, Earthquake perception. Considering all these factors the global risk perception total score can be derived. The complete questionnaire is accessible at: http://www.terremototest.it.

13.3 The Survey

The seismic risk perception survey began on 22 January 2013 and it is still underway. Compilation availability and accessibility has been spread through the social network, the web pages of regional, provincial, and municipal websites and on local online newspapers. The diffusion of the questionnaire was deliberately conducted through general interest locations, avoiding the specialized or official sites of the sector (Department of Civil Protection, INGV, OGS, universities, etc.) in order to limit the bias of educated/oriented samples.

The survey includes all the Italian regions; on 25 July 2013, 5,585 tests had been compiled, subdivided in Administrative units (Region) and seismic zones (hereinafter described) as shown in Table 13.1. Veneto Region represents over 1/3 of the sample, as a local newspaper in the Verona area advertised the initiative.

13.4 Data Processing

More than 5,000 questionnaires were compiled in few months, without any specific supporting initiative by the press or the mass-media. The examined sample consists of 5,585 people with the following characteristics.

The graph (Fig. 13.1) shows the differences in risk perception based on the characteristics of the sample: Education, Family, Age, Gender.

[1] Research on risk perception has identified a range of perception patterns that relate to key characteristics of the risk itself or the context in which the risk is taken. These patterns are called semantic risk images. The semantic images allow individuals to classify various risks on the basis of a few salient characteristics. Reducing complexity by creating classes of similar phenomena is certainly a major strategy for coping with information overload and uncertainty (for further details see Renn 2008).

Table 13.1 Distribution of the sample by regions and hazard zones

Regions	Seismic zones				
	Zone1	Zone2	Zone3	Zone4	Total
Abruzzo	58	83	50	0	191
Basilicata	41	50	5	0	96
Calabria	97	54	0	0	151
Campania	62	255	24	0	341
Emilia-Romagna	0	93	361	6	460
Friuli-Venezia Giulia	2	36	15	0	53
Lazio	16	215	34	0	265
Liguria	0	0	57	5	62
Lombardia	0	9	88	246	343
Marche	3	103	4	0	110
Molise	19	22	1	0	42
Piemonte	0	0	57	92	149
Puglia	0	47	60	39	146
Sardegna	0	0	0	23	23
Sicilia	35	178	2	5	220
Toscana	0	128	469	10	607
Trentino-Alto Adige	0	0	17	32	49
Umbria	1	51	3	0	55
Valle d'Aosta	0	0	1	4	5
Veneto	0	75	1,906	236	2,217
Total	334	1,399	3,154	698	5,585

Table 13.2 Characteristics of the sample

Gender (%)	Females	36
	Males	64
Age (%)	Young	15
	Adults (30–44 years)	40
	Adults (45–59 years)	35
	Seniors	10
Family (%)	With children	55
	Without children	45
Instruction level (%)	Primary or secondary school	8
	Graduate	46
	Degree	46

Looking at the sample of seismic risk perception found that some indications from previous studies on risk perception, in general, have been confirmed:

- Women tend to have a better perception of risk than men;
- Being parents (families with children) increases the perception of risk;
- A high level of education improves the perception of risk;
- Adults between 45–59 years old have a higher perception of risk than other age groups.

Furthermore, data are being analyzed comparing hazard perception scores with the so-called "hazard by law", i.e. the seismic hazard assessment assigned to a particular territory by experts (Gruppo di Lavoro 2004; Stucchi et al. 2011); we resorted to a simplistic subdivision of Italian municipalities in 4 seismic zones, as given by regional laws as they were in 2012 (http://www.protezionecivile.gov.it/resources/cms/documents/A3_class2012_03prov_.pdf).

To compare hazard perception with "hazard by law" we assumed for people living in seismic zone 1 (high hazard) scores between 6 and 7; in seismic zone 2 scores between 5 and 6 and so on. From this it follows that in seismic zone 1 is not possible to have overestimated values. The assigned scores are shown in Table 13.3.

The histogram in Fig. 13.2 shows that, in seismic zone 1, only 14 % of the sample has a good perception (green column) of the seismic hazard, while 86 % of surveyed people underestimate the hazard (39 % underestimated by 1 zone/class, 35 % underestimated by 2 zones/classes, 11 % underestimated by 3 zones/classes).

Figure 13.3 shows in detail the description for seismic zones obtained by hazard perception scales (HP). The colored lines represent the trend of scales in each seismic zone (all the samples considered). The zones appear well distinct for almost all the terms, and the descriptions well represent each seismic zone. Referred to our test interpretation (see Table 13.2), Zones 1 and 2 appear to be strongly undervalued with an average score of 5.12 in Zone 1 (whereas it should be included in the range from 6 to 7) and of 4.53 in Zone 2 (against 5–6 as expected). Hazard perception in Zone 3 with an average score of 4.15 is placed on the limit of appropriate range between 4 and 5. Only hazard perception in Zone 4 with an average score of 3.49 is a good match between hazard by law (HbL) and hazard perception (HP). It's worth mentioning that the scale "predictable-unpredictable" gets inversion in scores (the most hazardous the country is, the most predictable earthquakes are); it demonstrates that predictability of earthquakes is perceived independently form any other cognitive and expertise factors. This result is consistent with the outcome of debates recently underway in public opinion with respect to the possibility of predicting earthquakes.

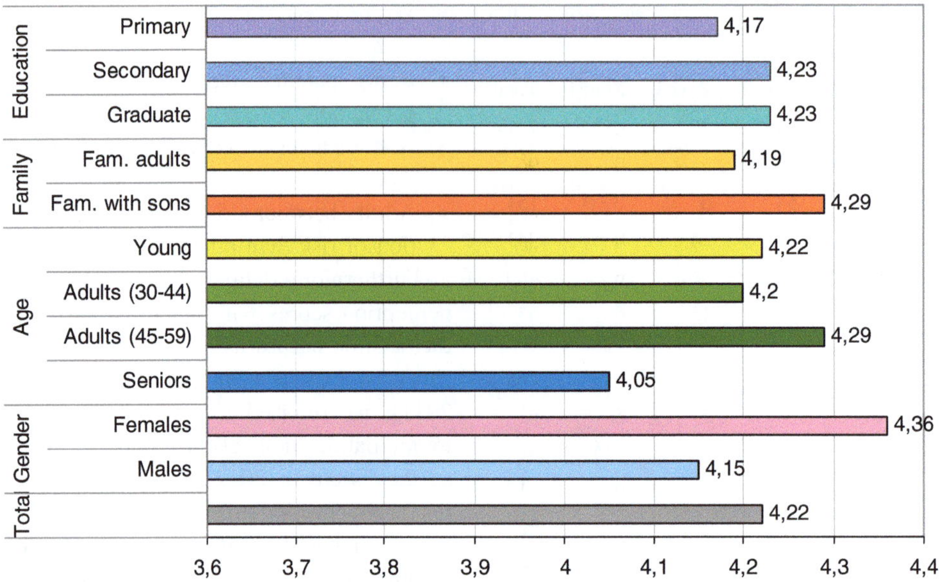

Fig. 13.1 Differences in Hazard perception for characteristics of the sample

Table 13.3 Interpretation of Hazard Perception scores (*HP*) respect to Hazard by Law (*HbL*)

Seismic zone	Semantic differential scores (7 points Likert's Scale)					
	1–2	2–3	3–4	4–5	5–6	6–7
Zone 1	−3 underestimated of 3 points and over			−2 underestimated of 2 points	−1underestimated of 1 point	0 good fitting
Zone 2	−3 underestimated of 3 points and over		−2 underestimated of 2 points	−1 underestimated of 1 point	0 good fitting	+1 overestimated of 1 point
Zone 3	−3 underestimate of 3 points	−2 underestimated of 2 points	−1 underestimated of 1 point	0 good fitting	+1 overestimated of 1 point	+2 overestimated of 2 points
Zone 4	−2 underestimated of 2 points	−1 underestimated of 1 point	0 good fitting	+1 overestimated of 1 point	+2 overestimated of 2 points	+3 overestimated of 3 points

Fig. 13.2 Frequency distribution of differences in Hazard perception for seismic zone

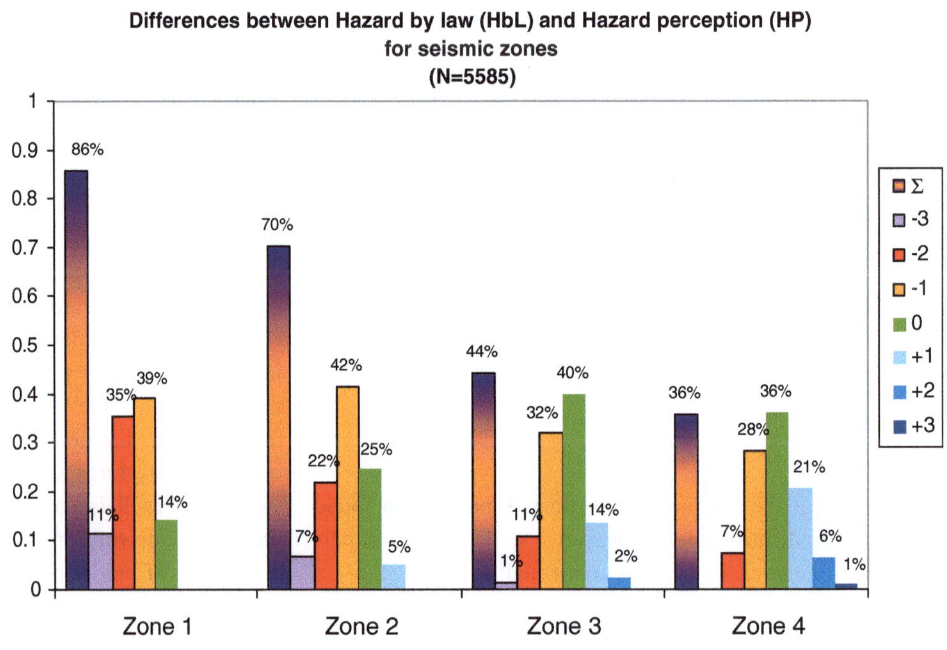

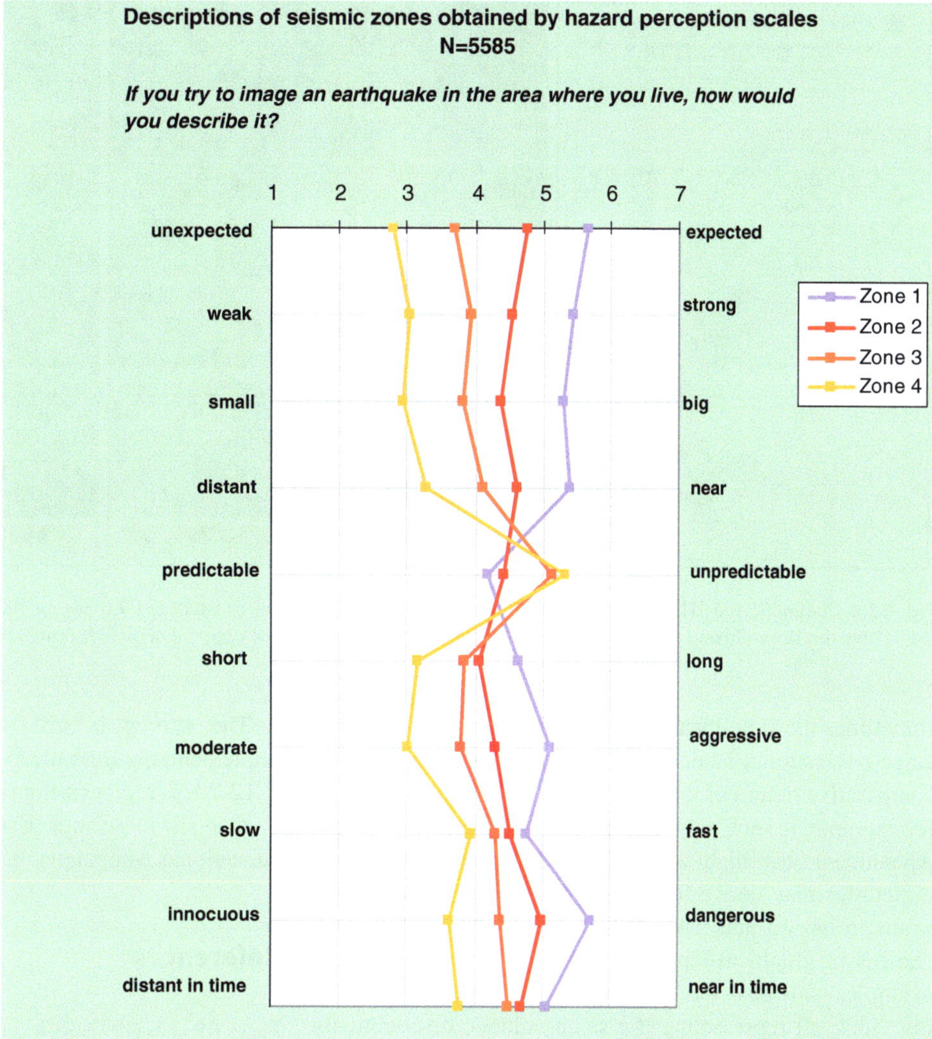

Fig. 13.3 Descriptions for seismic zone obtained by hazard perception scales (*HP*)

13.5 Conclusions

The international scientific community has many opportunities and tools to disseminate updated knowledge about seismic hazard and seismic risk. Conversely, the communication of these subjects to the public is not common, and surely not easy; in Italy it has been pursued by the classification of municipalities in seismic zones since the beginnings of the 20th century, and recently by the adoption of a national seismic hazard map. For historical and administrative reasons the two pieces of information are not fully coherent, causing confusion in the citizens.

Our study developed and spread a questionnaire on seismic risk perception in Italy; we performed a preliminary analysis on data collected during the first 6 months of the campaign. A strong underestimation of the seismic phenomenon in the most dangerous areas emerges: the problem of incorrect perception can not be simply attributed to a lack of information, as 61 % of the sample (N = 5,585) said to be "somewhat" (45 %) and "very" (16 %) informed about the earthquake, and only 8 % defined it's "not at all" informed. Furthermore, people declare to receive information about earthquakes by reliable sources: 30 % from the Department of Civil Protection, 15 % by Regions, Provinces and Municipalities, 13 % from research institutes, universities and schools; the authoritative sources are balanced by traditional media (television, newspapers, internet, books) that represent about 38 % of the information budget; unfortunately, the news are mainly driven by catastrophes and crimes, often representing negative information campaigns.

From the answers, we tried to examine if the main problems of seismic under-evaluation may be related to the matter of the communication or to pitfalls in the communication process. Some concepts such as the probabilistic nature of the seismic hazard analyses, stationary earthquake rates, random uncertainties in the earthquake waves propagation are definitely out of the common experience, and they have to be simplified in the messages for the public;

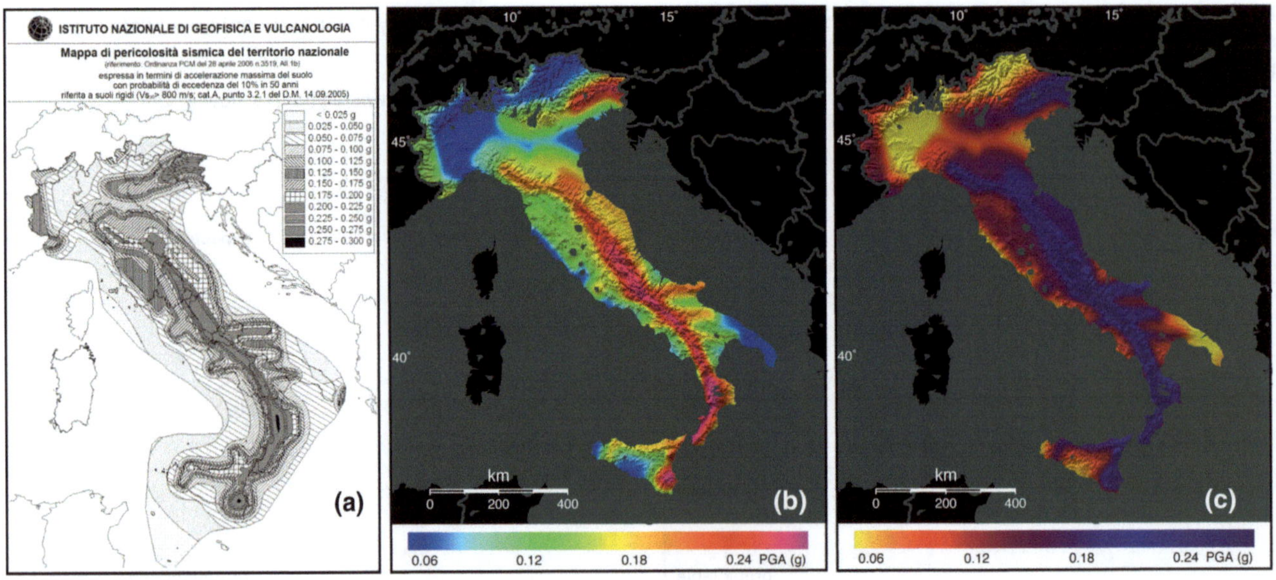

Fig. 13.4 Italian Seismic Hazard Map (MPS04, Stucchi et al. 2011): it displays the Peak Ground Acceleration (*PGA*) that in 50 years will not be exceeded 9 times on 10. Graphics have a strong impact on risk perception. On the left frame, the map published by the law 3519/2006

sometimes the simplification reduces hazard values to crude adjectives (strong, moderate, low, etc.), thus neglecting the comparative nature of their usage; the low hazard in Milan, for example, is such only if compared to Cosenza (Southern Apennines), but high with respect to Paris. Some other difficulties may derive from linguistic tricks (in Italian, for example hazard and risk are widely used like synonyms), choices in graphical representations (the color palette, for example, falsifies the perception, see the frames in Fig. 13.4, all representing the same values), or commonly accepted rules in ranking (the first is usually the best: with a reversal meaning seismic Zone 1 is assigned to most dangerous areas, but it should be perceived safer than Zone 4). All these considerations suggest that some improvement in the communication processes is needed, by spreading simple and direct messages, together with a larger effort in spreading the basic culture about earthquakes and the associated risk.

In the last decade, some coordinated educational campaigns have been undertaken under the umbrella of the Civil Defense Department (DPC). EDURISK, for example, is a school-oriented group of initiatives; since 2002 it trained about 3,000 teachers that reached about 60,000 students, in seismic and "less" seismic areas of the country. More recently, IONONRISCHIO, an awareness campaign for the reduction of seismic risk promoted by DPC, ANPAS (The Public Assistance Italian Association), INGV, OGS, ReLUIS (The Laboratories University Network of Seismic Engineering), which takes place every year in the squares of Italian municipalities; in the 2013 edition over 200 Italian cities have been reached.

The survey is still active and we are planning its implementation into the Framework Agreement DPC-INGV 2012-2021, reprocessing periodically the data, also to detect if changes in seismic risk perception can be assigned to educational campaigns, or seismic sequences.

References

Barke RP et al (1997) Risk perceptions of men and women scientists. Soc Sci Q 78:167–176

Crescimbene M (2008) Un test sulla percezione del rischio vulcanico. http://hdl.handle.net/2122/4678

Crescimbene M et al (2013) Seismic risk perception test. EGU 2013, Poster Session. http://hdl.handle.net/2122/8704

Dake K (1991) Orienting dispositions in the perception of risk: an analysis of contemporary worldviews and cultural biases. J Cross Cult Psychol 22:61–82

Dake K (1992) Myths of nature: cultural and social construction of risk. J Soc Issues 48(4):21–37

Douglas M, Wildavsky AB (1983) Risk and culture: an essay on the selection of technical and environmental dangers. University of California Press, Berkeley

Fischhoff B et al (1978a) Fault trees: sensitivity of assessed failure probabilities to problem representation. J Exp Psychol Hum Percept Perform 4:330–344

Fischhoff B et al (1978b), How safe is safe enough? A psychometric study of attitudes toward technological risks and benefits. Policy Sci 9:127–152

Greenberg MR, Schneider DF (1995) Gender Differences in Risk Perception: Effects Differ in Stressed vs non-stressed environments. Risk Anal 15:503–511. doi:10.1111/j.1539-6924.1995.tb00343.x

Gruppo di Lavoro MPS (2004) Redazione della mappa di pericolosità sismica prevista dall'Ordinanza PCM 3274 del 20 marzo 2003. Rapporto Conclusivo per il Dipartimento della Protezione Civile, INGV, Milano-Roma, pp 65 + 5 appendici

Gustafson PE (1998) Gender Differences in risk perception: theoretical and methodological erspectives. Risk Anal 18:805–811. doi:10.1111/j.1539-6924.1998.tb01123.x

Lave TR, Lave LB (1991) Risk Anal 11(2):255–267

Likert R (1932) Technique for the measure of attitudes. Arch Psycho 22(140):1–55

Lindell MK (1994) Int J Mass Emerg Disasters 12(3):303–326

Lindell MK, Perry RW (2000) Household adjustment to earthquake hazard: a review of research. Environ Behav 32:590–630

Osgood CE et al (1957) The measurement of meaning. University of Illinois Press, Urbana

Peters E, Slovic P (1996) The role of affect and worldviews as orienting dispositions in the perception and acceptance of nuclear power. J Appl Soc Psychol 26(16):1427–1453

Plapp T, Werner U (2006) Understanding risk perception from natural hazards: Examples from Germany by institute for finance, banking and insurance/postgraduate programme natural disasters. Universität Karlsruhe (TH), Karlsruhe

Renn O (2008) Risk governance: coping with uncertainty in a complex world. Earthscan, London, p 110

Rohrmann B (1999) Community-based fire preparedness programs: an empirical evaluation. Austral-Asian J Disaster and Trauma Studies, 1

Savage I (1993) Demographic influences on risk perceptions. Risk Anal 13:413–420. doi:10.1111/j.1539-6924.1993.tb00741.x

Siegrist M (2000a) The influence of trust and perceptions of risks and benefits on the acceptance of gene technology. Risk Anal 20:195–203

Siegrist M, Cvetkovich G (2000b) Perception of hazards: the role of social trust and knowledge. Risk Anal 20:713–719

Siegrist M et al (2000c) Salient value similarity, social trust, and risk/benefit perception. Risk Anal 20:353–362

Sjöberg L (2000a) Factors in risk perception. Risk Anal 20:1–11

Sjöberg L (2000b) The methodology of risk perception research. Qual Quant 34:407–418

Slovic P (1987) Perception of risk. Science 236:280–285

Slovic P (1992) Perceptions of risk: reflections on the psychometric paradigm. In: Krimsky S, Golding D (eds) Social theories of risk. Praeger, London, pp 117–152

Slovic P (1993) Perceived risk, trust, and democracy. Risk Anal 13:675–682

Stucchi M et al (2011) Seismic Hazard Assessment (2003–2009) for the Italian building code. Bull Seismol Soc Am 101(4):1885–1911. doi:10.1785/0120100130

Szalay LB, Deese J (1978) Subjective meaning and culture : an assessment through word associations. L. Erlbaum Associates; distributed by the Halsted Press Division, Wiley, New York

Thompson M et al (1990) Cultural theory. Westview Press, Boulder

Wachinger G, Renn O (2010) Risk perception and natural hazards. CapHaz-Net WP3 Report. In: DIALOGIK non-profit Institute for Communication and Cooperative Research. Stuttgart (available via http://caphaz-net.org/outcomes-results/CapHaz-Net_WP3_Risk-Perception.pdf)

Zacchi MC, Crescimbene M (2010) Mind quakes: visualizzare la percezione del rischio sismico per la prevenzione. http://hdl.handle.net/2122/6819

Long-Term Disaster-Prevention Strategies Based on Education

14

G. Musacchio, A.E. Bernhardsdóttir, M.A. Ferreira and S. Falsaperla and UPStrat-MAFA Outreach Working Group

Abstract

Although human behavior is the crucial factor in the degree of vulnerability and the likelihood of disasters taking place, preparedness and prevention programs are not mandatory in all countries around the world. Within the framework of UPStrat-MAFA (Urban disaster Prevention Strategies using MAcroseismic FAults), we have defined the disaster prevention strategies based on education management information and actions taken in Iceland, Portugal, Spain, and Italy. A detailed comparative study shows that compulsory school in these four participating countries is greatly unprepared with regard to hazard education, and these results are in line with worldwide studies. Moreover, when hazards are addressed, this is not done at an early age, which results in a missed chance to intervene in the noncognitive side of awareness, which decreases at later ages. To comply with the urge to take actions towards training and education at an early age, we used hands-on tools and learn-by-playing approaches in an informal learning environment. To reach the older population, the audio-visual media appears to be the best and lowest cost alternative to promote risk perception, awareness and education.

Keywords

Disaster prevention • Education • Seismic hazard • Information strategies

"UPStrat-MAFA Outreach Working Group": S. Thorvaldsdóttir, A. Carvalho, M. L. Sousa, S. Raposo, R. Nave, F. Sansivero, C. S. Oliveira, J. C. Nunes, R. Sigbjörnsson, and G. Zonno.

G. Musacchio (✉)
Istituto Nazionale di Geofisica e Vulcanologia, Rome, Italy
e-mail: gemma.musacchio@ingv.it

A.E. Bernhardsdóttir · UPStrat-MAFA Outreach Working Group
Earthquake Engineering Research Centre, Selfoss, Iceland

M.A. Ferreira · UPStrat-MAFA Outreach Working Group
Instituto Superior Tecnico, Lisbon, Portugal

S. Falsaperla
Istituto Nazionale di Geofisica e Vulcanologia, Catania, Italy

UPStrat-MAFA Outreach Working Group
Laboratorio Nacional de Engenharia, Lisbon, Portugal

UPStrat-MAFA Outreach Working Group
Istituto Nazionale di Geofisica e Vulcanologia, Napoli, Italy

UPStrat-MAFA Outreach Working Group
Istituto Nazionale di Geofisica e Vulcanologia, Milan, Italy

14.1 Introduction

There is a misconception about disasters, in terms of being Nature's curse, which has mostly to do with the mindset of people towards safety culture and the chance to live in areas prone to natural hazards in a sustainable way. Although seismicity is one of the most acknowledged causes of disasters, seismic hazard responses are far from being aligned with longer-term perspectives in terms of development, crucial socio-economic investments and infrastructure, and above all, preparedness (Horton 2011).

Risk awareness and correct perception are salient prerequisites for disaster risk reduction; a risk must be perceived before it triggers risk-reduction efforts. There is an absence of risk perception in the lives of many people, and thus in community and state development planning, in the educational curriculum, and in the media priorities.

G. Lollino et al. (eds.), *Engineering Geology for Society and Territory – Volume 7*,
DOI: 10.1007/978-3-319-09303-1_14,

UPStrat-MAFA (Urban disaster Prevention Strategies using MAcroseismic FAults) is a project founded by the European Union that is intended both to assess seismic hazard and risk at an urban scale, and to define disaster prevention strategies based on an education management information system that is linked to information about areas and population groups that are prone to particular kinds of emergencies (UPStrat-MAFA European project, 2013). Here, we present the actions that have been taken by the partners of this EU project in Iceland, Portugal, Spain, and Italy, concerning two major targets: children and the general public.

14.2 Information Strategies

To spread the necessary information about hazards is not an easy task. One might end up scaring people, or the approach might be too trivial. An additional challenge is intrinsic to this topic: disasters are infrequent in nature, and their recurrence is not predictable or linear. Often people have the idea that disasters will strike others, but not them. In part, this is connected with the education process itself: textbooks often present "horrible" cases from far away, compared to which, local disasters appear trivial. People might feel that scientists are presenting them with just another trouble to deal with.

Research has shown that the memory of a disaster remains preserved in the social sphere only for a certain period of time, unless it is kept vivid in the minds of the people, or they are reminded by the provision of information (media, web) and the socially active preservation of the memory (Wisner 2006; Biernacki et al. 2008; Komac 2009; Komac et al. 2013). People tend to deal with hazardous situations in the way they did in the past, no matter that they might have been trained otherwise. Strategies to spread knowledge and raise perception should either include repeated training until earlier incorrect behaviors are replaced, or start training and education at an early age when emotional intelligence is prevalent. The information needs to be balanced, easily understood, and accessible to all, and it should be linked to the areas and population groups that are prone to particular kinds of emergencies. It is important to distinguish between disaster prevention strategies in the field of education and those that are addressed to the general public.

14.2.1 Hazard and Risk Education and Schools

To be effective, long-term activities like education should be permanent and integrative, and should cut across all formal and informal educational efforts, while remaining in close contact with reality. Assessments of the educational curricula on natural hazards and accessibility to risk reduction information have highlighted how schools worldwide are greatly unprepared in their natural hazards education (Komac et al. 2013).

We have run a detailed comparative study on seismic and volcanic hazard education across four European countries: Iceland, Portugal, Spain, and Italy. The outcome of the study has highlighted in particular that no matter what the average level of education is, the risk of exposure to seismic and volcanic hazards are not mandatory subjects in the compulsory education (Bernhardsdóttir et al. 2012). We have analyzed standardized curricula and textbooks, and we have considered the age at which natural hazards are discussed.

None of these four countries provides a specific course that is devoted to the education of earthquake and volcano hazards and risks. Instead, standardized curricula include these aspects within different subjects, such as geography, geology, physics and history. Textbooks do not even thoroughly discuss all of the fundamental scientific and/or safety issues that are needed to understand a hazard. Furthermore, the core science textbooks on the market are riddled with errors. Our study highlights the gap that exists between the science world and that of education, which grows deeper as the misconceptions of the teachers themselves regarding the geosciences are passed on to their pupils (Bernhardsdóttir et al. 2012). Children do not approach hazard education at an early age, and the result is that the noncognitive side of awareness may never be triggered.

Bearing in mind what the major weak points of hazard education are in the four countries, we have developed tools to first take up informal learning, and then to plan to take actions on the formal curricula in the future. As we need to first address children, good approaches include the use of hands-on tools and learn-by-playing approaches.

Scientists need to challenge their communication skills to educate children through scientific role-play games and hands-on laboratory activities. In Italy, the National Institute of Geophysics and Volcanology (INGV) has built a portable shaking table, which allows children of every age to have direct experience of the motions related to an earthquake and its effects. In Portugal, the Advanced Technical Institute (IST) uses its digital educational small shaking table to demonstrate to students the dynamic performance of a building.

The IST has also been working on an interactive online game, named "*treme-treme*" (shake-shake) which allows children of 8 years to learn the skills and concepts for survival during an earthquake. The players are meant to build a survival kit, finding all of the safe and unsafe areas of their home, and to learn how to protect themselves (Fig. 14.1).

For younger children, as 5–10-year-olds, the Civil Protection of Lisbon Municipality promoted a space that is known as the "House of Tinoni", where each child learns to

Fig. 14.1 A snapshot from the *treme-treme* game

identify and respond to the risks involved in everyday life, as well as to exceptional events like earthquakes. Some pedagogical information is provided, and the experience culminates in an interactive game that summarizes the contents that are disseminated through the activities (http://www.tinoni.com/casa-do-tinoni.php).

In Iceland, the Earthquake Engineering Research Centre (EERC) provides awareness training for school children, and contributes to a permanent exhibition on seismicity in Hveragerði. Here, visitors can, for instance, enter a small "house" that shakes—an earthquake simulator—that allows them to experience powerful earthquake vibrations.

14.2.2 Hazard and Risk Education and the General Public

Hazard dissemination to the general public needs to be informative and balanced, and to be in a language form that is accessible to everybody.

The interactions between schools and the science world vary among these countries, from being formally defined in a curriculum, to being informally initiated by teachers and scientists. Both the formal and informal approaches are valuable for the educational system, but when the interaction depends on individuals, institutions and/or nongovernmental organizations, consistency can be lacking, due to the more unstable finances.

Science festivals and open days for research institutions are often environment hosting activities and exhibitions that are devoted to risk education and, in general, to Earth Sciences. However, these are spot initiatives that cannot reach the whole population. Audio-visual clips are, in our opinion, a more effective and long-lasting tool for the dissemination of information on hazards to a wider range of the population. Therefore, in the framework of the UPStrat-MAFA project, efforts have been devoted to the preparation of videos that can promote risk awareness and education. For example, in Portugal, the IST and the National Engineering Laboratory (LNEC) have produced a 16-min video titled, "Before it's too late", where the public perception and interviews with experts are presented, to bring people closer to the science (Fig. 14.2).

In Iceland, the EERC has produced the film "Hveragerði—in Compliance with Nature", which focuses on a small community coping with the risk of earthquakes. Similar actions have been taken in Italy, and together, Portugal, and Iceland have also contributed to the preparation of a video that provides an overview of the seismic hazard in these countries.

14.3 Discussions and Conclusions

We have presented the actions taken within the framework of the UPStrat-MAFA project. Comparing the education systems within countries participating to the project, namely Iceland, Portugal, Spain, and Italy, we have found that the compulsory school curricula are greatly unprepared in terms of hazard education. Our results are in line with worldwide studies concerning natural disaster education at each level (Komac et al. 2013).

To comply with the need of training and education at an early age, and to intervene in behavior that children, the future adults, will have in the case of hazards, we have

Fig. 14.2 Snapshots from the video "Before it's too late"

taken actions that include the education of children using hands-on tools and learn-by-playing approaches. To raise the perception and awareness of the older population, we have found that audio-visual media will be the best and lowest cost way to promote risk awareness and education.

Acknowledgements This study was co-financed by the EU—Civil Protection Financial Instrument (Urban disaster Prevention Strategies using MAcroseismic Fields and FAult Sources—UPStrat-MAFA, Grant Agreement N. 23031/2011/613486/SUB/A5).

References

Bernhardsdóttir AE, Thorvaldsdóttir S, Sigbjörnsson R, Musacchio G, Nave R, Falsaperla S, D'Adda S, Sansivero F, Zonno G, Sousa ML, Carvalho A, Raposo S, Ferreira MA, Nunes JC, Jimenez MJ (2012) Disaster prevention strategies based on an education information system. Paper presented at the 15th WCEE, Lisbon, 24–28 Sept 2012

Biernacki W, Bokwa A, Domanski B, Działek J, Janas K., Padło T (2008) Mass media as a source of information about extreme natural phenomena in southern Poland. In: Carvalho A (ed) Communicating climate change: discourses, mediations and perceptions. Centro de Estudos de Comunicacù ˜ao e Sociedade, Universidade do Minho, Braga, pp 190–200

Horton J (2011) Lancet 377:439

Komac B (2009) Acta Geogr Slov 49:199

Komac B, Zorn M, Cigli R (2013) Nat Hazards Earth Syst Sci Discuss 1:2255

UPStrat-MAFA European project (2013) http://upstrat-mafa.ov.ingv.it/UPStrat/

Wisner B (2006) Let our children teach us! a review of the role of education and knowledge in disaster risk reduction. ISDR system thematic cluster. Platform on knowledge and 20 education, Bangalore

Geoscientific Communication Problem with Communities for Disaster Prevention and Land Planning in Peru

15

Luisa Macedo, Sandra Villacorta, Shianny Vasquez, Jersy Mariño and Giuseppe Di Capua

Abstract

The communication process between the geoscientists and native communities in risk areas can significantly affect disaster prevention and land use planning. In Peru, the problem of disaster prevention is a fundamental policy due to unfamiliarity and deficiency of the associate information on the population. It is possible that talk of disaster prevention it will be an unlikely ideal in a country where most towns have settled on unplanned projects by the constant change and the lack of interest from the authorities in such topics. However, it is anachronistic that the rural communities and towns continue to live without a plan to enable them to improve their quality of life. The correct use of geoscience information in the mass media can help in this work. The characteristics of the enterprise in Peru require more training by professionals in the geosciences and support communication specialists. In this paper, we analyze the problem of communication for disaster prevention in Peru, with the aim of contributing to the articulation of a disaster prevention strategy.

Keywords

Communication process • Disaster prevention • Risk management • Peru

15.1 Introduction

In Peru there are many unknowns and little interest in the development of activities for disaster prevention, therefore awareness work is very important. The authorities, mainly the Central Government, act belatedly, only after the disaster that has occurred. They work in that direction at the time of the disaster and then forget the importance of prevention planning and sustainable reconstruction. For example, this can be seen in the performance of the authorities after the earthquake in Pisco (Ica, Perú) in 2007. Six years after the event, many affected people have not yet managed to regain the standard of life they had before the disaster. On the issue of land use, even though the regulations that should guide have been released, local authorities are not sufficiently prepared to use the information given by specialists; many times, norms are not understandable and so they cannot be applied. Properly designed geoscientific information would allow them to concentrate their actions on the development of alternatives initiatives, derived from management plans prepared by the specialists. By the above consideration, it is considered that disaster management and planning of land use are going tasks that must be managed from a multidisciplinary perspective.

L. Macedo · S. Villacorta (✉) · S. Vasquez · J. Mariño
Instituto Geológico, Minero y Metalúrgico, Lima, Peru
e-mail: svillacorta@ingemmet.gob.pe

L. Macedo
e-mail: lmacedo@ingemmet.gob.pe

G. Di Capua
Instituto Nazionale Geofisica e Vulcanologia, Rome, Italy

15.2 Background

The current regulation in Peru, although it seeks to contribute to disasters prevention and the management of land use, has not been properly prepared due to improper

G. Lollino et al. (eds.), *Engineering Geology for Society and Territory – Volume 7*,
DOI: 10.1007/978-3-319-09303-1_15, © Springer International Publishing Switzerland 2014

interpretation of the same. As an example, we cite recently approved Law No. 29869 (Law of population resettlement in very high risk areas not mitigable (CENEPRED 2012). This law states that local governments have the function to declare a "State of emergency" at the National Institute of Civil Defense (INDECI). This would allow them to access funding for relocation. However, this funding is often misused by local authorities that can take advantage of the economic support for appropriation of land and their subsequent sale. On the other hand, scientific institutions participate only as partners in the process of resettlement of populations located in areas classified with very high risk.

In many cases, it is left to scientific institutions to make the decision whether it is appropriate to relocate the population at a high risk level and the declaration of the so-called "IMMINENT DANGER", an inappropriate term that it is not based on a definition that is scientifically correct, because what it is being studied is the susceptibility and danger, and these don't end when it ceases to be imminent. The "imminence" is defined as the occurrence of a phenomenon in a "short-term" and it will be determined only when the processes will be monitored continuously; but in some cases this is not possible, as in case of earthquakes.

15.3 The Importance of Effective Communication to Ensure a Culture of Disaster Prevention

The communication by geoscientists to the authorities and population is a multidisciplinary, complex and dynamic work that involves a series of activities aimed at the application of scientific knowledge in the processes of human, social, territorial and sectorial development (Macedo et al. 2007).

The community, knowledge-generating institutions (entities proposing solutions) and the executing agencies (authorities, public and private institutions, etc.) have to be engaged in this task. The interaction with the community enables a collective and not individual production and it makes each person feel part of the project (Mucho et al. 2005).

A quality work is required and it has to be properly coordinated with all state institutions for the effective reduction of disaster risks. The primary objective should be to promote the transfer and application of the scientific research into the risk management. If the actors are not involved with other agencies and the community starting with the early stages of any study or research project, in many cases there is no continuity in the process of transforming into action the knowledge generated by scientific community and it would miss the opportunity to show the relevance of the Geosciences for society (UN/ISDR 2004; PMA: GCA 2005). This activity will ensure the strengthening and sustainability for proper risk management.

To consolidate citizen participation and action in disaster prevention, it is necessary that every person is aware of the importance of participation in the planning processes. Through grassroots organizations, roundtables, participatory plans, etc., people can work with the authorities. It is indispensable that people are well informed, so that their ideas are complemented by the scientific knowledge of the land.

15.4 Importance of Land Planning

Land planning is the process to organize spatial, social and economic development (Alfaro 2010). Its implementation is in the hands of specialists, but it is a task that must be engaged also by each of the country's inhabitants. Therefore communication of geoscientific information plays an important role in the development of the land.

Among the benefits of land-use we mention:

- The orderly growth of communities.
- Disaster prevention.
- Increased security for the private investments.
- The protection of natural areas and indigenous communities.
- The knowledge and sustainable use of resources.
- Conflict prevention.

Significantly, land-use or subsystems management present four types of approaches: natural, urban, economic, and socio-cultural. The ideal is to implement a comprehensive approach that reconciles the objectives of each of the above items.

15.5 Conclusions

- To develop an effective communication between authorities and people in disaster prevention, one should consider a multidisciplinary team, complemented with communication specialists.
- It must be a commitment that the geoscientist's knowledge dissemination to society will be in a simple language, and thus fulfill its purpose.
- The planning is no longer solely in the hands of specialists, it is a task that involves every citizen, so that they learn more of the potential disaster and that they better know their own territory.

- In Peru, authorities need to be trained to address the real needs of the people and to understand the true importance of land use and its planning. It also requires the support of competent professionals in national and local governments to work with the authorities.
- The authorities require large integrated projects in each region that must be considered essential in the land planning, the use of hazards maps and the use of technical reports that has to be explained by experts for their use.
- It is necessary for institutions to intervene in a "technical-scientific" way and seeks changes in the laws and regulations which have misused the terminology associated with risk prevention.
- It is important for scientific institutions to be involved in regional planning, as well as in emergency plans, and thus ensure better disaster risk management.

References

Inter-Agency Secretariat of the International Strategy for Disaster Reduction (UN/ISDR) (2004)—Living with risk: a global review of disaster reduction initiatives. Ginebra-Suiza

Mucho R, Page R, Alcántara P, Kempff O, Sougarret L, Villarruel J, Báez N, Ortegano O, Ellerbeck M, Muñoz F, Hickson C (2005) Papel de los Servicios Geológicos en el Ciclo de Conocimiento Aplicado a la Gestión de Riesgos. XII Congreso Latinoamericano de Geología Quito Ecuador. En web: http://www.geo.mtu.edu/~raman/papers2/Munoz2.doc

Macedo L, Mariño J, Fidel L, Luna R, Quispe R, Pareja H, Arguedas A, Siu A, Muñoz F, Ampuero F (2007) Documento Metodológico; Proceso de difusión, educación, sensibilización y acción frente a los peligros volcánicos del Misti en Alto Selva Alegre, Arequipa. INGEMMET, Especial Publication, Lima, p 60

Alfaro G (2010) La comunicación de la información geocientífica como estrategia para la planificación del territorio. XV Peruvian Geological Congress. Extended abstracts. Peruvian Geological Society, Pub. Esp. N° 9. Cusco, Peru, pp 460–463

Promoting Geo-awareness to Make Citizens the First Watchers of the Territory

16

Tiziana Lanza

Abstract

The earthquake occured in Tohoku (Japan) in 2011 showed that no matter how outstanding is the model of the mitigation of natural hazard, an earthquake can always surprise any studied attempt to contain a tsunami. I report this extreme example to draw the reader attention on how strong should be the commitment of the society in producing the optimal conditions to face natural hazard. Every country has a peculiar social context. If in the past we considered this as a secondary aspect, now we are obliged to recognize that the social context is of main importance when promoting mitigation of natural hazard. In this occasion I would like to refer to my experience as an educator in my country. Italy is a country where it is difficult to promote a dialogue between the Institutions and the citizens. The dialogue is difficult also among the communities of engineers, geologists and politicians. There is still a great confusion on the attribution of tasks and a lot of problems of difficult solution are related to corruption. In this context, education can play a crucial role and should be intensively addressed to promote geo-awareness in the citizens. People should learn about the geological site where their houses are built and get all the info necessary to understand if it has been done according to the law. Only in this way citizens become an active agent in promoting those changes without which an efficacious mitigation of natural hazard is impossible.

Keywords

Geo-awareness • Citizens • Risk • Natural hazard • Education

16.1 Introduction

As a science communicator, I have been involved now from several years in the mitigation of natural hazard. Since when I started working for the INGV in Rome, I witnessed two important earthquakes in my country in the last two decades and I have realized that few things have changed in the 13 years passing from the earthquake occurred in Colfiorito on September 1997 to that occurred in L'Aquila on April 2009, at least for what concern the mitigation of natural hazard. No matter the efforts by the scientific community, people keep on dying and houses and monuments are still at the mercy of natural hazard (Lanza et al. 2012). Thirteen years are a short time to expect some changes? I don't know. In any case, while so few has changed in our society, in the last decade communication has instead gained new opportunities for the birth of the social networks, a reality that has boosted the possibility of an interactive communication among the institutions and citizenship and among citizens themselves. In this paper I will consider a renewed approach to the manner in which modern societies organize in response to risk, starting from the idea that in this context a revised concept of seismic risk is necessary. Usually it is mainly based on the building vulnerability while other aspects remain in the background. Our country has indeed

T. Lanza (✉)
Istituto Nazionale di Geofisica e Vulcanologia (INGV), via di Vigna Murata, 605, 00143 Rome, Italy
e-mail: tiziana.lanza@ingv.it

G. Lollino et al. (eds.), *Engineering Geology for Society and Territory – Volume 7*,
DOI: 10.1007/978-3-319-09303-1_16,

shown how important is the social context in determining the inertia of a society confronting with natural hazard. In other words, the risk is strictly connected to the political asset, and eventually the grade of corruption, and the ability of the citizens to contrast it whenever is the case.

In this perspective, communication and in the specific education, assume a fundamental role in promoting awareness of the territory and a change starting from citizens when the institutions lack the sense of responsibility. A citizen becoming really aware of the hazard of the territory where he lives, is much more prone to solicit those changes necessary to mitigate natural risks.

16.2 Experiencing the Italian Social Context

In 2010, together with other INGV colleagues we implemented a pilot project in collaboration with the Castelli Romani Park to experiment the Eco-museum formula to educate students to the risks of the area. We involved students of three classes of a secondary school in Velletri, a country in the area of the Alban Hills, the same place where I live (Lanza et al. 2011). This is an area interested by volcanism for thousands of years. Volcanism is now thought to be quiescent but there is a general agreement among scientists that the Alban Hills may erupt again in the future as a recent study highlights (Carapezza et al. 2010). At present, the hazard comes from a moderate seismicity and from the release of carbon dioxide and other hazardous gases in the area. Also the nature of the soil has suffered from being interested by volcanism so that every engineering work should inevitably take care of it.

The evaluation of the Eco-museum pilot project confirmed us how poor was the students' knowledge of the area from the geological point of view, including a scarce

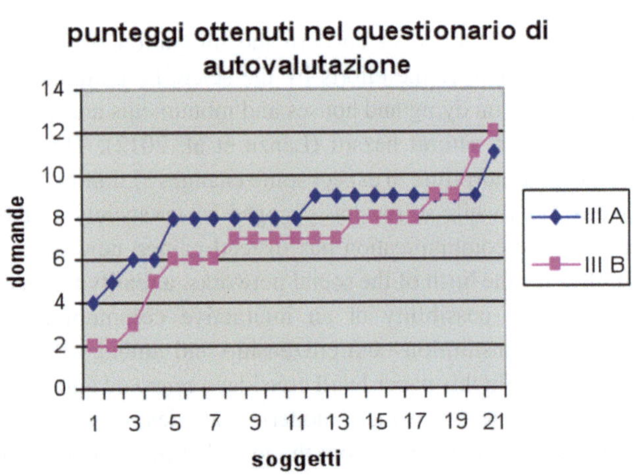

Fig. 16.1 The start up questionnaire provided data obtained by comparing experimental group (Class III-A) and the control group (Class III-B) (La Longa et al. 2013)

awareness of the risks (Fig. 16.1). Students do not generally have a prior knowledge of the territory around the lake of Nemi, one of the volcanic lake in this area, and the project resulted to be an opportunity and an exclusive means for acquiring that knowledge (La Longa et al. 2013).

I have also experienced in my own life what does it mean not having an appropriate education to the risks of the soil. I found myself involved in contrasting a dangerous project of opening a road in the valley in front of the building where I live, a valley that could be interested by landslides. In that occasion I confronted myself with the other citizens living in the same condominium to find out that very few of them were aware of the risk of such a project, because they lack the necessary education to understand it. So, it took a lot of efforts from my part and another couple of citizens in order to stop it. The project was strongly supported by the major who was making pressure on the builder. We fought against it in vain for a couple of years by creating a committee. Even if the Castelli Romani Park gave a negative technical advice, there was no way to obtain from the municipality the geological survey on which that project was based, even with the help of a lawyer. In the end, the project was stopped by nature itself! On the 17 March 2011 a landslide occurred starting from the exact point in which they were working to open the road (Fig. 16.2). It was just fortuitous if the landslide did not run over the houses immediately near. At present the valley has been abandoned to itself and up to now no measures have been undertaken to put it in security as we asked in our last letter addressed to all the authorities involved.

A recent study, that is still in progression, on seismic risk perception showed that in Italy, 9 of 10 citizens underestimate the risk coming from living in a seismic zone classified by the competent authorities with number 1, where major earthquake may occur. To go in details, in a sample of 5,585 people interviewed in a time span of 6 months (from January 2013 to June 2013), 86 % underestimate the hazard in zone 1, and 70 % in zone 2. Another important aspect of the research is how citizens get informed about earthquake risk in their area. A surprising result is that even if the source of information to which the majority of the population rely on is the Civil Protection Department, the citizens who get more information from the web seem to have a most appropriate risk perception (Crescimbene et al. 2013).

16.3 Experimenting the Right Educational Format

How to educate people to become aware of the territory where they live? A not less important question would be how to obtain this in a relatively short time period, since risk awareness is the first fundamental prerequisite to mitigate

Fig. 16.2 From *left* to *right* bulldozer at work for the preliminary operation to open a road in Rocca di Papa. In the second picture an image of the landslide occurred on the 17 March 2011

natural hazard. School is the place where people can acquire the tools necessary to make the right choices in the future. Even if a study highlighted that children have a "valuable and unique ability to conceptualize and analyze risk" (Tanner 2010), a recent European comparative research conducted on four countries (Italy, Portugal, Spain, Iceland) to investigate the education information system to the light of possible disaster prevention strategies, had to conclude that none of these countries provides a special course to educate students on earthquakes and volcano hazards and risks. Rather they include it within different subject such as geography, geology, physics and history, although most often within the natural sciences (Bernhardsdottir et al. 2012). The study then suggest that local awareness of seismic hazard can balance the lack of sufficient education in schools, by, for instance, training children living in high-risk areas to evacuate their school in case of an earthquake.

Seismic risk is just one of the risks. In a particular region of a country, as the case of the Alban hills in Italy, citizens can be subjected to different risks, maybe peculiar to the area, but not for this less important. It is in this perspective that we experimented the eco-museum formula. We conceived our Eco-museum pilot project in the widest context of experimenting science theatre as an efficacious tool to promote earth education and risk awareness. Starting with an experiment implemented in a primary school using theatre in a classical and traditional way (Lanza et al. 2014) we passed to experiment it in a secondary school in Velletri. In the second case we choose a cutting edge formula where theatre, in terms of scenic actions to be played in situ, is included in the widest context of an open-air museum. An eco-museum is a museum without walls that focuses on the identity of a place; it is based on local participation and refers to a new idea of it that involves the whole cultural heritage in opposition to the focus on specific items and objects, performed by traditional museums. In accordance with this philosophy, we have introduced in the classroom innovative teaching tools as the creation of an eco-museum itinerary.

Even if we realized how poor was the knowledge of the students about the territory around the Nemi Lake, the evaluation of the eco-museum experience conducted comparing a start up questionnaire with a final questionnaire gave encouraging results (La Longa et al. 2013). Since eco-museums are already well-established realities in our country (see http://www.ecomusei.net/index.php) we may wonder if in the future they can become the privileged path through which putting in contact the different realities such as municipalities, research institution, schools and citizens to enhance the territory knowledge and education to the risks.

16.4 Conclusion

The last two important seismic events occurred in my country have highlighted that the social context is of main importance in determining the inertia of a society confronting with natural hazard. My experience as an educator has confirmed my belief of how important is to make grow the geo-awareness of the citizens to solicit the institutions to undertake the necessary changes to promote an even better risk preparedness. Taking into account that at present the education information system at school is not enough to promote appropriate risk mitigation, as an educator working for a scientific institution, I do believe that experimenting different educational format is the first important step to reach the goal. In particular, experimenting the eco-museum format in a secondary school has triggered new opportunities to promote the culture of the territory as a local and environmental heritage to be shared and safeguarded by those who inhabit it.

References

Bernhardsdóttir AE, Thorvaldsdóttir S, Sigbjörnsson R, Musacchio G, Nave R, Falsaperla S, D'Adda S, Sansivero F, Zonno G, Sousa ML, Carvalho A, Raposo S, Ferreira MA, Nunes JC, Jimenez MJ (2012) Disaster prevention strategies based on an education information system. 15WCEE, Lisbona. http://hdl.handle.net/2122/8148

Carapezza ML, Barberi F, Tarchini L, Ranaldi M, Ricci T (2010) Volcanic hazards of the Colli Albani. In Funiciello R, Giordano G (eds) The Colli Albani volcano. Special Publications of IAVCEI #3. Geological Society, London, pp 279–297

Crescimbene M, La Longa F, Camassi R, Pino NA (2013) Report on risk perception DPC-INGV—S2 Project https://sites.google.com/site/ingvdpc2012progettos2/deliverables/d2_6

La Longa F, Lanza T, Crescimbene M (2013) Ecomuseum itineraries, a path of rediscovery and enhancement of the territory EGU 2013 Poster session. http://www.earthprints.org/bitstream/2122/8703/1/Posterecomuseo_EGU2013.pdf

Lanza T, Crescimbene M, La Longa F, Pizzicannella E, Tortorici G, Pizzino L, Frepoli A, D'Addezio G (2011) Geophysics on stage: bringing Earth into scene. The INGV science theatre experiences. The Pantaneto Forum. http://www.pantaneto.co.uk/issue42/lanza.htm

Lanza T, Crescimbene M, La Longa F (2012) From Colfiorito to L'Aquila earthquake: learning from the past to communicating the risk of the present EGU2012 Poster session. http://www.earth-prints.org/bitstream/2122/8288/1/EGU2012.pdf

Lanza T, Crescimbene M, La Longa F, D'Addezio G (2014) Bringing earth into the scene of a primary school: a science theatre experience. Sci Commun 36(I):131–139

Tanner T (2010) Shifting the narrative: child-led responses to climate change and disaster in El Salvador and the Philippines. Child Soc 24:339–351

Laboratory Activities for Seismic and Volcanic Hazards Education: A Challenge for Scientists

17

Gemma Musacchio and Nicola Alessandro Pino

Abstract

Balanced information and education are fundamental prerequisites for risk prevention and preparedness. Among others, children embody our best chance to implant appropriate behaviors that will be recalled during hazardous situations and to involve adults according to a knowledge chain reaction. In this work, scientists challenge their communication skills to built a set of hands-on and learn-by-play based laboratory activities, for primary and secondary schools, addressing three major issues: (1) the location of earthquakes and volcanoes on Earth; (2) earthquakes and eruptions mechanisms; (3) earthquakes unpredictability. Students are asked to place volcanoes and earthquakes epicenters (issue 1) on a wooden plate puzzle according to Plate dynamics. To addresses eruption mechanisms (issue 2) and related hazard, we use backing soda forced blowing out from a volcano vent and suggest that a pyroclastic flow is fast, it can spread over a large area and raise high up to the stratosphere. Earthquake mechanisms (issue 2) are discussed describing the energy buildup, release, and transfer, using a wooden sticks bend-and-break analogy. The display of acoustic waves caused by the breakage in different situations allows understanding of both the rupture energy and the wave attenuation. Earthquakes occurrence (issue 3) is addressed using steadily pulled blocks sliding on a frictional surface, where pins simulate asperities. These activities were tested, involving thousands of students. Discussions with students and teachers and the analysis of the answers to specific questionnaires gave us confidence that we proposed proper tools to raise risk awareness.

Keywords

Seismic and volcanic hazards • Outreach • Education

17.1 Introduction

No matter how disasters overwhelmingly crash our lives, it seems we all are reluctant to preparedness and prevention. Since disasters are infrequent in nature (compared to human life span) and memories are short (Komac 2009), there is an urge to promote awareness and proper perception of natural hazards and risks.

The science underlying natural hazards is complex, the impact is potentially devastating and there are no quick fix solutions. Balanced information and proper addressing of the unavoidable uncertainties in natural hazard assessments require that public must have a basic understanding of the phenomena. Science has the fundamental duty of communication and the meeting point between school and science

G. Musacchio (✉)
Istituto Nazionale di Geofisica e Vulcanologia, via di Vigna Murata 605, 00143 Rome, Italy
e-mail: musacchio@mi.ingv.it

N.A. Pino
Istituto Nazionale di Geofisica e Vulcanologia, Osservatorio Vesuviano, via Diocleziano, 328, 80124 Naples, Italy
e-mail: alessandro.pino@ov.ingv.it

G. Lollino et al. (eds.), *Engineering Geology for Society and Territory – Volume 7*,
DOI: 10.1007/978-3-319-09303-1_17,

world can be the key to start the new age of risk prevention and preparedness. Although being fundamental, science communication is not an easy task, as it requires a deep knowledge of the topic as well as the ability to use a proper language for each public, so that misunderstanding is minimized while raising interest.

In this paper we challenge science education and outreach to explore the rationale behind seismic and volcanic hazards, fully relying on the definition and the unraveling of basic issues, being able to communicate them in the proper way.

17.2 Seismic and Volcanic Hazard Education

There is a stringent paradox between the pressing demand by the general public towards seismology and volcanology—often challenging scientists communication skills—and the fact that these subjects are worldwide neglected in today schoolbooks and school programs (King 2010 and references therein; Bernhardsdóttir et al. 2012). Research suggests that risk perception may be built upon experience of local disasters, combined with information gained from external information sources, such as the media, training sessions and school curricula (Wachinger and Renn 2013). If training and education start at early age there is a better chance that appropriate behaviors can be placed in the mind and recalled during hazardous situations. Having children as chief target we can rely on the chain reaction spreading of knowledge: children will pass information to adults within their families.

Learn-by-playing based laboratory activities can provide children with appropriate education, training and, in a way, strengthen their ability to participate in disaster risk reduction efforts. The creativity, the knowledge, the experimentation and the discovery triggered by playing activities also might allow overcome the natural fear of hazard and trigger emotional learning that most efficiently influence behaviors.

17.3 Laboratory Activities

We set up activities for primary and secondary schools students, addressing three basic issues: (1) what cause the Earth to have earthquakes and volcanoes; (2) how earthquakes and volcanoes work, and (3) why these natural phenomena are not predictable. Each issue is tuned to specific learning processes—emotional or cognitive—capable to trigger proper behavior towards hazards. Activities highlight in a simple way just a few basic aspects of the phenomena so that the audience might independently repeat experiments.

17.3.1 What Cause the Earth to Have Earthquakes and Volcanoes?

No matter the age of a student or general public, they are always extremely curious about what causes disasters on Earth. We challenge children understanding that the stress related to Earth dynamic is prevalently released (1) at plate boundaries and (2) within the lithosphere. These are two basic concepts that can be explained to almost every age students, although with rather different approach.

Primary school children have only a little knowledge on worldwide geography and extremely simple demonstration of phenomena are most likely to cut cross and provide basic answers to questions. We found that a good way to address the topic is to mimic Plate Tectonics: four children will pretend to have their foot glued on four different Plates; they are placed on sliders either moving apart or one against the other. Diverging sliders will drag two children moving apart and mimic the breaking up of the lithosphere. Converging sliders will lead the two children pushing one against the other.

Once plate motion is worked out, we can use the puzzle concept to allow comprehension of seismicity and volcanism occurrence. In a wooden plate puzzle individual pieces represent lithospheric Plates; the stiffness of the wooden plate will highlight that the lithosphere is the most rigid layer of the Earth, then capable of breakage and where earthquake hypocenters are confined. The puzzle we use is a build up information tool (Fig. 17.1), where only the plate motion is a given data. Students are asked to place Plate names labels according with geographic hints, and guess the occurrence of earthquakes and volcanoes according to Plate Tectonics while placing little cardboard circles and playdough symbols on the puzzle. Pop-up wooden blocks showing cross sections of the major plate margins are used to trigger discussion on Plates motion.

17.3.2 How Do Earthquakes and Volcanoes Work?

Among concepts to be communicated in a simple way and relevant for understanding hazards are the mechanisms behind earthquakes and volcanic eruptions.

We take children into earthquakes' world by exploiting the simple and frequently used analogy with bending and breaking of wooden sticks. In this activity, we use sticks of different thickness and broken sticks that have been previously glued and recomposed. One of the many available computer programs designed to visualize acoustic waves is used to allow the display of the energy produced by the breakages and the description of basic concepts such as energy buildup, release, and transfer. We used GarageBand,

Fig. 17.1 The plate tectonic puzzle. Volcanoes and earthquake epicenters are placed according to plate dynamics

a commercial program for Apple supports designed for music composition, but well suitable for our purpose. A microphone will record the acoustic wave propagating by the steak rupture and displayed on a screen. The attenuation of energy is simply demonstrated by breaking sticks at different distances from the microphone. Letting children play with the sticks gives us the chance to introduce them to the occurrence of earthquakes with distinct magnitude, their different frequency of occurrence, the correspondence with preexisting faults, the distribution of damages with distance from the epicenter. Children participate with great enthusiasm, happy to freely break something and their comprehension greatly benefit from the possibility of literally seeing the energy they are putting in the game. In this way, they easily get familiar with basic ideas of the physics of earthquakes.

Relevant to understand volcanic hazard is the difference between lava and pyroclastic flow, as to the phase (liquid, solid or gas) that mostly characterize the eruption and its dynamic. We use the well known backing soda and vinegar model for the lava flow simulation to model magma resurging from the volcano conduit. The pyroclastic flow simulation relies on an extremely light and thinned particle material (i.e. backing soda powder) pumped out from the volcano conduit. We highlight that a pyroclastic flow derives from an explosion, is mostly composed by gas and extremely light particles spreading over a large area and rising up to the high troposphere. The baking soda powder will raise a maximum height; it will eventually be blown away by wind and collapse all around the volcano and further away, almost as just as during real eruptions. Students will taste the soda in their mouth even though they are standing a bit far away from the vent. We found this to be crucial in the understanding the effects of the eruption on a large area.

17.3.3 Why Earthquakes Are Not Predictable?

The QuakeCaster is a hands-on teaching model, simulating earthquakes and their interactions to help students explore and test the four leading hypotheses for earthquake occurrence. The experiment is meant for secondary school students, but might catch attention to younger students. Here we present a modified version of the QuakeCaster (Linton and Stein 2012) that allows additional inferences on earthquakes occurrence.

The idea is to demonstrate that earthquakes are neither time- nor slip-predictable so that the best way to reduce risks is prevention and preparedness.

A reel steadily pulls and simulates the steady plate tectonic motions far from the plate boundaries, wooden blocks having different thickness simulates different shear zones of the fault which are connected with a rubber band that furnishes the elastic property of the crust. Friction is simulated by sandpaper stacked on a wooden plank using pins slightly rising from the board: when the blocks simulating the fault get stuck on the pin we can also discuss asperity of the fault. The use of wooden blocks and sandpaper (Fig. 17.2) instead of the rocky blocks and the porcelain surface suggested by Linton and Stein (2012) is meant to emphasize the

Fig. 17.2 The QuakeCaster. A demonstration of faults dynamic

possibility to repeat the experiment independently in an extremely, yet efficient way.

The quake caster experiment can be run hiding away the whole tool while just observing the display of the shaking measured by an accelerometer placed on the plank. This allow emphasize the relationship between available data and complexity of natural phenomena: we record earthquake shaking while we never have a comprehensive knowledge of the fault, nor we know where the asperities might be located and we cannot predict when the stress will be released. We found that at this point the public really understands unpredictability.

17.4 Conclusions

We tested our activities reaching more than a thousand students in different environments such as in school classrooms, research institutions open doors, science festivals and other scientific venues. The public reactions were in general of wonder on how they could get in a simple way topics, not necessarily trivial or intuitive; teachers were enthusiastic in getting new educational ideas and gave us positive feedbacks. Activities were set to allow students to pose questions on major issues and give themselves a proper answer, so we could directly test the effectiveness of our approach. However we hand out questionnaires to students, teachers and general public to assess the communication efficacy (e.g. D'Addezio et al. 2014). They gave us confidence that the activities we proposed were a proper tool to trigger discussions on risk mitigation, and therefore to raise risk perception.

Acknowledgments The activities described in this paper are part of a long-term outreach program supported by INGV.

References

Bernhardsdóttir AE, Thorvaldsdóttir S, Sigbjörnsson R, Musacchio G, Nave R, Falsaperla S, D'Adda S, Sansivero F, Zonno G, Sousa ML, Carvalho A, Raposo S, Ferreira MA, Nunes JC, Jimenez MJ (2012) Disaster prevention strategies based on an education information system. In: Paper presented at the 15th WCEE, Lisbon, 24–28 Sept 2012

D'Addezio G, Rubbia G, Musacchio G, Lanza T (2014) INGV Laboratorio Didattica e Divulgazione Scientifica Team. In: Abstract of the EGU General Assembly, Vienna, 27 April–02 May 2014

Horton J (2011) Lancet 377:439

Komac B (2009) Acta Geogr Slov 49:199

King CJ (2010) Int J S. Edu 32:565

Linton K, Stein RS (2012) Seis Res Let 83:150

Wachinger G, Renn O (2013) Risk perception and natural hazards. CapHaz-Net WP3 Report. http//caphaz-net.org/outcome-results/CapHaz-Net_WP3_Risk-Perception.pdf. Accessed 10 Sept2013

18 Geological Engineers from the University of Salamanca: The Experience of Ten Years from Education to the Professional Life

J. Nespereira, M. Yenes and J.A. Cabezas

Abstract

The results of a survey distributed to students who have completed the degree of Geological Engineer (GE) at the University of Salamanca (USAL) are presented. Surveýs purpose was to know their integration into the labour market, their satisfaction with the degree and what is the perception they have about societal point of view towards GE. The GEs from the USAL work mainly in the field of construction and believe that GE is not properly valued in the labour market mainly because of the ignorance that still exists in relation to their training and capabilities. Overall, they are satisfied with the education received, and they would choose again the same university studies. Respondents suggest a broader training, mainly focussed in the exploitation of natural resources and geotechnics, more practical content, and they also claim for a mandatory work in industry during the study years.

Keywords

Geological engineer • Profession • Salamanca • Education • University

18.1 Introduction

In Spain Geological Engineering (GE) education begins at the Complutense University of Madrid, with a postgraduate course taught between 1980 and 1990, becoming an M.Sc. degree in Geological Engineering later on. Also in 1990 the joint implementation of their own degree in Geological Engineering at the University of Barcelona and Catalonia Polytechnic began. Years later, during 1997–1998 the University of Alicante did the same, and finally in 1999, with the publication of Real Decreto 666/1999 (Ministry of Education. Government of Spain 1999), the degree of GE is officially recognised. At the USAL, comparable studies began in 2001/2002, structured according to the curriculum published in January 2002 (University of Salamanca 2002) and based in turn on the Real Decreto, which divided the formation in two cycles of three and two courses respectively. Thereby, the titulation consisted of five courses, but the effective time spent to fulfill the second cycle used to increase because a final project must also be presented.

Annual reports about the international status of geological engineering showed the evolution of the professiońs activity from different points of view, and among them how university education it is also dealt with (Hatheway and Reeves 1997, 1999; Hatheway and Kanaori 2002; Hatheway et al. 2005). Focussing on the geological engineering degree in Spain, Tomás et al. (2005) presented the universities where it was taught and the program contents but it analyzed a period of time too short to show how the graduates were joining the labour market.

J. Nespereira (✉) · M. Yenes
Department of Geology, University of Salamanca, 37008 Salamanca, Spain
e-mail: jnj@usal.es

M. Yenes
e-mail: myo@usal.es

J.A. Cabezas
Department of Mechanical Engineering, University of Salamanca, 37008 Salamanca, Spain
e-mail: jacf@usal.es

G. Lollino et al. (eds.), *Engineering Geology for Society and Territory – Volume 7*,
DOI: 10.1007/978-3-319-09303-1_18, © Springer International Publishing Switzerland 2014

This paper has three main objectives, which are to provide an image of the GE from the USAL, get close to the social perception that the GE observed in their environment in relation to their activity and training, and create a platform for teachers to receive those contributions or improvements that according to the GE's criteria should be included in training. With these purposes in mind we created and distributed by e-mail a survey with thirty questions. The potential population of the survey was 110 students, which were the total course enrollments between 2001/2002 and 2007/2008 that have not abandoned the degree. Finally, twenty-four questionnaires were returned, based upon which a statistical analysis has been carried out looking at the different issues questioned.

18.2 Results

From 2001/2002 to 2009/2010 one hundred and sixty six students began the GE degree at the USAL, of which some 28 % abandoned the effort before conclusion (Fig. 18.1). More than half of the students accessed the GE degree after failing to get into other options: GE ended up being the final alternative for students with very different original preferences (medicine, architect, chemical engineering…).

Of the 110 students who join the degree between courses 2001/2002 and 2007/2008 and that did not quit, there are sixty-seven that have already finished. Considering the first 6 years, rate of success, defined as the ratio between the number of students graduated until December 2012 and the number of students who have not abandoned the degree and could have completed it on that date, stands at 67 %.

Some 88 % of respondents are already GEs, after spending an average of 7 years for completing the degree, that is 1 year above the generally accepted minimum period of time for 5 years degrees that include a final Project Defense. This average matches with the requirements of European convergence process of University studies, which points out that the time spent for the student getting their degree should suit the expected duration of the studies. The final majority income stands at the cradle of five to seven points out of a maximum of ten.

The annual dropout rate is one of the criteria taken into account when assessing the universities. Between 2002/2002 and 2010/2011 the average for the GE studies was 28 %. Even though this rate varies among different universities, the average in Spain for technical studies is 40 % (Cabrera et al. 2006), so that the rate of the USAL is clearly lower.

About 71 % of GEs respondents consider they are working on tasks related to their training. In most cases, the lapse of time between the completion of their studies and their first day of working life does not exceed a year. Nevertheless, it is representative of the current economic crisis that since 2011 this interval of time is increasing.

The GEs trained at the USAL have found work in the areas of construction (72 %), research (12 %) and the exploitation of natural resources. We have also observed a substantial mobility in the profession, because up to 60 % persons have already been in at least two different companies.

High youth unemployment in Spain is causing an important phenomenon of emigration. Is it happening to the GEs from USAL? The answer is yes, with 26 % of them working or have being working abroad: recipient countries are Ireland, UK, Australia, Norway and Burkina Faso.

The degree in GE has a serious problem related with the scarce penetration of his profile in the society, and so, in Spain no one doubts about the formation of a Civil Engineering (CE), a Industrial Engineering (IE), a Mining Engineering (ME), a Geologist (G)… But we can not say the same for GE, and this fact certainly hampers their professional development, as was warned by Knill (2003) as one of the weaknesses of their profession. This point is corroborated by the results of the survey, with 75 % of respondents considering the degree is still unknown to society and that GE is undervalued in Spain; curiously, among those who have worked or are working outside the country, the majority opinion is that the GE is much better

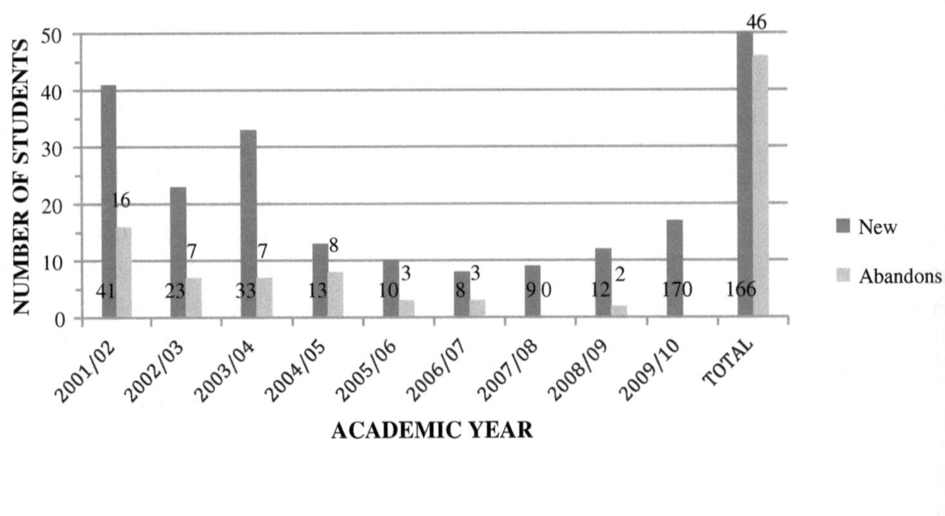

Fig. 18.1 Evolution of the number of students during the years considered in this study

considered outside our borders. A reason that partially helps to explain this situation is that in Spain there are some degrees that have associated professional responsibilities officially established and regulated by law, among which are those of some co-workers of GE's. They thus use their more advantageous legal situation for progressing inside the companys at the detriment of GEs. This is opposite to what happens in other countries like Canada where, after an initial dispute, the degree is accredited as engineering at several universities (Hatheway and Kanaori 2001). In Spain the future in this area is still in the air, and will be linked to the implementation of a new law that will regulate Professional Services.

With respect to the adequacy of the tasks performed in the company and the tasks for which they feel qualified, the graduates general opinion is that they are suited mostly or always (61 %), and do not fit in 11% of the cases. The GE's share office with IE (46 %) and G (31 %), ME (15 %) and to a lesser extent, CI (8 %). Faced with II, IM and CI, the GE believes they have an advantage because of their greater geotechnical training, but they also feel they lack knowledge of mechanics, electricity (IE), the operation of a mine (MI) and structures (CI). Faced with the G, the advantage is the increased ability to manage projects, and as a disadvantage less knowledge of geology.

Soil Mechanics and Applied Geology are the most useful subjects for the GE jobs, calling more attention than other areas that close to the geological engineering profile such as hydrogeology, Earthquake Engineering, and Environmental Engineering, that have not been within the most cited. Probably this is so because society does not perceive or do not know the possibilities that a GE has in these fields.

Again 77 % of respondents are satisfied with their training, and 71 % would complete the same degree again; lack of careers and lack of recognition within the industry are the causes signaled for the ones that would not follow the same degree again. Improvement proposals made by the respondents are focussed on increasing the practical training through more field classes and site visits, and also introducing the use of some of the most used computer tools in geological engineering. There are also some opinions that claims to include curricular practical training as compulsory, but that polled also suggest that some subjects should be refocus from a closer perspective to the geological engineering professional life. In relation to this matter, Hatheway and Reeves (1997) saw a problem in the fact that a lot of teachers come from areas far away from the activity of Engineering Geology; referring to the situation in the UK, Griffiths and Culshaw (2004) noted not only the lack of GEs within the lectures, but also the fact that many teachers had never been in direct contact with the subject. No doubt the suggestion can and should be taken into account, but the teacher profile drawbacks pointed out by previous authors are also present in the USAL, where there are teachers with professional experience outside the University, and therefore can easily give a more practical approach to their teaching, but also teachers outside of what society should and can request to a GE, to whom this task will be more difficult to address.

18.3 Conclusions

The degree in Geological Engineering has lasted for 10 years since its introduction as an officially recognized title in the USAL. Students consulted by us show high satisfaction with their studies, and most of them have found a place in the labor market. The degree has a dropout rate of 28 %, below the national average for technical careers.

The socioeconomic crisis in Spain is being reflected in the working life of GEs of the USAL, with an increase in the time spent to get a first job. Clearly, this is due to the fact that until 2007 in Spain there was an intense activity in construction—infrastructure projects and new houses—which favour the hiring of GEs. Today, with a drastic reduction in budgets for new infrastructures, the GEs should turn their eyes to new horizons, without ruling out the job search outside our borders that some have already successfully tackled.

Socially, GEs think they play a second class role within the areas or departments in which they participate, perhaps motivated by the lack of knowledge about the possibilities of a GE, and because another graduates working together with GEs benefit themselves from professional responsibilities that in Spain are assigned to them by law, which is not the case today with GEs. The future in this area is still in the air, and will be linked to the implementation of a new law regulating Professionals.

The suggestions in teaching aimed towards strengthening the practical training of the degree. The authors of this study believe that the organization of an annual journey dedicated to exposing the alumni professional activity may be a mechanism to bring the day to day of the profession both to teachers and students. We also believe that the degree would get a big boost if curricular practical training turns to be considered as mandatory, as it happens in other engineering degrees in Spain.

References

Cabrera L, Bethencourt JT, Pérez PA, Afonso MG (2006) El problema del abandono de los estudios universitarios. RELIEVE 2:171–203
Griffiths JS, Culshaw MG (2004) Seeking the research frontiers for UK engineering geology. Q J Eng GeolHydrogeol 37(4):317–325
Hatheway AW, Kanaori Y (2001) 6th annual report on the international status of engineering geology—year 2000 encompassing

hydrogeology, environmental geology and the applied geosciences. Eng Geol 61(1):1–27

Hatheway AW, Kanaori Y (2002) 7th annual report on the international status of engineering geology—year 2001: encompassing hydrogeology, environmental geology and the applied geosciences. Eng Geol 65(1):47–79

Hatheway AW, Reeves GM (1997) Status of engineering geology in North America and Europe1. Eng Geol 47(3):191–215

Hatheway AW, Reeves GM (1999) A second review of the international status of engineering geology—encompassing hydrogeology, environmental geology and the applied geosciences. Eng Geol 53(3):259–296

Hatheway AW, Kanaori Y, Cheema T, Griffiths J, Promma K (2005) 10th annual report on the international status of engineering geology—year 2004–2005; encompassing hydrogeology, environmental geology and the applied geosciences. Eng Geol 81(2):99–130

Knill J (2003) Core values: the first Hans-Cloos lecture. Bull Eng Geol Environ 62(1):1–34

Ministerio de Educación y Cultura. Gobierno de España (1999) Real Decreto 66/1999. Boletín Oficial del Estado n° 109, pp 17210–17213

Tomás R, Bañón L, Cano M (2005) Ingeniería Geológica: una carrera moderna como respuesta a la demanda de especialistas en el terreno. Documentos del XIII Simposio sobre Enseñanza de la Geología, pp 1–18

de Salamanca Universidad (2002) Resolución de 7 de enero de 2002. Boletín Oficial del Estado 21:3201–3215

Rumours Related to the 2012 Emilia Seismic Sequence

19

Milena Bertacchini, Doriano Castaldini and Giovanni Tosatti

Abstract

In the aftermath of an earthquake, pseudo-scientific, scaremongering rumours spread very quickly and with greater effect than correct scientific information. The aim of this article is to describe the authors' observations and examine how rumours linked to seismic shocks spread in the community following the strong 2012 seismic sequence in Emilia (northern Italy). During the two mainshocks, 27 people lost their lives, over 400 persons were injured and 14,000 families were evacuated. In the weeks following the mainshocks, in collaboration with other universities and research groups, we carried out field surveys, organized interviews and public meetings with the local population and held direct contacts with journalists and reporters. The mainshocks of this sequence ($M_L = 5.9$ on 20th May and $M_L = 5.8$ on 29th May) found the local population culturally and psychologically unprepared. As a consequence, there were attitudes of suspicion and lack of trust towards the authorities and the scientific community. Many people considered these earthquakes induced by human activities such as exploitation of subsurface resources or fracking. Moreover, in spite of the experts explaining in all possible ways the origin of earthquakes and emphasizing their unpredictability, the population has given credit to various groundless alarms on the basis of gas emissions, bubbling water and ground fractures. In order to effectively counter the spreading of wrong convictions about earthquakes, a constant, updated dialogue must be kept up between the population and the scientific community.

Keywords

Rumours • Earthquake • Environmental effects • Emilia • Northern Italy

19.1 Introduction

Every time there is an earthquake in Italy, and probably not only in this country, pseudo-scientific information spreads more quickly and to greater effect than correct scientific facts. Wrong or irrational information spread as *rumours* can have an economic, political, social or cultural origin. Typically, this misguided information is not supported by objective facts but rather by the fears and uncertainties of most of the population (Crescimbene et al. 2012).

In this era of global communication and social networks, there are preferential ways which facilitate and amplify the spreading of rumours. This happened, for example, with news appearing on the web concerning the purported prediction of an earthquake in Rome on the 11th May 2011 (Georgiadis and Pescerelli Lagorio 2012; Nostro et al. 2012), not to mention the Mayan prophecy of disasters potentially resulting from the alignment of celestial bodies in December 2012.

The aim of this article is to examine how rumours linked to seismic shocks develop in the community, as observed by the authors following the strong seismic sequence in Emilia

M. Bertacchini · D. Castaldini (✉) · G. Tosatti
Department of Chemistry and Earth Sciences, Modena and Reggio Emilia University, Modena, Italy
e-mail: doriano.castaldini@unimore.it

G. Lollino et al. (eds.), *Engineering Geology for Society and Territory – Volume 7*,
DOI: 10.1007/978-3-319-09303-1_19, © Springer International Publishing Switzerland 2014

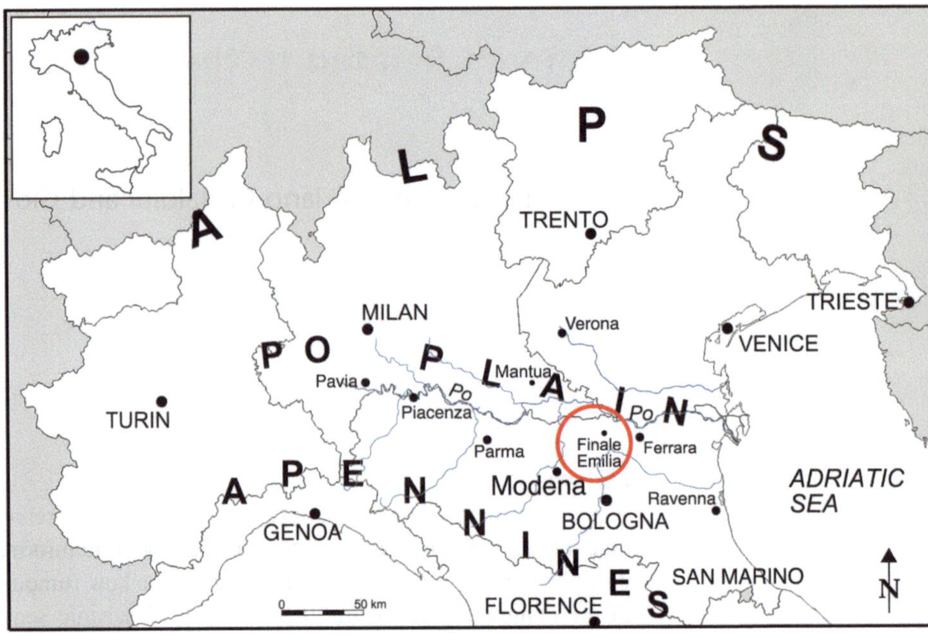

Fig. 19.1 Epicentral area of the 2012 Emilia seismic sequence

(northern Italy). In collaboration with researchers from other universities and Italian research agencies, the authors have carried out field surveys for detecting environmental effects and held meetings with the local population, together with interviews and direct contacts with journalists and reporters. In addition, further elements for reflection are suggested in order to deal with the problem of correct information on seismic events.

19.2 The 2012 Emilia Seismic Sequence and Its Environmental Effects

The Emilia seismic sequence struck the lower central part of the Po Plain, in northern Italy. All together, 27 lives were lost, over 400 persons injured and 14,000 families evacuated (Fig. 19.1).

The mainshock occurred on 20th May 2012 ($M_L = 5.9$; focal depth ca. 6 km), with the epicentre located in the low Modena plain. A second strong shock took place on 29th May 2012 ($M_L = 5.8$; focal depth ca. 10 km), about 12 km west of the first earthquake. The seismic sequence, caused by buried Apennine faulted folds, known as the "Ferrara Folds", was characterized by another five $M_L \geq 5$ events, and more than 2,500 aftershocks of lower magnitude were recorded from 19th May 2012 to 19th May 2013 (ISIDe 2013).

According to Di Manna et al. (2012), more than 500 ground effects were recognised, spread over an area of about 700 km^2, whereas, according to the Emergeo Working Group (2013), more than 1,000 environmental effects, spread over an area of about 1,200 km^2 were detected. They occurred during and immediately after the shocks and mainly consisted of soil liquefaction phenomena, ground ruptures, and water table fluctuations. About 80 % of the coseismic effects were induced by the mainshock. Many liquefaction cases triggered by the 20th May 2012 earthquake were reactivated by the 29th May 2012 event (Di Manna et al. 2012). This seismic sequence has been analysed from various viewpoints (seismological, seismotectonic, geomorphological, historical etc.) as shown by many articles (among them, worthy of note is the volume edited by Anzidei et al. 2012).

19.3 Impact of the Environmental Effects on the Population

19.3.1 Rumours Attributing the Causes of Earthquakes to Human Activities

This seismic sequence and, in particular, the two strongest quakes found the local population culturally and psychologically unprepared, generating attitudes of suspicion and lack of trust towards the authorities and the scientific community. One of the main causes of this attitude is lack of knowledge of the geological features of the territory and, consequently, a widespread, wrong conviction that the lower Po Plain is a low-seismicity area. This is also related to the fact that this sector of Italy is classified as "low seismic hazard", although this definition does not mean that strong earthquakes cannot take place in this area but, rather, are a rare occurrence.

The main cause of damage is due to the fact that the norms concerning the planning of new buildings were not implemented in the affected area before 2003 and in some cases, not even afterwards. Therefore, what was really underestimated in non-scientific circles was the need to reduce seismic risk (Stucchi et al. 2012).

According to this widespread view resulting from poor knowledge, the deposits of alluvial soil should contribute to reduce the effects of seismic shocks, when, in fact, the opposite is true. One reason for this common mistaken opinion is the loss of historical memory about strong seismic events of the past. For example, the M_L 5.5 Ferrara earthquake of 1570, characterised by a seismic sequence lasting 4 years (Solerti 1889), was recalled to mind only after the shock of 20th May 2012.

As a consequence of loss of historical memory, many people have attributed the causes of the Emilia earthquakes not to tectonic causes linked to the dynamics of the Northern Apennines but to actions carried out by man. Part of the local population was convinced that the seismic shocks were triggered by exploitation of subsurface resources (such as the Cavone oil wells in the epicentral area) or by fracking activities (i.e. hydraulic fracturing of subsurface low-permeability rocks). Another element connected to these mistaken beliefs was the planning of a gas storage facility in Rivara, near the epicentre areas, although this project was rejected by the regional and local authorities and therefore was never implemented. Since all these wrong convictions are widespread among the population, in November 2012 the Emilia-Romagna Region established an international "Technical-scientific Commission for assessing the possible relationships between hydrocarbon exploration activities and increase of seismic activity in the Emilia area struck by the 2012 earthquakes". For the time being, the Emilia-Romagna Region has decided to suspend future programmes of hydrocarbon exploration in all the territory struck by this seismic sequence until the appointed Commission expresses an opinion. In addition, late in July 2013, the regional Geological and Seismic Survey wrote a report clarifying the real causes of these quakes (Regione Emilia-Romagna—Servizio Geologico Sismico e dei Suoli 2013). This report clearly states that there is no relationship between hydrocarbon research and exploitation and origin of these earthquakes, also because fracking techniques have never been authorised. Furthermore, the presence of hydrocarbons on the ground surface has been a well-known phenomenon all over Emilia since ancient times and cannot be considered a premonitory sign of an imminent earthquake.

In addition, some people even believed that extreme meteorological events could generate earthquakes: the M_L 3.8 quake generated by the Ferrara Folds in the lower Modena plain on 4th May 2013 was linked to a tornado which, in the late afternoon of the previous day, had hit the same area causing considerable damage.

19.3.2 Rumours About Earthquake Prediction

In any case, the most important and delicate problem following the Emilia seismic sequence has concerned earthquake prediction. Although experts—including the authors—have repeatedly stressed in all possible ways the unforeseeable nature of earthquakes, the population has given credit to various groundless alarms which were spread around. In fact, a few months after the main seismic shocks of 2012, many people who had gone back to their homes abandoned them again out of fear, following the unjustified prediction of future quakes.

The tragic experiences of the earthquake-struck population, have induced some people in the epicentral area and its surroundings to pay particular attention to natural phenomena occurring in their own territory, wrongly linking them to premonitory signs of earthquakes.

For example, in mid-February 2013, in the Ferrara plain, a jet of methane and hot water gushing out from the soil aroused concern among local inhabitants since it was interpreted as a precursory sign of a seismic event. Actually, this vent was coming from a poorly sealed old methane well and the alert ceased. A similar phenomenon, accompanied by gurgling of water and gas (Fig. 19.2a), took place in May 2013 in Mantua province. In this case, it can be explained as the emission of atmospheric gas (mainly nitrogen) due to the rising of the water table (Fig. 19.2b). In mid-August 2013, at the outskirts of Ferrara, a ground fracture (Fig. 19.2a) similar to those occurring in the 2012 seismic sequence (Fig. 19.2b), was observed and reported sensationally by the local media. However, this crack was caused entirely by soil settlement. Furthermore, in the epicentral area, many people observed a marked increase—up to 400 cm—of the water table inside phreatic wells, confirmed also by instrumental recording, usually occurring some days before and during the mainshocks. In some cases an increase in water temperature was also observed from irrigation wells, but it was later verified that this phenomenon was related to the overheating of water pumps clogged with sand (Marcaccio and Martinelli 2012). The memory of these anomalies in the people of Emilia created alarm and fear of a new earthquake when similar phenomena were observed during a heavy rain period in spring 2013 (further information on the problem of earthquake prediction can be found at INGV-DPC Project S3 2012–2013).

19.4 Final Remarks

The meetings and conferences between the authors and the earthquake-struck population of Emilia have helped in understanding the importance of a constant, updated

Fig. 19.2 **a** Bubbling water in a field near San Giovanni del Dosso (Mantua Province) on 16th May 2013, **b** Soil liquefaction in progress on 29th May 2012 triggered by the M_L 5.8 quake at Moglia di Gonzaga (Mantua Province). *Sources* **a** A. Zibordi; **b** L. Righi

Fig. 19.3 **a** Ground crack noticed in mid-August 2013 at the outskirts of Ferrara, **b** ground crack formed by the 20th May 2012 M_L 5.9 quake across the football pitch of Mirabello (Ferrara Province). *Sources* **a** and **b** D. Castaldini

dialogue between the population and the scientific community in order to wipe out irrational fears and discredit tales and wrong ideas about earthquakes (Fig. 19.3).

The dialogue is continuing by means of various activities organized by the authors, among which a course directed to high-school teachers and a collaboration with the Municipality of Rovereto s/Secchia for setting up an earthquake museum project. Only constant dialogue and comparison will be able to provide people with correct information and advice on the most appropriate behaviour in case of seismic shocks. Scientists should also remember the role of local populations as keepers of their own territory and as irreplaceable witnesses of any change occurring on their land. Therefore, the dissemination of scientific culture should be promoted by taking into account not only technical criteria and scientific arguments but also the cultural and traditional aspects of a given territory (De Marchi 1996).

Acknowledgements Thanks are due to Dr. G. Martinelli from the Geological Survey of the Emilia-Romagna Region and to Ms. A. Zibordi for the information on the phenomena observed in the territory of San Giovanni del Dosso.

References

Anzidei M, Maramai A, Montone P (eds) (2012) The Emilia (northern Italy) seismic sequence of May–June, 2012: preliminary data and results. Ann Geophys 55:4, 842

Crescimbene M, La Longa F, Lanza T (2012) The science of rumors. Ann Geophys 55(3):1–7

De Marchi B (1996) Il terremoto come metafora, la ricostruzione come sogno. Quaderno 96 (6), ISIG, 1–10

Di Manna P, Guerrieri L, Piccardi L, Vittori E, Castaldini D, Berlusconi A, Bonadeo L, Comerci V, Ferrario F, Gambillara R, Livio F, Lucarini M, Michetti AM (2012) Ground effects induced by the 2012 seismic sequence in Emilia: implications for seismic hazard assessment in the Po Plain. Ann Geophys 55(4):697–703

Emergeo Working Group (2013) Liquefaction phenomena associated with the Emilia earthquake sequence of May–June 2012 (Italy). Nat Hazards Earth Syst Sci 13:935–947

Georgiadis T, Pescerelli Lagorio P (2012) The communication of science as an ethical issue: the case of Raffaele Bendandi. Ann Geophys 55:3

INGV-DPC Project S3 (2012–2013) Short term earthquake prediction and preparation. http://sites.google.com/site/ingvdpc2012progettos3/

ISIDe (2013) Italian seismic instrumental and parametric data-base. http://iside.rm.ingv.it/iside/standard/index.jsp

Marcaccio M, Martinelli G (2012) Effects on the groundwater levels of the May–June 2012 Emilia seismic sequence. Ann Geophys 55(4):811–814

Nostro C, Amato A, Cultrera G, Margheriti L, Selvaggi G, Arcoraci L, Casarotti E, Di Stefano R, Cerrato S, the May 11 Team (2012) Turning the rumor of the May 11, 2011, earthquake prediction in Rome, Italy, into an information day on earthquake hazard. Ann Geophys 55:3

Regione Emilia-Romagna—Servizio Geologico Sismico e dei Suoli (2013) Terremoti emiliani 2012, tra certezze storiche e indagini scientifiche. http://ambiente.regione.emilia-romagna.it/primo-piano/2013/terremoti-emiliani-2012

Solerti A (1889) Il terremoto di Ferrara nel 1570. Rassegna emiliana di storia, letteratura ed arte, Modena 2:517–528

Stucchi M, Meletti C, Bazzurro P, Camassi R, Crowley H, Pagani M, Pinho R, Calvi GM (2012) I terremoti del maggio 2012 e la pericolosità sismica dell'area: che cosa è stato sottostimato? Progettazione Sismica 4(3):63–73

The Experience of ScienzAperta, a Week of Scientific Information and Dissemination

20

Giuliana D'Addezio, Giuliana Rubbia, and Antonella Marsili and The Laboratorio Didattica e Divulgazione Scientifica Team

Abstract

ScienzAperta, i.e. Open Science, is an outreach week conceived and promoted by the Laboratorio Didattica e Divulgazione Scientifica of the Istituto Nazionale di Geofisica e Vulcanologia (INGV). Every year ScienzAperta is an opportunity to open to the public and to share with the community the places where research is performed, through events distributed in different times and places, with scientific exhibitions, hands-on laboratories for kids, meetings and seminars with researchers and guided tours to scientific laboratories. All the activities have the common idea to intrigue, interest and stimulate audiences of all ages. Some of these initiatives have been organized jointly with other institutions and research institutes, cross-pollinating disciplines. On the occasion of the 2013 edition, questionnaires have been distributed among adults and children during the Open Day held in INGV Rome headquarters. Features of initiatives performed during the three first editions as well as hints from appreciation surveys are illustrated.

Keywords

Scienzaperta • Science and society • Hands-on laboratories • Outreach

20.1 Introduction: An Open Science Week

ScienzAperta is an outreach week promoted by the Laboratorio di Didattica e Divulgazione Scientifica of the Istituto Nazionale di Geofisica e Vulcanologia (INGV). This initiative was born in 2011 and it stems from the original "Week of Scientific and Technological Culture" promoted by the Ministry of Education, University and Research (MIUR) of Italy. It is intended to respond to the needs and the requests of the community for more information on issues regarding our planet. Since Italy is a place prone to seismic and volcanic activities education and outreach programs play an important role to convey simple, clear and complete information able to contribute to a better understanding of the territory features and the related hazard and risk. This approach derived from the consciousness that preparedness is the best way to live with and to mitigate natural hazards. The common goal is to engage the community in a correct, straightforward and efficient communication on scientific research and technological innovations. In a world that request citizens to be more informed, aware and able to make crucial decisions about their own health and safety, the knowledge is crucial to handle doubts and to know how to choose with consciousness.

G. D'Addezio (✉) · G. Rubbia · A. Marsili · The Laboratorio Didattica e Divulgazione Scientifica Team
Istituto Nazionale di Geofisica e Vulcanologia, via di Vigna Murata 605, Rome, Italy
e-mail: giuliana.daddezio@ingv.it

G. Rubbia
e-mail: giuliana.rubbia@ingv.it

A. Marsili
e-mail: antonella.marsili@ingv.it

G. Lollino et al. (eds.), *Engineering Geology for Society and Territory – Volume 7*,
DOI: 10.1007/978-3-319-09303-1_20,

ScienzAperta weeks include events of different kinds, which are conceived for both general public and schools. Events are held in INGV headquarters in the Italian territory as well as in museums and cultural centres. In the three past editions, programs have been organized in the INGV headquarters of Rome, Catania, Naples, Genoa, Arezzo, Pisa, Portovenere and Grottaminarda. Events are scheduled over 5–6 days in Spring. Content covers the different sectors of INGV scientific mission. Event types include: seminars held by researchers, guided tours of laboratories and of seismic surveillance rooms, educational tours for students, exhibitions, educational games, movies and debates, scientific theatre performances and concerts. Programs, photos and press reviews are available on the dedicated website http://www.scienzapertaingv.it. Aims of the paper are to illustrate into details features of ScienzAperta and to comment feedbacks from public.

20.2 ScienzAperta in Rome

We describe the most noteworthy events organized specifically for the INGV headquarters in Rome. In each of the three editions we welcomed 1,000–1,200 visitors of all ages. The first edition was co-organized with another research institute, Istituto Nazionale di Ricerca per gli Alimenti (INRAN). On this occasion, geophysics was linked to food and health, for example presenting geodynamic evolution of the Mediterranean and evolution of the Mediterranean diet.

In all three editions, events that efficiently link scientific information to music were also organized, engaging the emotions of the audience in the transfer of knowledge. These events where conceived as evening happenings for the general public, as concerts which follow the scientific seminars such as: "Seismic waves, sound waves: from earthquake to music" (Hunstad et al. 2013); "A musical journey among the Italian earthquakes", aims to raise the awareness of the great historical earthquakes of the Italian peninsula, making a historical and musical journey through the regions hit and using the local folk music to revive the popular traditions (Marsili et al. 2013). "Waves sympathy and music", experiments that help in understanding the features of seismic waves and reducing damages upon buildings. "Landscapes, territory and wine" was conceived as a new geo-enological event, presenting a journey in the Italian peninsula to understand the link between geodynamic phenomena, landscapes and earth products, followed by a wine degustation.

20.2.1 ScienzAperta: Seminars and Debates with General Public

According to European and Italian citizens, scientists who work in public research centers are the most authoritative professionals to explain social impact of science and technology (EC 2010). In this line, we proposed seminars regarding topics of general, current and sometimes crucial interest, implying a continuous interaction with researchers. For example, for the 2011 edition we proposed seminars on Fukushima, Japan, earthquake and on outstanding aspects related to the perception of seismic risk in Rome (i.e. Rome: not only earthquakes). The latter format were used also for the Open Day devoted to the presumed prediction of a devastating earthquake in Rome on May 11 which concerned Roman inhabitants and tourists (Amato et al. 2011).

20.2.2 ScienzAperta: Educational Games, Hands-on Laboratories and School Calendars for Kids

In all programs special attention has been devoted to initiatives dedicated to schools and teachers, with hands-on laboratories on earthquakes, volcanoes, of course, and also on INGV researches in Antarctica.

During educational path students meet up geosciences researchers that works in the field of the natural hazard.

The ScienzAperta week includes also both the promotion and the award ceremony of a drawing competition for children of primary schools. The intent is to provide a pleasant stimulus of discussion both for teachers and pupils about earth sciences arguments. Schools participate with enthusiasm and send drawings made by children on a specified theme, different each year. During the 2011 edition, the drawing competition for 2012 calendar was inspired by the International Year of Sustainable Energy and the calendar "Mission possible: let's save the world" was launched; in the same occasion, the calendar of the previous year "I am a scientist too! Science and scientist from the children point of view" was awarded. The analysis upon hundreds of collected drawings revealed precious elements of image of scientists in kids (D'Addezio et al. 2013). The 2013 calendar "In the heart of earth" was intended to illustrate all those phenomena which occur under our feet, while the 2014 one "The magic of water" to raise water resources awareness.

Fig. 20.1 Pupils and their parents attending a laboratory about volcanoes held by a researcher in INGV Rome headquarters' courtyard on the occasion of ScienzAperta Open Day in April 2013

20.2.3 Scienzaperta: The 2013 Open Day: Laboratories, Games and Theatre

On the occasion of the 2013 edition, the Open Day occurred on Saturday April 20th in Rome INGV headquarters (Fig. 20.1). It included visits to laboratories for adults, educational games and hands-on laboratories for kids, an exhibition and a performance of scientific theater, suggested from a previous experience of narrative (Lanza et al. 2013).

Questionnaires have been distributed during the second part of the afternoon. We collected 33 questionnaires compiled by children, while 42 by adults.

Although in small percentage in comparison with the number of visitors of the whole Open Day, that summed up to hundreds, compiled questionnaires revealed some precious hints about users themselves, appreciation and margins of improvement of the initiatives, both in organization and in content.

Here some of the questions addressed in the questionnaire:

- Did you find the visit/lab/seminar/game/show: interesting-boring; easy-difficult; useless-useful; unorganized-organized?
- Does your knowledge of these arguments come from: studies, work, outreach books, documentaries and TV programs about scientific arguments?
- Do you think that those information you received were appropriate to deep your knowledge? Yes, no. If yes, which of them?
- Was there something unclear or that you did not appreciate? Yes, no. If yes, which item?
- Do you have any opinion or suggestion about the visit?

Adults of all ages, 25–80 years old, as well as pupils, 4–11 years old, participated to the different visits. Adults, mainly women (62 %), have all a degree or an university degree, and half of them have already visited a scientific exhibition or a research center. Visitors were employees, retirees, teachers and professors, students and researchers, housekeepers and also physicians, lawyers, craftsmen and a musician. For most people background knowledge derives from personal interests in documentaries and TV programs about science, for a fewer from educational books, specific studies and professional related reasons.

75 % of visitors took part to at least three events in the program, such as the visit to the seismic surveillance room, the scientific laboratories, as well as the performance of scientific theater, 25 % attended them all (Fig. 20.2).

Information acquired during the visits were perceived as useful to get more in depth with the topics by ¾ visitors. Nothing was perceived as not clear, and a few visitors wrote also appreciation comments and invited to repeat such events more frequently. An university student in geology, male, 30 years old wrote "*It has been a wonderful experience, and very useful with regards to the faculty I attend. In my opinion such Open Days should be organized more frequently*". A craftsman suggested that "*a ticket would be appropriate. Public research funding are not enough*". We choose these comments because they are not only

Fig. 20.2 Percentages of visits according to adult responders of the questionnaire

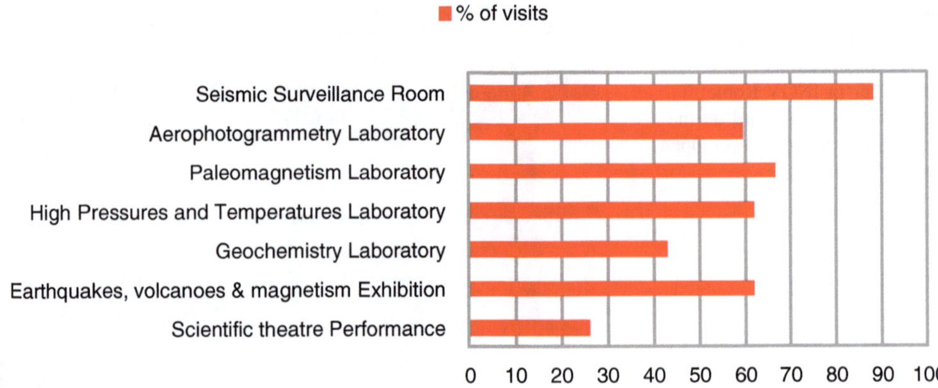

	N. of visitors	% of visits
Scientific theatre Performance	11	26.19
Earthquakes, volcanoes & magnetism Exhibition	26	61.9
Geochemistry Laboratory	18	42.86
High Pressures and Temperatures Laboratory	26	61.9
Paleomagnetism Laboratory	28	66.67
Aerophotogrammetry Laboratory	25	59.52
Seismic Surveillance Room	37	88.1

Fig. 20.3 Perceptions about the educational games. Mean values on a 1–4 Likert scale are plotted for each game: **1** What (rock) am I touching?, **2** The plate tectonic puzzle, **3** What shall I do in case of an earthquake?, **4** What shall I take in case of an earthquake? Respondents are 18 girls and 11 boys, 4–11 years old

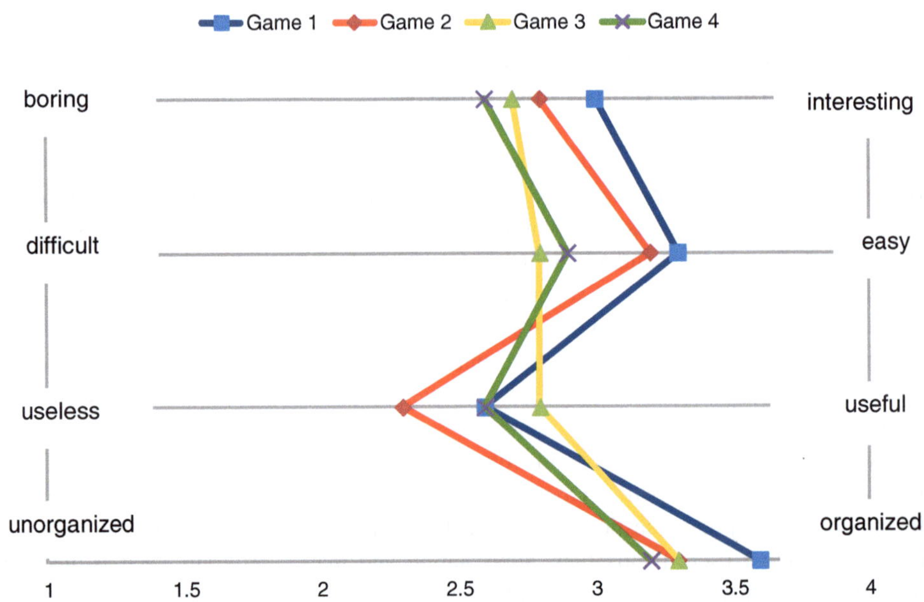

		Game 1	Game 2	Game 3	Game 4
Interesting-boring	1	3.6	3.3	3.3	3.2
Easy-difficult	2	2.6	2.3	2.8	2.6
Useful-useless	3	3.3	3.2	2.8	2.9
well organized badly organized	4	3	2.8	2.7	2.6

interesting but also much more grounded in the context than we can expect from general public.

A woman retiree appreciated courtesy of the staff. A couple of students suggested to separate children from adults.

On a four degree scale, children perceived the activities performed during the games as very interesting, very useful, well organized, but in some cases not so easy to be understood. In all, information acquired during the visits were perceived as useful to get more in depth with the topics and half of children explicitly wrote the arguments of the lesson learned: plates and their positions, volcanoes and volcanic rocks, how to behave in case of an earthquake. (Fig. 20.3).

20.3 Conclusive Remarks

As stated by the European Charter for Researchers, researchers should ensure that their research activities are made known to society at large in such a way that they can be understood by non-specialists, thereby improving the public's understanding of science (EC 2005). ScienzAperta at INGV is an outreach week which is intended to go in this direction through a wide range of educational and dissemination events for people of all ages; after three years, it would be close to become a routine institutional appointment, provided that resources are available; feedbacks from public appear encouraging, and visitors ask for repetition of these outreach programs. In this light and based on the gained experience, new formats and new ideas will be designed for future editions.

Science communication can be considered a tool in risk mitigation strategies, but also a tool to promote science study, and science as possible student's choice for their future carriers. A long-term effort in educational programs is essential. But, how can we measure long term effect on public? How can we measure and increase society appreciation and recognition of our research center? What about evaluation of careers for researchers involved in these "third mission" activities? Careful considerations about performance indicators of such initiatives are desirable, in order to overcome the "goodwill exercise with no reward" stereotype of public engagement activities; it seems that both institutions and individual researchers have to put more efforts upon this issue, for example referring to rating models described in Neresini and Bucchi, (2011). Ethical perspectives are also to be kept in mind, like those invocated by Matteucci et al. (2012), involving responsibility of geologists as professionists and communicators in geohazards prone countries, like Italy.

References

Amato A, Cultrera G, Margheriti L, Nostro C, Selvaggi G, 11 Maggio Team (2011) How to turn the rumor of the May, 11 earthquake in Rome (Italy) into a day of information on earthquakes: the Open Day at INGV. Geoitalia 2011, VIII Forum Italiano di Scienze della Terra, Torino, 19–23 settembre 2011. http://hdl.handle.net/2122/7650

D'Addezio G, Marsili A, Rubbia G, Carosi A (2013) Science and scientists from the children point of view, an overlook from drawings. Geophys Res Abstr 15:EGU2013-13070. http://hdl.handle.net/2122/8728

European Commission (2005) The European charter for researchers. The code of conduct for the recruitment. Office for Official Publications of the European Commission, Luxembourg. http://ec.europa.eu/euraxess/index.cfm/rights/europeanCharter

European Commission (2010) Special eurobarometer 340. Science and technology. Office for Official Publications of the European Commission, Luxembourg. http://ec.europa.eu/public_opinion/archives/ebs/ebs_340_en.pdf

Hunstad I, Marsili A, Casale P, Vallocchia M, Burrato P (2013) Seismic waves and sound waves: from earthquakes to music. Seismol Res Lett 84(3). doi:10.1785/0220120095

Lanza T, Crescimbene M, La Longa F, D'Addezio G (2014) Bringing earth into the scene of a primary school: a science theatre experience. Sci Commun 36(I):131–139. doi:10.1177/1075547012473841

Marsili A, Hunstad I, Burrato P, Casale P, Vallocchia M, D'Addezio (2013) New methods and dissemination models for risk education at INGV (Italy). Geophys Res Abstr 15:EGU2013-7064

Matteucci R, Gosso G, Peppoloni S, Piacente S, Wasowski J (2012) A Hippocratic oath for geologists? Spec Issue Geoethics Ann Geophys 55(3):365–369. http://www.iapg.geoethics.org/pub/6030.pdf

Neresini F, Bucchi M (2011) Which indicators for the new public engagement activities? An exploratory study of European research institutions. Public Understanding Sci 20(1):64–79

Hazardscape, Territorial and Individual Resilience in an Interdisciplinary Study: The Case of Pollino, Southern Italy

21

Francesco De Pascale, Marcello Bernardo and Francesco Muto

Abstract

In this work, certain aspects of the Pollino territory's hazardscape are analysed. Pollino spans the border area between the Italian regions of Calabria and Basilicata which has been the centre of a seismic swarm (up to Mw 5.6) over the last 3 years. The factors which have been taken into consideration in this research have been true risk, perceived risk and local resilience. Therefore, the study focuses upon the role of seismic risk perception in strengthening, or not, local resilience and individuals' behaviour. The research develops through the use of direct and indirect sources. Interviews were used to discover what level of knowledge people have and how it relates to their age, experience and where they are from. Field work was carried out focusing on 'mental map' and questionnaires. The relation between hazardscape and community response highlights the awareness, perception, past experience, response culture and trust in various response measures and agencies. This helps to identify the areas and causes of weakness and shortcomings that contribute to the social vulnerability and break down in the equilibrium between the population, environment and resources, the most frequent reason for the triggering of catastrophes.

Keywords

Geoethics • Geography of perception • Hazardscape • Resilience • Seismic risk

21.1 Introduction

The recent seismic activity in the Calabro-Basilicata border area (Fig. 21.1) has exhibited certain relevant sequences, the most important of which occurred in the Mercure structural depression in September, 1998 (Mw 5.6, Brozzetti et al. 2009). The present seismic sequence, which began in early 2010, has given rise to great anxiety in the local population and authorities because of its duration over time and numerous peaks in activity, with many events being clearly felt. The south-western part of the Pollino Massif has average seismicity, at least with regards the information that we currently have (INGV 2012). The occurrence of several earthquakes along N–S striking faults, in the northern Calabrian sector suggest their still ongoing activity. The northernmost part of the Crati basin, bordered by the crustal Pollino fault, is characterised by an important seismic gap; this contrasts with the geological evidences of NNW–SSE striking faults outcropping in the Castrovillari area, active during the Quaternary. North to the Pollino fault, a series of fault segments shows historical and present activity. The municipalities affected by the sequence are classified in zone 1 or 2 on the map of seismic danger.

F. De Pascale (✉) · M. Bernardo · F. Muto
University of Calabria, Street Pietro Bucci, 87036 Rende, Italy
e-mail: francesco.depascale@unical.it

M. Bernardo
e-mail: bernardo@unical.it

F. Muto
e-mail: francesco.muto@unical.it

G. Lollino et al. (eds.), *Engineering Geology for Society and Territory – Volume 7*,
DOI: 10.1007/978-3-319-09303-1_21, © Springer International Publishing Switzerland 2014

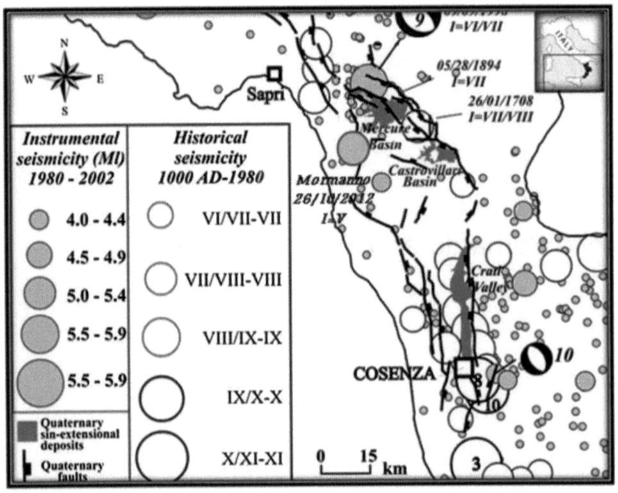

Fig. 21.1 Major faults and seismicity in the Pollino area and Northern Calabria, from Brozzetti et al. 2009. The municipalities involved in the present study are classified in zone 2 on the map of seismic hazard (from INGV)

21.1.1 The Role of Geoethics and the Geography of Risk

In collaboration with geological analyses and the geography of risk, geoethics (Peppoloni and Di Capua 2012) forms an important component to inform the population and to further develop integrated risk management approaches that can enhance the resilience of communities. For example, there is an OCSE report for the 2009 L'Aquila earthquake which makes specific political recommendations and presents concrete options with the aim of increasing the resilience of the Abruzzo region to future shocks. The report is inspired by experiences in other OCSE regions, both those where natural catastrophes brought about a revision of the development model and those long term decline brought about a revision. The sharing of such experiences can provide examples of the political options and governance available for the reconstruction of a region so that it will be stronger, more resistent and more prosperous in the long term. Community involvement in the improvement of the region's post disaster decisional process was also examined with reference to the difficulties encountered in L'Aquila following the 2009 earthquake and the evaluation some of the recent initiatives taken in that city to recreate social cohesion and collective involvement (OECD 2013). Given the premise that earthquake prediction which is useful for emergency management is not possible at present, it is essential to recognise that different forms of intervention which might mitigate and prevent seismic risk are possible. Indeed, it is clear that the effects of seismic events (in terms of economic, social and political activity) are always a direct consequence of physical damage and the impact it has upon a community's capacity to react. Over recent years, earthquakes in Italy have demonstrated that the damage done to buildings (both public and private) has been the main cause of loss, while community resilience, i.e. the system's capacity to adapt to new, negative conditions, has been generally non-existent. In the long run, a return to normality has generally been due to substantial economic investment, often with no consideration of long-term planning strategy. If these resources had been employed in preventative mitigation policies, they might have produced significant economic growth, increased (or, rather, created) resilience and, as a consequence, avoided many deaths (Vona et al. 2012).

21.2 A Questionnaire in the Schools of Pollino Based on New Assumptions in the Geography of Perception

This research included giving a questionnaire to primary (4th and 5th year, 9–10 years of age) and lower secondary school pupils (between 11 and 13 years of age) in some of the villages effected by the Pollino earthquakes so as to evaluate pupils' understanding and how it correlates with age, experience, locality and perception of seismic phenomena. The sample was made up of 542 students. The questions referred to direct earthquake experience, reactions, behaviour before and after the earthquake, safety of the school and the home. Emergency plans and the terminology specific to seismic risk, awareness of the risks within the municipal area, territorial organisation with regards earthquakes and how they are perceived of. The final question asked the pupil to design a "mental map" regarding what to do should an earthquake occur while they are in the classroom with classmates and the teacher. Perceiving one's living space is the equivalent of symbolically representing reality through a mental image which lies between the individual and his/her environment. Before giving out the questionnaire, a lesson was given, in collaboration with young graduates in geology and geographic and humanistic disciplines from the University of Calabria, on earthquakes and the correct behaviour to adopt in the event of one. The sample taken into consideration covered Pollino geographically by taking in municipalities from Basilicata and Calabria; Basilicata: Terranova del Pollino (44 students), Noepoli (24 students), Cersosimo (18 students) and San Costantino Albanese (20 students); Calabria: Mormanno (99 students), Morano Calabro (134 students), Castrovillari (95 students), Laino Borgo (85 students) and Laino Castello (23 students).

21.2.1 Individual and Community Resilience and Pupil's Behavioural Attitudes

The first studies of resilience (Anthony 1974; Garmezy and Rutter 1983) were aimed at understanding the process through which children who have had difficult experiences activate strategies which can help them reach a positive resolution of the stressful event. The longitudinal study carried out by Werner (1989) on the island of Kauai allowed certain typical characteristics of resilient children to be identified. These children appear to be active, autonomous, are of a "good" character and are able to establish positive relationships with others, look after the weak and make use of their own abilities. In the case of Pollino too, students help colleagues in difficulty during the seismic event. On the other hand, it is the fear of earthquakes which has brought about this de facto solidarity around a common object and is the starting point for socialisation. Firstly fear and then confusion are the most widespread reactions among the pupils. As suggested by Boschi (Minciaroni 2012), fear is a justified and positive sentiment when buildings are poorly constructed. The fear which true danger triggers causes us to behave in such a way as to avoid that danger, while panic, even when generated by true danger, leads to dysfunctional and, sometimes, dangerous behaviour. With repetitive seismic activity, habit slowly takes over from fear during the intervals between events.

In the context of the Pollino seismic swarm, pupils tended to grow accustomed and, in some cases, became indifferent. However, the fear remained. Only 26 % of the Basilicata pupils managed to keep calm during an event. This is another positive element which renders the work of the Basilicata teachers easier, but is in contrast, as we shall see shortly, with the situation for Calabrian teachers who have to deal with their pupil's agitation and anxiety. Another positive element is the trust in their homes which the vast majority of pupils of both regions expressed. This is due to the fact that they are often recently built according to anti-seismic regulations. These aspects can be included among the protection factors which are of influence in the context of individual and community resilient behaviour (Table 21.1). An impulsive character trait emerges from pupils' perceptions which brings to mind the Calabrian impulse for catastrophe (Teti 2004). This emerges from the dangerous tendency to rush outside during and after the earthquake. The feeling that one's school is unsafe which the Basilicata children manifest and the perception of a "disorganised" territory on the part of a significant percentage of students are alarming facts which should make local institutions reflect. These fundamental facts are factors of risk with regard resilient behaviour, again within the context of individual and community resilience (Table 21.1). In her direct testimony regarding earthquakes, a thirteen-year-old schoolgirl from Morano Calabro affirmed, "the last earth tremor was on 26th October, 2012. I was sleeping at that precise moment. I was very afraid and stayed in bed, immobile. I live in the old town and, although our house was destroyed, many old houses were not. My family and I went to sleep at my aunt's for a few nights. Her house is anti-seismic". The perception that many of the Pollino pupils have of their territory as being "ancient" brings back historical ideas and recollections of earthquakes and the presence of ancient settlements and sites which were destroyed by past seismic activity. Indeed, it is common in Calabria to come across "doubles", where old settlements have been abandoned and new ones built nearby, or sometimes, even, in the same position. There are some historical episodes which are rationally inexplicable, such as villages which, although completely destroyed by earthquakes on different occasions, have always been rebuilt in the same place. It is difficult to say why this happens, although ignorance certainly plays a part. However, ignorance is not simply a not-knowing of things (with respect to not knowing, it is very important that the culture of the territory is a part of the population's social education from the start of schooling, and that geological faults, for instance, are talked about at primary school in certain seismic areas). As psychoanalysis indicates (Freud 1976), ignorance can also be the removing of a trauma. Indeed, unless it is an action of collective removal, it is difficult to understand how earthquakes are forgotten about, sometimes within so few years that the younger generations of today can still have a direct conversation with those who were present. This makes us reflect upon a facet of reality which is of great weight at times of important decision making, but can not be grasped by the networks of rationality, i.e. logical deductive discourse: a removal of such importance is a warning, a clue to something prevalent and profound, an instinct which is stronger than others, which is stronger even than the intention to reason itself. This instinct involves the affective identification of the inhabitants with the place where they live (Mazzoleni 2005). In some ways, this constitutes a negative factor for territorial resilience because emotive, affective ties with a place of birth contribute to the increase of exposed value and vulnerability through the presence of people and buildings in areas of high seismic danger. In other ways though, ties to the territory also constitute a positive factor in that they act as an important driver to continue and survive in one's native community.

Table 21.1 Protection and risk factors in the context of individual and community resilient behaviour

Protection factors in the context of individual and community resilient behaviour	Factors of risk in the context of individual and community resilient behaviour
26 % of the Basilicata pupils managed to keep calm during a seismic event	33.2 % of the Calabrians immediately ran outside
The most common reaction was fear (58 % in Basilicata and 71.1 % in Calabria). The fear which true danger triggers causes us to behave in such a way as to avoid that danger	17 % of the Calabrians declared that one should get out of the building immediately, during a seismic event. This answer, confirmed elsewhere, shows how the unstable, impatient, impulsive and bilious Calabrian nature, a recurrent theme in many descriptions of southern Italian regional characters
After the event 68 % (Basilicata) and 74.3 % (Calabria) declared said they followed the teacher and 70 % (Basilicata) and 72 % (Calabria) offered help to colleagues in difficulty	After the event, the Calabrian tendency to run outside is confirmed (49 %). Only 32 % of the Basilicata pupils have this inclination
Most pupils (93 % Basilicata, 91.2 % Calabria) gave correct answers with regard the behaviour to adopt in the event of a seismic movement, i.e. get under your desk or a door architrave. Moreover, 88 % in Basilicata and 80.5 % in Calabria said that they should keep away from windows, cupboards and the blackboard	Another very common reaction to the earthquake was confusion (24 % Basilicata and 40 % Calabria). Anxiety was less common (7 % Basilicata, 13 % Calabria). A significant number of students also "froze" during the earthquake (15 % Basilicata, 18.34 % Calabria). If these symptoms are not attended to, they may represent the prelude to more serious and invalidating psychological disturbances
In Calabria, 51.38 % of the secondary school pupils and 50.67 % of the primary school pupils considered their school to be a safe place	59.15 % of the Basilicata secondary school pupils declared that their school building was not safe in terms of seismic risk
A clear majority of pupils think that their home is safe with regard seismic risk (Basilicata: 68.57 % primary school and 54.92 % secondary; Calabria: 80.4 % primary and 73.95 % secondary)	76 % of the Basilicata pupils stated that the Fire Department was responsible for the emergency plan in their replies
79 % of the students in Calabria and Basilicata know that it is impossible to predict when and where an earthquake will occur, but that the dangerousness of an area can be evaluated	58 % of the Basilicata pupils didn't know whether there was an emergency plan in their municipality of residence
An increasing number of pupils consider human factors to be influential with regard catastrophic events (8 % Basilicata, 12.84 % Calabria)	The perception of a "disorganised", "not looked after" and "ancient" territory on the part of a significant percentage of Basilicata students (29 % "disorganised", 39 % "not looked after" and 35 % "ancient") is an alarming fact which should make local institutions reflect

21.3 Conclusions

The pupils in Basilicata involved in the research showed a level of feeling and response to the seismic risk which was no more than average. The situation in the Calabrian schools was better, due principally to the work of teachers who periodically conducted lessons in class on the correct behaviour to adopt in the event of an earthquake. The teachers complained, though, about the lack of initiatives by competent institutions from outside the school to encourage lessons for deeper study of the Pollino seismic activity and of the correct rules of behaviour in the case of a seismic event. Pupils showed a good level of intellectual curiosity during the lesson in class, where some slides were shown, before the questionnaire was given out, demonstrating interest in the argument and asking us questions. Besides certain behaviour which needs correcting and other signs of alarm with regard the lack of safety of their schools and territory which should be studied at a later stage and indicated to families, schools and competent institutional organs, most of the pupils gave correct answers to the questions on the questionnaire. This was without forgetting man's fundamental role as it can be decisive in triggering catastrophes and influencing their impact. Indeed, as a large group of pupils indicated, a correct environmental planning might help limit damage to the territory. Within the context of spatial perception, the most advanced "map" was drawn by a pupil from Basilicata. This "map" was of the 5th stage (with a more geometrical graphical layout), objectively correct from a Euclidean point of view, and the product of "cognitive mapping" (Downs and Stea 1973) that was a true process of mental cartography, which evolved from simple, approximate maps to something that was more complete and correct, and was closely related to development of the capacity for orientation and the reference system adopted. Most of the pupils involved in the study declared that they had first heard about earthquakes from the television. Indeed, the media constitutes one of the most important vehicles of risk communication. In the case of this research, a comparison between real and perceived risk (initially considering pupils at primary and lower secondary schools and then involving adults) is useful to help us understand how the future of our society depends upon social decisions. It emerges from our results that, at different levels, pupils only have an average awareness of risk with respect to the area's high level of danger. However, the road is still long towards achieving a resilient community of young people who should be rendered even further aware through the promotion of risk knowledge, the transmission of historical and ecological memory, encouragement to acquire positive individual and social behaviour for the reduction of risk, and initiatives through which the territory and its inhabitants can be placed at the centre of an analysis of risk.

References

Anthony EJ (1974) The syndrome of the psychologically invulnerable child. In: Anthony EJ, Koupernik C (eds) The child in his family: children at psychiatric risk. Wiley, New York, pp 529–545

Brozzetti F, Lavecchia G, Mancini G, Milana G, Cardinali M (2009) Analysis of the 9 September 1998 Mw 5.6 Mercure earthquake sequence (Southern Apennines, Italy): a multidisciplinary approach. Tectonophysics 476:210–225

Downs RM, Stea D (1973) Cognitive maps and spatial behavior: process and products. In: Downs RM, Stea D (eds) Image and environments. Aldine Publishing, Chicago, pp 8–26

Freud S (1976) La Rimozione. In: Metapsicologia. Opere, Torino, Boringhieri, vol 8

Garmezy N, Rutter M (1983) Stress, coping, and development in children. Johns Hopkins University press, Baltimore

INGV (2012) Relazione su attività sismica nell'area del confine calabro-lucano (massiccio del Pollino) per la riunione della Commissione Nazionale per la previsione e la prevenzione dei grandi rischi, Roma, 4 ottobre

Mazzoleni D (2005) Identità ambientale e terremoto del 1980 nella ricerca universitaria: il lavoro del Comitato Interdisciplinare Universitario (1980-81). In: Mazzoleni D, Sepe M (eds) Rischio sismico, paesaggio, architettura: l'Irpinia, contributi per un progetto, Napoli, CRdC, AMRA, pp 119–129

Minciaroni S (2012) Terremoto, Boschi: "la paura? Non l'ho mai provata"—Santarnecchi "è come un cobra". Intervista ad Enzo Boschi. http://tuttoggi.info/articolo/49154/

OECD (2013) Lezioni per l'elaborazione delle politiche nelle regioni post disastro. In: OECD, L'azione delle politiche a seguito di disastri naturali: Aiutare le regioni a sviluppare resilienza—Il caso dell'Abruzzo post terremoto, OECD Publishing. doi:10.1787/9789264189621-11-it

Peppoloni S, Di Capua G (eds) (2012) Geoethics and geological culture. In: Reflections from the geoitalia conference 2011. Ann Geophys 55(3). ISSN 2037-416X

Teti V (2004) Il senso dei luoghi. Paesi abbandonati di Calabria, Roma, Donzelli editore

Vona M, Harabaglia P, Murgante B, Attolico A (2012) Rischio sismico e resilienza delle città. Riassunti estesi del 31° Convegno Nazionale del Gruppo Nazionale di Geofisica della Terra Solida (GNGTS). Potenza, 20–22 novembre, pp 402–409

Werner EE (1989) Vulnerability and resiliency: a longitudinal perspective. In: Brambring M, Lösel F, Skowronek H (eds) Children at risk: assessment, longitudinal research and intervention. De Gruyter, New York

The Role of Earth Science and Landscape Approach in the Ethic Geology: Communication and Divulgation for the Prevention and Reduction of Geological Hazard

22

Piero Farabollini, Francesca Romana Lugeri, Barbara Aldighieri and Vittorio Amadio

Abstract

Territorial and environmental problems demand, now more than ever, and particularly in Italy where the situation is dramatic, the constant presence of the geologist in all social, cultural, economic and territorial contexts. It is necessary to act with greater authority and to make the public aware of how the territory in which they live can be both a source of risk and a resource. Knowledge is the key tool; the diffusion of scientific heritage, using well known and appreciated topics that are more accessible to the public, may represent one of the goals for the communication and sensitivity versus natural disasters. From the latest dramatic events in Italy emerged the need to create a new kind of communication that can activate a wider and conscious target, providing society with correct and clear information about the geo-environmental conditions in Italy. There is an obvious need for a new approach to the problems related to the complex context that now shows us a planet going beyond the critical point. A holistic approach is imperative, that is a one that considers the environmental and social ecosystem in its entirety, providing all policy makers with a realistic view of the situation and of possible developments. In fact, a big problem for scientists as well as for society, is the prediction of natural disasters. Knowing what is going to happen, as accurately as possible, is the key to be able to provide an effective warning of the population and plan action to safeguard people and properties.

Keywords

Earth science • Landscape • Geoheritage • Geological hazard • Italy

P. Farabollini (✉)
Scuola di Scienze Ambientali - Università di Camerino, 62032 Camerino, Italy
e-mail: piero.farabollini@unicam.it

P. Farabollini
Consiglio Nazionale dei Geologi, Roma, Italy

F.R. Lugeri (✉)
ISPRA Istituto Superiore per la Protezione e la Ricerca Ambientale, Roma, Italy
e-mail: francesca.lugeri@isprambiente.it

B. Aldighieri
IDPA Istituto per la Dinamica dei Processi Ambientali–CNR, Milano, Italy

V. Amadio
Università Mediterranea di Reggio Calabria, 89124 Reggio, Calabria, Italy

22.1 Earth Science, Landscape and Environment

The many territorial and environmental problems, which increasingly end in dramatic circumstances, demand the constant presence of geologists in all social, cultural, economic, and territorial contexts. Today more than ever, efficient and timely participation in actions to protect the environment as well as making the public aware of the importance of preventative measures especially in regards to natural hazards, must be a priority for geologists. The geological landscape is an excellent vehicle for education about the environment and about Earth Sciences; it is a result of endogenous and exogenous activity that create and form the Earth's surface while at the same time it can be

G. Lollino et al. (eds.), *Engineering Geology for Society and Territory – Volume 7*,
DOI: 10.1007/978-3-319-09303-1_22, © Springer International Publishing Switzerland 2014

considered the result of the interaction of many natural and cultural components (Turner et al. 2001). Awareness, therefore, that the Geosciences can be a powerful tool for achieving and sharing a "sense of environmental identity" deriving from an awareness of being part of an ecosystem, is obtained through knowledge and experience of and in the environment (Lugeri 2011).

We will illustrate various ways of studying and researching the landscape that open up new interpretations and ways of understanding landscape as an element of a system in which geology plays a primary role and acts as a catalyst to arouse emotions that otherwise would not be felt or appreciated. In this way it is possible to communicate and spread information about natural hazards and raise the public awareness about natural hazards and risks.

The relationship between the components of landscape and the environment it- self is a very close one, such as the cause and effect of natural phenomena and the biotic component, and they are in turn, reciprocal. The concept of "environmental diversity" can be understood as the integration of geodiversity and biodiversity and its components can be subdivided in hierarchical levels. Conservation must therefore consider the integrated studies about living organisms, habitats, ecosystems, and geological heritage.

Landscape is everywhere, but it must be understood and appreciated as heritage, and, at the same time, it must be protected in order to become a resource. The quality of landscape affects social and individual well-being, as underlined by the statement of the European Convention on Landscape (2000). The key role of landscape is its perceptive and symbolic value; from our first contact with the environment we perceive landscape as a source of emotions that translates into an awareness that landscape is also a source of resources and of risks, which should be identified and understood" (Farabollini et al. 2013a).

The aims of a systemic approach to landscape in the divulgation of geosciences are the following:

- establish the principles and criteria of theoretical reference and methodologies for the study of landscape and for the communication of its components;
- provide an environmental diagnosis and assessment, including specific indicators and qualitative and quantitive models;
- provide synthetic models for forecasting the evolution of landscape;
- influence choices about conservation and land management;
- increase the culture of prevention through knowledge of natural hazards and of associated risks;
- allow monitoring of planned action.

Cartography, digital cartography and GIS in particular are instruments that permit the representation of the connection between nature and culture. Maps are the most comprehensive tool to arrive at an in-depth understanding and a clear image of the areas in question. Thematic maps provide the identification and the visualization of the natural environment and constitute a necessary basis for the evaluation of its condition, and, at the same time, they represent a powerful medium for the public's use. The overlaying of thematic maps (for example geological, geomorphological, and physiographic maps) and the integration of collected data permit the identification and definition of the areas of greater natural and cultural value. This approach is particularly suitable for the Italian territory, where the distinctive features of the landscape add an ulterior element of cultural diversity, which over millennium has been highlighted in painting and literature.

22.2 Dissemination of the Geosciences Through Geotourism

The study of landscape must be addressed as a complex process. Only through the integration of all components of the system being studied, its geomorphological and ecological features and its cultural aspects, is it possible to arrive at a complete understanding of the landscape. In recent years a new theoretical approach to Geosciences that integrates nature and culture has been developing, and this has introduced new powerful educational tools and has stimulated greater communication between researchers and local land managers. This approach has proved essential in creating a new type of dissemination of the Geosciences and of the natural heritage in order to involve the whole society in common actions for sustainable land management (Angelini et al. 2012). A good example of the link between nature and culture is the connection between Earth Sciences, natural landscape, and oenogastronomy.

In this context, a journey through landscape is like a journey through the Earth Sciences. For the modern geologist, the integration of the complex aspects of landscape and the historical development of certain areas of particular interest, with the main aim of creating public participation, can be achieved by utilizing nature, culture, and sport. The discipline of Geosciences seeks to understand the history and the evolution of our planet. By way of geological and geomorphological sites it is possible to make the places known where the evolutionary processes are most evident, giving rise to that which is commonly called "Geotourism" (Lugeri et al. 2012).

Talking about landscape, its forms and its origins, using clear language and models that aid comprehension, would allow people to understand how many of the places of great scenic beauty and geological interest were formed and would result in an awareness that these geosites should be protected (Larwood and Prosser 1998).

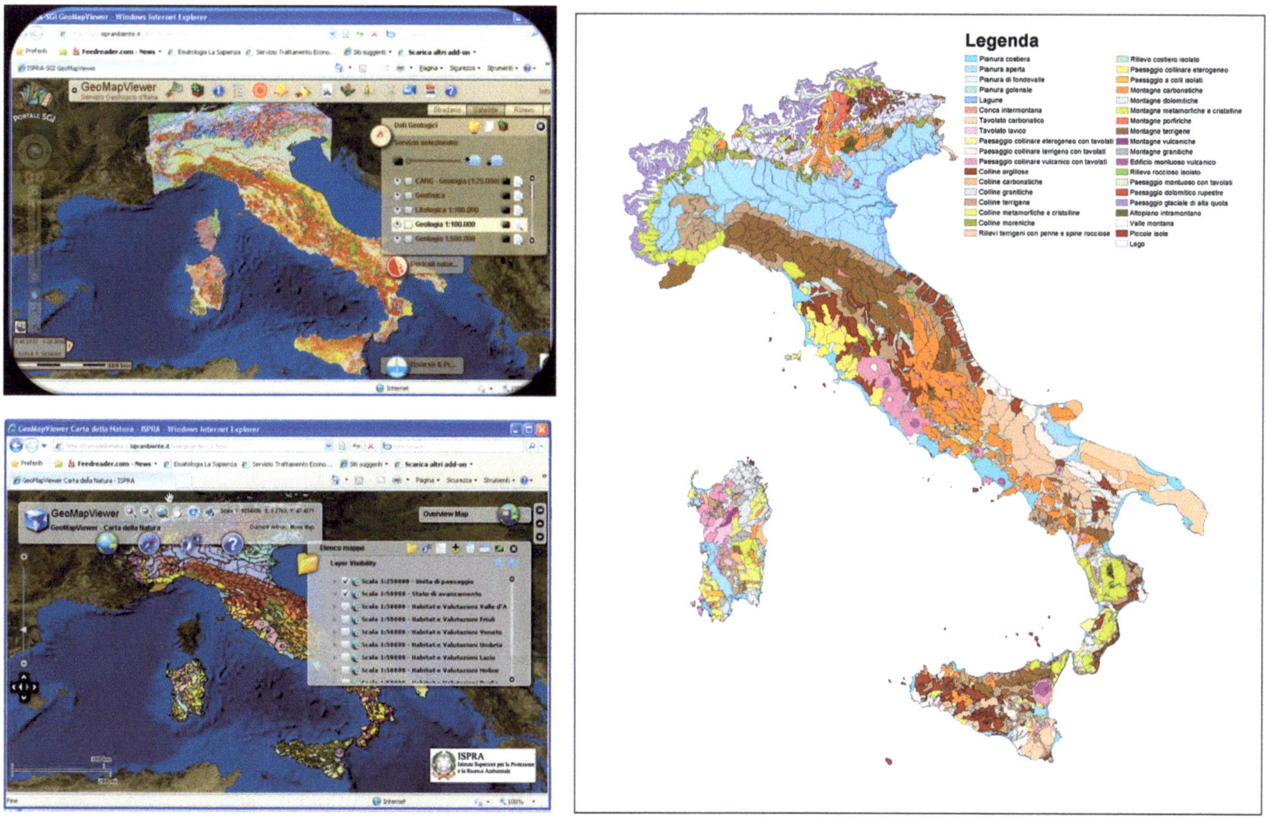

Fig. 22.1 The physiographic units of Italy map (*right*) from geological map (in the *upper left*) and "Carta della Natura" map (*bottom left*)

A suitable instrument for the spreading and popularization of Geosciences through geosites and geotourism, and thus making it possible to create itineraries that give a conscious understanding of abiotic aspects of a territory, is the GIS and connected databases (Fig. 22.1). The use of methodologies such as those that present and make known the geotouristic elements of a territory (cards, pictures, charts, maps, panels, audiovisuals etc.), derives from the accurate identification of geosites according to their uniqueness, their ability to be representative teaching models and that show, by their physicality together with strong elements of scientific knowledge, the stages of the geological evolution of the area (Fig. 22.2).

22.3 Earth Science, Natural and Cultural Heritage in the Fiction and Sport

A well-known Italian television series, "Il Commissario Montalbano", adapted from the novels of Sicilian writer Camilleri, used as backdrops some of the most interesting and important natural sites of international importance, some of which are included among the sites declared by UNESCO World Heritage. In fact, the stories are set in landscapes and archaeological sites in Sicily, spectacular and significant components of the Italian natural and cultural heritage: in particular, many of them were shot in the Val di Noto, Ragusa, Modica, Scicli, Agrigento, in Syracuse and the Aeolian Islands.

The landscape, in particular, as a major component of the set of the film, offers new communication codes to spread scientific knowledge: in many episodes of the series, the arrangement of the geological landscape is an important component for the dramatization. The natural setting gives it a special significance, where the local population, as well as the many tourists who visit the island, they recognize the area and therefore can more appreciate its value (Magazzino and Mantovani 2012). Communicate the natural and cultural heritage through film and fiction could represent a new way to engage the public to appreciate the landscape and to promote eco-tourism and sustainable development in Italy (Lugeri et al. 2012).

Cycling, popular sport par excellence, is a spatial-temporal relationship between individuals, communities and societies. The Tour of Italy (it has recently been de- fined as "the toughest race in the world in the most beautiful country in the world") is the most popular race in our country and is a means of great communicative potential and multifunctional.

The possibility to show to the general public (and even athletes) the geological and geomorphological components of the places visited by the various stages of the race, describing the main characteristics and peculiar geological

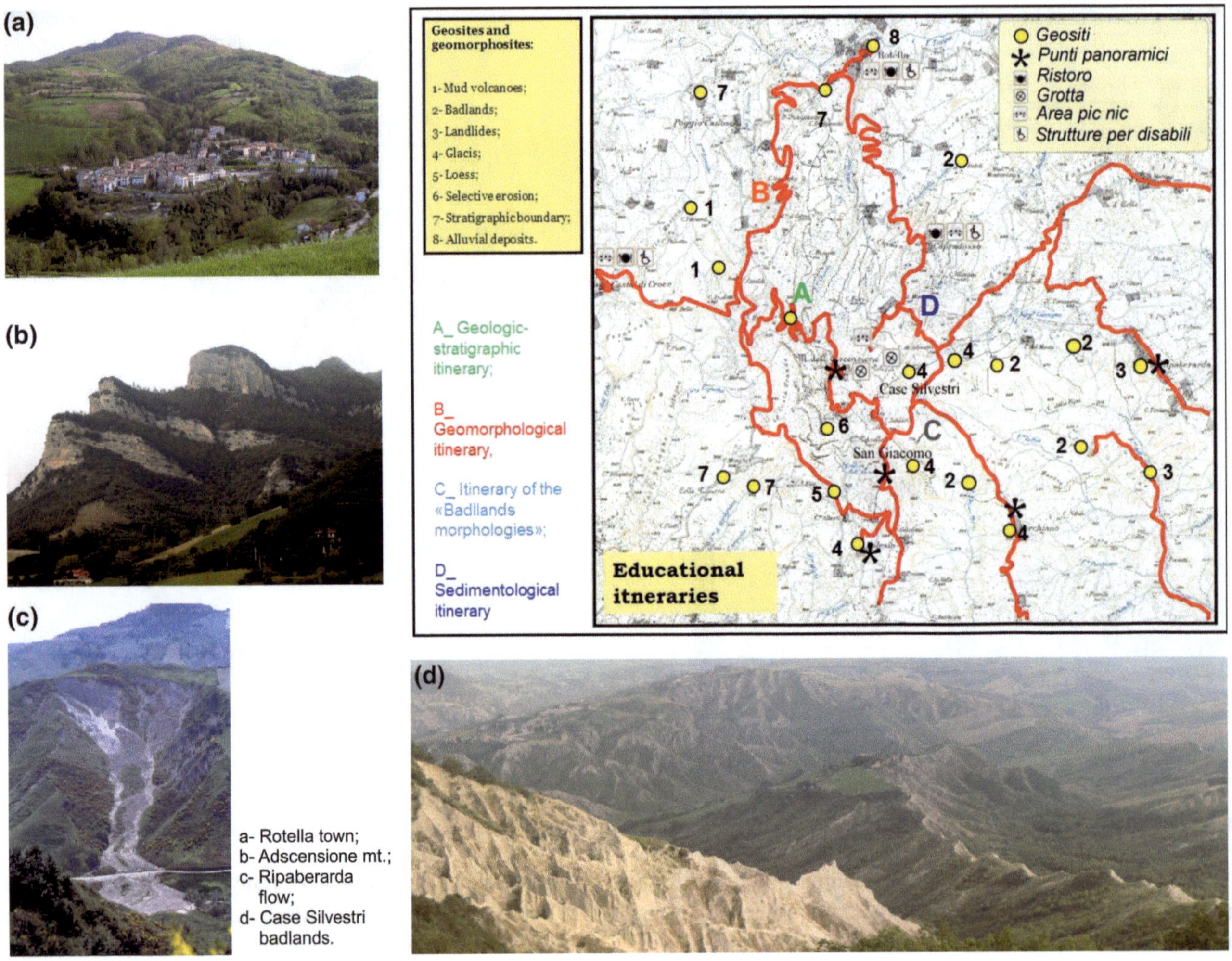

Fig. 22.2 Example of earth science divulgation and geoturistic map

settings as well as the natural and environmental areas, can be a great conveyance of Geosciences and especially represent a new way of reading its own territory, knowing and valuing its cultural components and perceiving the potential socio-economic conditions. The characteristics elevations of the various stages (such as running up and down), are essentially linked to the morphology which depends on the geological characteristics of the territory, linked to endogenous processes, and subsequent exogenous agents and morphodynamic processes that have changed. GIS processing and images (also 3D; Fig. 22.3) can help to show the "geo-environmental" path, making it easier for the public to recognize the landscape (Farabollini et al. 2013b). The multi-scale approach gives an added value in terms of instruments, the study of landscape: in parallel, the new focus is in the communication and dissemination of earth science, the nature and the territory recognized as important factors for a holistic approach to the territorial analysis, integrating the natural with the cultural aspects of the landscape itself, facilitating the understanding of what are natural disasters, and which are human-induced.

22.4 From the 1963 Vajont Dam Disaster (1963) to Earthquake in Emilia (2012): The Communication of Risks and the Risk of Communication

Natural disasters are events that Italy has always had to deal with and although there is a common belief that there is little one can do when faced with these events, providing the public with updated scientific information for an in-depth knowledge of territory and of the phenomenon is the best way to create strategies for prevention and reduction of natural risks.

From an epistemological point of view, the popularization of science through the teaching of the Geosciences, represents the starting point for risk education. Experience

Fig. 22.3 Earth Science and Giro d'Italia: geological 3D Map of the Ischia island and images during the different stages of the Giro d'Italia, 2012 (*upper right* and *bottom left* Stelvio; *bottom right* Vajont) (modified from Farabollini et al. 2013b)

gained in this regards, in particular from the INGV or other research institutions and universities (La Longa et al. 2011), compares educational methodologies carried out in "peacetime" far from seismic events (Risk and Regulation Advisory Council 2009; Progetto EDURISK 2011), with those used during and following emergencies such as the 2009 earthquake in Abruzzo region (Camassi and Peruzza 2011). The communication of risk is still a problem to resolve, both in terms of the quality and the content of the communication. Scientific divulgation, should have the objective of raising the public awareness of issues related to geology, thus strengthening the role of Geosciences in society.

Spreading awareness through dialogue with the community, creating interest in the Geosciences, describing natural hazards, creating a link between science and society, means creating scientific awareness and teaching the culture of hazards and therefore that of prevention. Formal and informal knowledge are part of a system that guides ones understanding and action; data transformed into information that, when properly organized, becomes awareness. This awareness, if placed in a context through action and/or inaction, becomes wisdom and as such enables educational strategies, in particular if used for the communication of risk (Tobin and Montz 1997; La Longa et al. 2011; Tola 2011; Farabollini et al. 2013a).

Awareness of natural hazards, of prevention and of risk management, has a particular importance in modern society; knowing how to recognize dangerous situations and having the competence to evaluate and face them in a reasoned manner, making conscious choices, should be the minimum knowledge for all citizens, especially in times outside emergencies. Communication of risk should not respond to training and informational needs with the intent of providing a response in order to contain anxiety levels generated by the uncertainty of an emergency situation, nor should it be aimed at explaining what took place and providing re- sources in order to overcome the crisis phase (all of which are only necessary during an emergency), rather it should always be focused on the culture of prevention.

Today, more than ever, we refer to prevention as the principal strategy of the defense against risks. Choices that impact the reduction of risk are those that we make or we do not make long before an event takes place and are tied to our perception of risk. Very often the responsibility of taking action falls on others—the government, institutions, local authority—and the citizen considers him or herself impotent in the face of the inevitability of natural events and shifts the responsibility of realizing effective politics of prevention to those in power. The aim of the dissemination of the Geosciences is to develop sufficient competencies for responsible and aware individuals. The concept of risk is often used as a synonym for hazard. This seemingly harmless semantic confusion generates serious consequences. If people perceive risk as the hazard of an area, an earthquake becomes an expression of the force of nature thus affirming the impotence of human action.

The culture of prevention is therefore the result of a long process of learning that is not merely the simple utilization of information: it must be capable of developing risk awareness, through the acquisition of knowledge and values, in order to activate choices and actions directed at reducing the risk. Ultimately, prevention is about motivating people's life choices before an event happens and therefore putting more effective strategies in motion that reduce risk.

22.5 Conclusion

The Italian landscape is composed of a spectacular alternation of abiotic peculiarities, a dynamic combination of geosites and geomorphosites. Additional value is represented by the local institutions that should be the principal protagonists in establishing programs for monitoring and protecting territory, for promoting projects aimed at renewing, protecting and enhancing, in a sustainable way, the areas of interest. At the same time, however, special attention must be dedicated to social involvement: communication, transmission and education.

A proper technical and cultural approach to the issue of managing and protecting the environment, and of knowing the level of hazard in order to know how to defend oneself from the associated risks, demands a multilayered and multidisciplinary approach that aims at a balance between use and respect for the planet, so as to have a development that is sustainable for natural and social ecosystems.

New environmental problems require new solutions and new creative strategies in collaboration with citizens and responsible politicians. This objective can be reached only if it becomes a common one, moving the politics of development to- wards an actual integration of ecological and socio-economic needs.

The socio-economic situation of a territory is closely tied to its geo-naturalistic scenery; even on different scales, the endogenous and exogenous processes, the physiographic features, such as elements of landscape and of evolution of the environment, form a basis for spatial-temporal development of a region. It is for this reason that raising the public's awareness about the natural hazard that can affect a certain area must be done together with the proper education. The aim of Geosciences to transmit and popularize its knowledge in many different ways is driven by the necessity to guarantee a conscious and enjoyable use of the landscape in all its forms for future generations. The development of a territory follows the same path as eco-tourism. Transmission of knowledge of the natural and cultural heritage through new methods of communication as we have suggested (by ways, for example by film, cycling, wine) represents a new strategy for a balanced and sustainable management of territory.

References

Angelini S, Farabollini P, Lugeri FR, Scalella G (2012) GIS technologies as a tool for earth science learning through geotouristic itineraries: application in the Marche. Rendiconti Online SGI 19:61–63

Camassi R, Peruzza L (2011) EDURISK, 10 anni di lavoro. L'approccio, il percorso, a che punto siamo. Miscellanea INGV, 8–12

Farabollini P, Graziano G, Lugeri FR (2013a) Paesaggio, Ambiente natu- rale e culturale in un approccio sistemico per la divulgazione delle Geoscienze. Geologia Tecnica ed Ambientale 1:62–73

Farabollini P, Graziano G, Lugeri FR, Lugeri N, Lugeri M (2013b) Earth Sciences divulgation, geoheritage and landscape approach: the project of the Geologiro d'Italia. In: Proceedings international congress Geoparks, Ascea

La Longa F, Camassi R, Crescimbene M (2011) Strategie per la riduzione del rischio sismico: divulgare per diffondere conoscenza scientifica o educare per promuovere consapevolezza? Abstracts Geoitalia 2011, Torino

Larwood J, Prosser C (1998) Geotourism, conservation and society. Geo- logica Balcanica 28:97–100

Lugeri FR (2011) Il landscape approach nello studio del territorio e nella di- vulgazione. Geologia Tecnica & Ambientale 2–3:54–65

Lugeri FR., Farabollini P, Graziano G, Amadio V (2012) Geoheritage: nature and culture in a landscape approach. EFG Eur Geol Mag Geoheritage 34:23–28

Magazzino C, Mantovani M (2012) L'impatto delle produzioni cinemato- grafiche sul turismo: il caso de "il Commissario Montalbano" per la Provincia di Ragusa. Rivista di Scienze del Turismo 1:29–42

Risk and Regulation Advisory Council (2009) A practical guide to public risk communication. The five essentials of good practice. UK Government office for Science and Economic and Social Research Council, p 12

Tobin G, Montz B (1997) Natural hazards: explanation and integration. The Guilford Press, New York

Tola E (2011) La comunicazione. Miscellanea INGV, pp 58–64

Turner MG, Gardner RH, O'Neill RV (2001) Landscape ecology in theory and practice. Pattern and process. Springer, New York

Piovono idee! (Cloudy with a Chance of Ideas!): An Interactive Learning Experience on Hydrogeological Risk and Climate Change

23

Giovanna Lucia Piangiamore, Gemma Musacchio and Maurizio Bocchia

Abstract

In 2011, on October the 25th a devastating flood hit the Vara and Magra valleys in Italy and left an unforgettable scratch in the inhabitants' minds. *Piovono idee!* (*Cloudy with a Chance of ideas!*) is an active journey of discovery and training on hydrogeological risk, and climate change. Land preservation and safety of people living on it are issues which we would like to help citizens get perception about, in order to instill awareness on the actions that can be taken towards risk mitigation. *Piovono idee!* stemmed from this belief and it is the result of a collaborative planning, in which primary and secondary school students, living within cities heavily hit by the flood, took actively part. Children were helped by experts and scientists to build an exhibition devoted to hydrogeological risk. Here interactive workspaces, games and educational laboratories allow visitors explore concepts, phenomena and their consequences on land and inhabitants. Issues are addressed from a daily actions perspective, where everybody might make the difference towards sustainability and trigger good practices on natural hazards risk reduction.

Keywords

Hydrogeological risk • Climate change • Prevention • Environmental impact • Territory

23.1 Introduction

In recent years the combined effect of global warming and changes in land use has led to higher flood risk, worsened by the rising probability of events of considerable intensity. The state of abandon of cultivated land and the neglect of undergrowth are aggravating factors to the exceptional nature of climatic variables, as they provoke the reduction of soil geotechnical resistance, debris formation and waterways obstruction. The hydrogeological structure and land excessive cementation worsen the scenario and highlight that disasters are the result of natural and social processes (Okazaki et al. 2008). Unlike the natural conditions that have a strong unpredictability component, social dimension of the disaster risk relates to the way human being interacts with natures (Luciani 2007).

The flood of the 25th October 2011 hit many municipalities in Liguria (Italy) and followed a very heavy rain (542 mm of water fell in just 6 h). This caused the Vara and Magra rivers to overflow, streams of water to gush and the valleys to be flooded. This event marked the peak of three catastrophic years in which every rainfall above average caused disasters, ranging from flooding to landslides.

G.L. Piangiamore (✉)
Istituto Nazionale di Geofisica e Vulcanologia (INGV), Sede di Portovenere, Via Pezzino Basso, 2, Fezzano di Portovenere (SP), Italy
e-mail: giovanna.piangiamore@ingv.it

G. Musacchio
Istituto Nazionale di Geofisica e Vulcanologia (INGV), Sede di Roma, Via di Vigna Murata 605, Roma, Italy

M. Bocchia
Provincia della Spezia, Settore Difesa del Suolo, Servizio di Protezione Civile, Via XXIV Maggio, 3, La Spezia, Italy

G. Lollino et al. (eds.), *Engineering Geology for Society and Territory – Volume 7*,
DOI: 10.1007/978-3-319-09303-1_23, © Springer International Publishing Switzerland 2014

In this paper we describe *Piovono idee!* an active journey of discovery and training on hydrogeological risk and climate change. It is an exhibition resulting from a project run together with primary and secondary schools students, scientists and experts. *Piovono idee!* is meant to preserve the memory of a disaster and ultimately promotes the raise in awareness on risk reduction (Wisner 2006; Komac 2009).

It was firstly held within 'ScienzaAperta', an Italian national science venue run by the Istituto Nazionale di Geofisica e Vulcanologia (INGV) and aimed at spreading scientific knowledge, augment curiosity on the world of research and natural phenomena surrounding us.

23.2 *Piovono idee!*: The Project

The *Piovono idee!* project was run together with 378 children, aged 8–14, from 15 primary and secondary schools in Borghetto Vara, Brugnato, Monterosso, Vernazza, Riomaggiore, Pignone, Riccò del Golfo, Aulla and Ragnaia (Aulla), the areas strongly affected by the flood of the 25th October 2011, as well as a sample of 3 secondary schools from the city of La Spezia. It had many phases. The *laboratory planning phase* and *laboratory creative phase* were run together with the schools and meant to build the exhibits. The work done with the schools was then finished and perfected by ConUnGioco, an educational association experienced on ineractive learning projects and the models were then displayed to public explaining hydrogeological risk and its impact on the Spezzino territory during the *exhibition phase* in March 2013. The personnel of the educational association ConUnGioco and INGV ran for 10 days free guided tours lasting an hour and a half for groups of a maximum of 26 participants at a time.

During the planning process we run laboratories activities for each class participating to the project. Laboratories were targeted to the exploration and understanding of processes causing hazard and risk. We used models to explore the key issues on hydrogeological risk. The understanding process involved young children with games and narration. All the activities were to inspire the creation of objects that could help express their own experiences of the flood and to work out questioning and problem solving on difficult past and present.

23.3 The Learning Path

Within *Piovono idee!* (Fig. 23.1) the participants, who are more than just simply visitors, improve their specific scientific knowledge with hands-on activities, while exploring feelings and emotions triggered by the experience of a flood. Moreover the learn-by playing approach was used to instil appropriate behaviours.

The stands were structured within a 'scientific'- and 'experience-' session. The former was to explain the geological phenomenon; while the latter was meant towards actions to be taken towards risk mitigation. They were made with simple materials (mainly cardboard) and are completely interactive, to encourage visitors to add-on, leave a trace of their own experience (Brasini et al. 2006). The workshop approached with games topics on the industrialization effects on the environment and on behaviors 'burdening' the environment.

In the first exhibit we ask the question *'Is nature scary?'* Students had made out of play dough little objects representing gifts that nature might give us. They are flowers, fruits, trees, animals, the sun, the moon, stars, a rainbow. Objects are hidden inside little boxes that are meant to be search through whilst a cardboard face smiles. However, the cardboard face can turn around, showing an angry look, and this might be turn on just blowing or poking on it. The message we wanted to convey is that we need to think on every actions we take, make up a choice on those that pay respect towards nature, and be aware that even little actions can turn into negative or positive.

The exhibit equipped with several experiments focused on the understanding of hydrogeological risk. Here key concepts, such as (1) the water cycle, (2) the effect of the rain on different rocks, (3) the underwater world, (4) the greenhouse effect, are discussed.

The totem of *the comparison between natural and anthropic environment* shows two empty landscapes which visitors are invited to fill in using the elements we made available. For the natural environment it is possible to pick out some dangerous elements to be stuck on the picture (ferocious animals, lightning, forces of nature, etc.). For the anthropic environment, man's 'transformations' are valued with pros and cons (industry, breeding farms, woods, galleries, streets, skyscrapers, etc.). Here usually discussions concerning the environmental impact of every single person's daily habits are run.

In the following exhibit we ask the question *'The territory in which we live belongs to us. Are we responsible for it?'* Here a model portrays river banks, from springs to lowlands, where several elements such as houses, bridges, cars, people and trees can be placed. Swamped areas safe behaviors in the case of river flooding are discussed.

The river exhibit focus is on the consequences of foolish urbanization. Little houses, dams, bridges, trees and decorative architectonic elements can be added on while the flood comes as cascade of little balls, until the river bursts its banks and the balls will destroy those elements mostly exposed at risk.

Fig. 23.1 *Piovono idee! Is nature scary?* totem (topmost *left*); an overview of *Piovono idee!* active path, *the risk* totem (topmost *right*), *the greenhouse effect* exhibit (*right*), another overview of *Piovono idee!* (*bottom*), *the river* exhibit (*left*), the model portrays river banks from *the territory* totem (*center*)

Can we avoid all risks? And how would you behave when faced with a risk? This is the theme of *'the risk'* totem, which asks visitors to pick possible options for a specific illustrated scene and the first and second choice's box to find possible options to finish the story. This triggers interesting debates, which focus on assessing the risks and danger surrounding us. These are assessments, which we often make without realizing we are doing so.

23.4 Learning with Games

The combination play and learning can be a valuable tool in science education. Games often have a fantasy element that engages players in a learning activity through a storyline. Research about new ways of teaching and learning (Lepper and Cordova 1992) demonstrates that a story context often included within games, combined with a challenge for the student to overcome (in other words, making it into a game) significantly improves the learning performance of children. Here we discuss the games included within the exhibition.

The society of yesterday and today: the same pollution? is a role-play game confronting head-to-head characters such as the housewife, the publisher, the animal breeder, the shopkeeper, … on the amount of carbon dioxide and methane their usual activities produced within two main type of societies: one before the nineteenth century and the other after industrialization.

How much do you affect the environment? is to show how much we affect the environment with our daily actions (food choices, cleaning our teeth, going to school, turning on the AC, …). Players are given out stones according to the choices made from different options, while a special scale shows if the weight of the team falls into the green quarter of good practices, the grey quarter of improving practices or that belonging to pollution. The scores could be changed after awareness: heavy stones can be relieved if the team members promise behavioral changes towards sustainability.

23.5 Conclusions

Piovono idee! is the result of research following a stimulating didactic/communicative model which emotionally involves the participants. It is aimed at raising awareness on the social dimension of the disaster risk, pointing out that human behavior is the crucial factor in the degree of vulnerability and the likelihood of disasters taking place. *Piovono idee!* is a pleasant and effective way to disseminate that little effort are needed to gain significant improvement fast positive response. Our behavior affects the community, we urge promote good habits too. All exhibits and scientific games of *Piovono idee!* rely on emotional intelligence. As knowledge is clearly connected with understandings, we argue that emotional learning can prompt behaviors best imprinted in the mind and therefore act on hazard perception.

The scientific venue at his first edition involved 700 children and teachers who followed with great appreciation the insights given by scientists.

Acknowledgments *Piovono idee!* was founded by Province of Spezia within the framework of the 'Laboratories of shared and participatory citizenship- IV edition' project promoted by the Directorate-General for the Third Sector and Social Formations- Italian Ministry of Labour.We would like to thank D. Modenesi and F. Brasini from ConUnGioco Onlus for the creation of the exhibition, R. Camassi from INGV in Bologna for the part on climate change, M. Miconi from INGV in Rome for logistic and technical support, P. V. De Leonardis and G. Iardella, MARIDIPART and MARICOMMI for kindly allowing us to use CRDD locations and all the school heads, teachers and children of the primary and secondary schools of Borghetto Vara, Brugnato, Monterosso, Vernazza, Riomaggiore, Pignone, Riccò del Golfo, Aulla and Ragnaia (Aulla) and S.M. "V. Alfieri", "U. Mazzini" and "J. Piaget" from La Spezia who contributed to the initiative. Many thanks to M. Casarino, P. Milano and O. Zocco from La Spezia Provincial Europe Service—Coordination of Municipalities, G. Mancini, M. Biso and G. Savoldi from La Spezia Provincial Civil Protection Service, G. Forlani, M. S. Ariodante, A. M. Bimbi from La Spezia Prefect's Office's Civil Protection Department, the voluntary Civil Protection associations and the GEV of the Province of La Spezia.

References

Brasini F, Modonesi D, Camassi, Sidoti B (2006) Tutti giù per terra (all fall down)—an active course to discover earthquakes topics. In: Proceedings of the first european conference on earthquake engineering and seismology, Geneve (CH), SS 3, 3–8 Sept 2006

Komac B (2009) Social memory and geographical memory of natural disasters. Acta Geogr Slov 49:199–226. doi:10.3986/AGS49107

Lepper MR, Cordova DI (1992) A desire to be taught: instructional consequences of intrinsic motivation. Motiv Emot 16(3):187–208

Luciani R (2007) Il clima che cambia, cosa possiamo fare noi. INGV-Giunti Progetti Educativi, Firenze

Okazaki S, David EJR, Abelmann N (2008) Colonialism and psychology of culture. Soc Pers Psychol Compass 2(1):90–106

Wisner B (2006) Let our children teach us! A review of the role of education and knowledge in disaster risk reduction. ISDR System Thematic Cluster, Platform on Knowledge and 20 Education, Bangalore

Interactive, 3D Simulation of Natural Instability Processes for Civil Protection Purposes

24

Furio Dutto, Massimo Arattano, Marco Bacenetti, Marta Chiarle, Concetta Contrafatto, Marco Giardino, Fabrizio Longo, Luigi Perotti, Filippo Racca, Lorenzo Rocci and Laura Turconi

Abstract

A computer simulation of a debris flow in motion, a landslide in evolution and a snow avalanche is here presented that was realized in the context of a series of risk education activities developed within the European project RISKNAT. The pc simulation allows to view in 3D and in their natural context of occurrence the three types of mass movements mentioned above. The peculiarity of the simulation is that the user is able to directly interact with the processes with the help of a special joystick that allows to vary the main parameters governing the process and also its sight. It is possible, for example, to choose an aerial view, or to get a view closer to the ground to observe in more detail the phenomenon in its progress. The computer simulation is based on real data and intends to spread among the local population, primarily exposed to risks, instructional and educational bases by offering a virtual experience of some recurrent natural processes occurring in alpine valleys. In fact it allows the user to reach a greater awareness of these latter, learning about their aspect, their mode of propagation, their velocity and their interaction with the natural and urbanized landscape. The simulation is therefore a tool to support the communication activities devoted

F. Dutto · C. Contrafatto · F. Longo · L. Turconi
Provincia di Torino, Civil Protection Department, Via A. Sordi 13, Grugliasco, Turin, Italy
e-mail: furio.dutto@provincia.torino.it

C. Contrafatto
e-mail: contrafa@provincia.torino.it

F. Longo
e-mail: fabrizio.longo@provincia.torino.it

L. Turconi
e-mail: laura.turconi@irpi.cnr.it

M. Arattano · M. Chiarle (✉)
Istituto di Ricerca per la Protezione Idrogeologica, Consiglio Nazionale delle Ricerche, Strada delle Cacce 73, 10135 Turin, Italy
e-mail: marta.chiarle@irpi.cnr.it

M. Arattano
e-mail: massimo.arattano@irpi.cnr.it

M. Bacenetti · M. Giardino · L. Perotti
Dipartimento di Scienze della Terra - GeoSITLab, Università degli Studi di Torino, Via Valperga Caluso, 35, 10125 Turin, Italy
e-mail: marco.bacenetti@unito.it

M. Giardino
e-mail: marco.giardino@unito.it

L. Perotti
e-mail: luigi.perotti@unito.it

F. Racca · L. Rocci
ThalesAlenia Space, Strada Antica di Collegno, 253, 10146 Turin, Italy
e-mail: evilastharoth@gmail.com

L. Rocci
e-mail: lorenzo.rocci@thalesaleniaspace.com

G. Lollino et al. (eds.), *Engineering Geology for Society and Territory – Volume 7*,
DOI: 10.1007/978-3-319-09303-1_24, © Springer International Publishing Switzerland 2014

to explain to the general public the civil protection basics, with the belief that the direct involvement, albeit virtual, in a situation of danger can be an important stimulus to better understand the forces of nature and to assimilate the basic behaviors needed for self defense.

Keywords
Debris flow • Landslide • Snow avalanche • 3-D stereo simulation • Virtual reality

24.1 Introduction

The risk posed by landslides and snow avalanches is very well known worldwide to scientists, administrators, practitioners and all those professionals who deal with environmental issues. The same cannot be told for the general public who is often poorly aware of the environmental risks that might affect and threaten his life and possessions. For this reason in 2011, in the framework of the project RISKNAT (program ALCOTRA, Alpes Latines Cooperation TRAnsfrontalière, 2007–2013, Projet stratégique I RiskNat "Gestion en sécurité des territoires de montagne transfrontaliers") a consortium of different institutions led by the Civil Protection Department of the Turin Province decided to realize a software that allowed a 3D stereo simulation of three different types of natural instability phenomena: a debris flow, a landslide and a snow avalanche. The idea was to present the simulation of these three phenomena in their natural context of occurrence and provide people the chance to gain a personal experience about them. The consortium included the CNR IRPI, the University of Turin and ThalesAlenia Space (ThalesAlenia Space is a joint subsidiary of Thales (67 %) and Finmecca-nica (33 %), and a partner in the Space Alliance along with Telespazio).

24.2 The Computer Simulation

As previously mentioned the software realized by the consortium allows to view a 3D simulation of a debris flow in motion, a landslide in its multitemporal evolution and a snow avalanche, all of them in their natural context of occurrence.

24.2.1 The Debris Flow Video Simulation

Debris flows are a type of mass movement that may occur in steep mountain torrents and consist of a highly concentrated dispersion of poorly sorted sediment in water that commonly move at very high speeds and have great destructive power. They generally appear as waves (surges) that have steep fronts consisting mostly of boulders that can be surprisingly large. Debris flows cause damages and casualties worldwide every year, due to their destructive power (Arattano et al. 2010). Many efforts have been carried out so far by researchers of CNR-IRPI to monitor their behavior (Arattano 2003; Arattano et al. 2012). The results allowed to simulate numerically their propagation, to investigate their rheology, to better understand their complex dynamics and to compare the different sediment processes involved (Arattano and Franzi 2003, 2004; Lin et al. 2005; Arattano et al. 2006; Mao et al. 2009). However all these efforts, that may lead to a better hazard zone delineation (Tsai et al. 2011), could be worthless if an adequate awareness of this natural phenomenon and its dangerous characteristics were not instilled in the general public.

The 3D video that has been realized simulates the propagation of a debris flow wave from its inception to the deposition on the fan and allows a direct, albeit virtual, involvement of the public in a situation of danger. This might be an important stimulus to better understand the examined natural phenomenon and to assimilate the behavior basics needed for self defense. The 3D video is based on the case study of the Champeyron Creek, a 3 km^2 mountain basin located on the right side of the Susa Valley (Fig. 24.1). The debris flows that develop along the Champeyron Creek pose a hazard to the Beaulard village, and in particular to the camping site located on the right side of the alluvial fan. During the night of August 7, 1981 a thunderstorm caused the initiation of a debris flow which reached the alluvial fan and flooded the campsite.

Fig. 24.1 View of the Champeyron basin obtained by overlaying an orthophoto image on the digital terrain model

The propagation of this debris flow along the Champeyron Creek was simulated taking advantage of the 20-years experience on debris flow characteristics and behavior gained by the researchers of CNR-IRPI. Field data were examined and used in the simulation that had been collected in several other catchments. These included stage, volume and velocity data that had been obtained through the different measurements and monitoring methodologies developed by CNR IRPI during the years. These methods included, for instance, the analysis of seismic data and hydrograph curves through techniques of cross-correlation, which are nowadays widely used in engineering geology (Lollino et al. 2002, 2006) or the use of image processing techniques (Arattano and Grattoni 2000; Arattano and Marchi 2000).

The Champeyron Creek was chosen as a case study because of the availability of documentation on past debris flow events, of clearly visible tracks of past debris flow processes, and because of the presence of elements at risk. A geomorphological analysis of the basin, carried out through a multi-temporal photo interpretation, allowed to reconstruct the development of the debris flow in 3D. For this purpose also some photos of the period were used, together with interviews of witnesses and some specific field surveys. In particular the areas of debris supply for the 1981 event were identified. These latter corresponded to the extended screes visible in Fig. 24.2 and to some other areas prone to periodic rockfalls. The available DTM was adjusted to obtain the needed scale that allowed a simulation of the process. The characteristics of the debris flow mixture (solid fraction in % and dimensions of the largest boulders) were faithfully reconstructed, and this allowed to appreciate the dynamics of the process evolution and the mass transport capacity. Furthermore the expansion areas of the debris flow on the fan and the flow directions have been defined, which are essential components for the interpretation and numerical simulation of the 1981 event (Fig. 24.3).

24.2.2 The Landslide Simulation

The Cassas landslide is located on the right side of the middle Susa Valley, in the "Gran Bosco" natural park (Turin, Piedmont, Italy). The Cassas slope is deeply affected by instability phenomena; the landslide top is located near a minor watershed at about 2,000 m a.s.l., while the landslide foot is placed at about 1,000 m in the wide Salbertrand valley bottom.

Between June 12th and 14th, 1957 the western area of the Cassas landslide had a paroxystic phase: some millions of m^3 of rock mass slid and triggered a debris flow that reached the valley bottom. Re-activations of the landslide were recorded until the middle Sixties (Peretti 1967; Brovero et al. 1996).

Fig. 24.2 Catchment head of the Champeyron Creek; extended screes can be recognized that have depths of several meters at the bottom of the rocky slopes of the Grand Hoche Peak (2,760 m a.s.l.)

Fig. 24.3 Detail of the debris deposits of August 1981 in the Beaulard camping area

The estimation of the risk for the Cassas landslide is difficult, because of the slow and continuous deformation that occur in depth. The different studies carried out over the last years have supplied important information for a correct understanding of the kinematics and dynamics of the landslide. These studies left behind, nevertheless, several unsolved problems. The landslide has been divided into five different zones for investigation purposes (Amatruda et al. 2004):

1. the main scarp zone: intensely fractured and released calcschists, with unstable blocks of variable volumes; in the crow zone many deep fractures are apparent;
2. the active landslide accumulation zone: made up of rocks blocks of variable size floating in a sandy-muddy matrix;
3. the complex landslide accumulation zone: the kinematics are complex because of the combination of sliding and slow-to-rapid flow movements;
4. the alluvial fan area: area of debris flow expansion and accumulation originating in the middle and upper parts of the slope;
5. the landslide accumulation zone in meta-stable equilibrium: block accumulation and mega-blocks of older landslide.

The 3D simulation of the landslide evolution (Fig. 24.4) has been realized following the steps listed below:

- collection of outcropping data on the area of interest (DTM, orthophotos)
- acquisition of multi-temporal aerial images
- orthorectification of photogrammetric data (years 1954–1963–1994–2000, Fig. 24.5)
- creation of levels of multi-temporal coregistered orthophotos
- 3D perspective visualization and animation of the landslide evolution.

24.2.3 The Snow Avalanche Simulation

During the heavy snowfalls occurred between December 14th and 17th, 2008 (Cordola et al. 2009), the Western Alps were affected by major avalanches that resulted in substantial damage to infrastructures and houses with the involvement of areas that for decades had not been affected by avalanches.

Fig. 24.4 *Left* Cassas landslide case study. An example of 3D perspective view (Susa Valley). *Center* and *right* example of hardware installation at 2012 European Researchers' Night. TV 3D video with case studies visualization by stereo view capabilities (3D passive glasses)

Fig. 24.5 Multitemporal evolution of Cassas landslide: ortophoto (year 1957) and real panoramic views (years 1965, 2001, 2005, 2011)

Fig. 24.6 Snow avalanche case study. Avalanche occurrence area frapped on satellite high resolution image

In the Venaus Municipality on December 15th, 2008 a medium size avalanche affected a large area of forest, damaging an important road (namely the SS 25 Mont Cenis) and obstructing the flow channel at about 200 m from houses located just below. The main damage caused by the event was the complete removal of a deciduous forest, that covered an area of over 20 ha. The avalanche events on the mountain side above the town of Venaus show a historical recurrence of about 20 years. The most serious event ever documented occurred on January 18th, 1885: it affected some villages and caused the destruction of several houses and the burial of 23 people, 6 of which deceased. After the December 2008 event, the City Council decided (November 2009) to setup an Avalanche Emergency Plan (AEP) prepared by the Civil Protection Service of the Province of Torino in collaboration with the Oulx Upper Valley Susa Forest Consortium, and the ARPA Piemonte Department of Systems Projections (Bertea et al. 2010). The AEP goal is to safeguard people and villages and to ensure the safety of some sections of the road SS 25.

In the computer simulation, a 3D view of the slope affected by the avalanche events above the village of Venaus (TO) is diplayed. This 3D view (Fig. 24.6) allows to show three different simulations with different avalanche scenarios. These latter are produced by three input snow packs of different thickness at the top of the slope. On the valley bottom, two different limits can be viewed which are used to establish the number of people that need to be evacuated from the town during the emergency.

24.3 Conclusions

A computer simulation has been realized in 2011, by a consortium of different institutions leaded by the Civil Protection Department of the Turin Province, that allows a 3D view of three different types of natural instability phenomena: a debris flow, a landslide and a snow avalanche. The computer simulation has been shown to the public during several occasions: at the "European Researchers' Night 2012" (University of Torino—Geositlab), at Liceo "M. Curie" (Pinerolo—Turin, April 2012), at Liceo "Des Ambrois" (Oulx—Turin, June 2012), during civic protection drills (Ivrea—Turin, October 2012; Bussoleno—Turin, June 2013) at the 2013 Science Festival (Genova, October 2013). It has repeatedly revealed to be useful to attract the public attention and so to produce an increase of awareness of the environmental risks produced by the natural phenomena that have been simulated.

Acknowledgments The authors wish to thank Dr. Giovanni Mortara for his valuable advices and the support given in retrieving and analyzing the historical documentation.

References

Amatruda G, Bonnard CH, Castelli M, Forlati F, Giacomelli L, Morelli M, Paro L, Piana F, Pirulli M, Polino R, Prat P, Ramasco M, Scavia C, Bellardone G, Campus S, Durville J-L, Poisel R, Preh A, Roth W, Tentschert EH (2004) A key approach: the IMIRILAND project method. In: Bonnard C, Forlati F, Scavia C (eds) Identification and mitigation of large landslide risks in Europe. Advances in risk assessment. European Commision, fifth framework programme, IMIRILAND project. A.A. Balkema, Taylor & Francis, London, pp 13–43. ISBN: 90 5809 598 3

Arattano M, Grattoni P (2000) Using a fixed video camera to measure debris-flow surface velocity. In: Wieczorek G, Naeser N (eds) Proceedings, second international conference on debris-flow hazard mitigation: mechanics, prediction, and assessment. Taipei, Taiwan, 16–18 Agosto, 2000. A.A. Balkema, Rotterdam, pp 273–281. ISBN: 90-5809-149-X

Arattano M, Marchi L (2000) Video-derived velocity distribution along a debris flow surge. Phys Chem Earth 25(9):781–784 (Elsevier)

Arattano M (2003) Monitoring the presence of the debris flow front and its velocity through ground vibration detectors. In: Rickenmann D, Chen CL (eds) Proceedings 3rd international conference on debris-flow hazard mitigation: mechanics, prediction, and assessment, pp 719–730. Mill Press, Rotterdam

Arattano M, Franzi L (2003) On the evaluation of debris flows dynamics by means of mathematical models. Nat Hazards Earth Syst Sci 3(6):539–544

Arattano M, Franzi L (2004) Analysis of different water-sediment flow processes in a mountain torrent. Nat Hazards Earth Syst Sci 4:783–791

Arattano M, Franzi L, Marchi L (2006) Influence of rheology on debris flow simulation. Nat Hazards Earth Syst Sci 6:519–528

Arattano M, Conte R, Franzi L, Giordan D, Lazzari A, Luino F (2010) Risk management on an alluvial fan: a case study of the 2008 debris-flow event at Villar Pellice (Piedmont, N-W Italy). Nat Hazards Earth Syst Sci 10:999–1008

Arattano M, Marchi L, Cavalli M (2012) Analysis of debris flow recordings in an instrumented basin: confirmations and new findings. Nat Hazards Earth Syst Sci 12:679–686

Bertea A, Cordola M, Dutto F, Longo F, Dotta A, Caffo L, Vangelista Z, Fontan D (Agosto 2010) Il piano di protezione civile per emergenza valanghe del comune di Venaus (TO). Neve e Valanghe, n. 70, pp 36–43

Brovero M, Campus S, Forlati F, Ramasco M, Susella G (1996) La Frana del "Cassas", Salbertrand, Val di Susa. In Regione Piemonte & Université J Fourrier (eds), Rischi Generati da Grandi Movimenti Franosi, Programma Interreg I Italia-Francia, 143–177

Cordola M, Turroni E, Prola MC, Bertea A, Zaccagnino M, Turco M, Martorina S (2009) Piogge e nevicate intense del 14–17 Dicembre 2008. Neve e Valanghe, n. 67, pp 28–37

Lin ML, Wang KL, Huang JJ (2005) Debris flow run off simulation and verification–case study of Chen-You-Lan Watershed, Taiwan. Nat Hazards Earth Syst Sci 5(3):439–445

Lollino G, Arattano M, Cuccureddu M (2002) The use of the automatic inclinometric system for landslide early warning: the case of Cabella Ligure (North-Western Italy). Phys Chem Earth 27(36):1545–1550. doi:10.1016/S1474-7065(02)00175-4

Lollino G, Arattano M, Allasia P, Giordan D (2006) Time response of a landslide to meteorological events. Nat Hazards Earth Syst Sci 6(2):179–184

Mao L, Cavalli M, Comiti F, Marchi L, Lenzi MA, Arattano M (2009) Sediment transfer processes in two Alpine catchments of contrasting morphological settings. J Hydrol 364(1–2):88–98

Peretti L (1967) Collegamento autostradale del traforo del Frejus con Torino. La Rivista della Strada 36:310

Tsai MP, Hsu YC, Li HC, Shu HM, Liu KF (2011) Application of simulation technique on debris flow hazard zone delineation: a case study in the Daniao tribe, Eastern Taiwan. Nat Hazards Earth Syst Sci 11(11):3053–3062

The Roles and Responsibilities of Engineering Geologists and Other Geoscientists in Serving Society and Protecting the Public—an Overview of International Approaches to Ensuring Effective and Ethical Professional Practice

25

Ruth Allington and Isabel Fernandez-Fuentes

Abstract

This paper is about approaches to ensuring effective and ethical professional practice in the application of geoscience in the interest both of protection of the public and a better informed and more resilient society. It starts by providing an overview of the aims and objectives of the Task Group on Global Geoscience Professionalism (TGGGP) formed in 2012 by the International Union of Geological Societies (IUGS), with the aim of supporting and promoting professionalism in all areas of geoscience practice and developing shared perspectives internationally about how this should be done most effectively. The authors will highlight the importance of achieving a shared understanding of, and commitment to, ethical professional practice across the entire geoscience community in the interests of serving society and protecting the public. To illustrate this, it will propose that continuing dislocation and separation of the academic community from the applied community in geoscience is not in the interests of anyone—least of all the public—and present a framework for promoting closer and more collaborative relationships between geoscience professionals who work as educators, researchers, in government and for 'industry'.

Keywords

Professional geoscience practice, professionalism • Ethical practice • Public engagement • CPD

25.1 Introduction

In August 2012 the IUGS approved the formation of a Task Group on Global Geoscience Professionalism (TGGGP). The Task Group was formed by a group of international professional geoscience organisations comprising: the European Federation of Geologists (EFG); Geoscientists Canada; American Institute of Professional Geologists (AIPG); Australian Institute of Geoscientists (AIG); Geological Society of South Africa (GSSA); and El Colegio de Geólogos de Bolivia (College of Geologists of Bolivia).

A key objective of TGGGP is to support and promote professionalism so as to ensure excellence and best practice in the application of geoscience in the interests of both protection of the public and a better informed and more resilient society. To meet this objective, it aims to help the whole geoscience community embrace professionalism and work together towards breaking down the barriers that currently exist between academic, and applied geoscience.

The authors will highlight the importance of achieving a shared understanding of, and commitment to, ethical professional practice and participation in professional geoscience organisations across the entire geoscience community in the interests of serving society and protecting the public.

R. Allington (✉)
GWP Consultants LLP, IUGS Task Group on Global Geoscience Professionalism, European Federation of Geologists, Oxford, UK
e-mail: rutha@gwp.uk.com

I. Fernandez-Fuentes
European Federation of Geologists, Brussels, Belgium
e-mail: isabel.fernandez@eurogeologists.eu

G. Lollino et al. (eds.), *Engineering Geology for Society and Territory – Volume 7*,
DOI: 10.1007/978-3-319-09303-1_25, © Springer International Publishing Switzerland 2014

25.2 Professional Geoscience Organisations

Professional geoscience organisations around the world support governments, industry and academic institutions to assist with the setting of standards for:

- Communication of geoscience information relevant to public safety;
- Responsible use of geoscience information in relation to natural hazard mitigation, sustainable development and capital investment; and
- Standards and content of academic training, experience and continuing professional development (CPD) required for safe, ethical and competent geoscience practice.

Some professional geoscience organisations focus solely on professionalism, whilst some combine this with more traditional roles of scientific societies and institutions. A common feature of all such organisations is that they have codes of professional conduct/professional ethics binding on all their members (http://tg-ggp.org/professionalism-in-geoscience/codes-of-ethics/); they also operate disciplinary codes. Some have a role in setting standards for or administration of statutory regulation of geoscience professionals where licensing or registration is a legal requirement. Others award professional titles to those judged by their peers to have sufficient qualifications, experience and competence to meet particular criteria—these titles are voluntary and are maintained through continuing professional development (CPD).

Professional geoscience organisations and the regulatory frameworks they promote and operate (whether statutory or voluntary) have traditionally focussed primarily on supporting and regulating the professional practice of geoscientists working in applied fields with very little (and sometimes no) involvement of the academic community and/or geoscientists employed by government. TGGGP wishes to promote the idea that professional skills are essential for effective and ethical practice alongside technical and scientific expertise for anyone who makes their living as a geoscientist, whether as an educator, researcher or applied/industry practitioner. It therefore uses as the starting point for is mission a wider definition of "professional geoscientist" than is commonly understood, encouraging both applied/industry practitioners on the one hand and educators and researchers on the other to see themselves as "professionals".

25.3 Regulatory Frameworks

The primary stated purpose of regulatory frameworks applied to professional practice (including in geoscience) is to avoid harm to people (health, safety, economic wellbeing) or the environment caused by malpractice or faulty products. In general, regulation of professional practice is applied to services that are considered as public goods (and where errors or wilful malpractice have the potential to cause harm).

As indicated above, there are two distinct models of regulation of the practice and profession of geoscience:

- Mandatory registration or licensure (legally required and enforceable by law); and
- Systems based on the award of professional titles and voluntary registration (operated by professional geoscience organisations and with assessment by peers).

The different approaches taken in different jurisdictions reflect cultural and legal differences in approaches to regulation and the extent to which the practice of geoscience (or aspects of it) is considered in that jurisdiction to have the potential for significant harm.

Sometimes the two models of professional regulation of geoscience co-exist. For example, in some States of the USA licensure is mandatory, but geoscientists may also hold the voluntary professional title "Certified Professional Geologist" (CPG) awarded by the AIPG. Similarly, European geologists working in countries where the profession is regulated by law may also choose to hold the voluntary professional title "European Geologist" (EurGeol) awarded by the EFG. The table produced on the website of the European Federation of Geologists (http://www.eurogeologists.eu/index.php?page=1141) illustrates the range of geoscience regulatory models that exist within Europe, and the TGGGP is currently working on a similar table summarising the situation world-wide.

25.3.1 Compulsory Registration or Licensure

In many jurisdictions, the geoscience professions are regulated by law and registration or award of a licence to practise is compulsory before geoscientists are allowed to work as a geoscientist (other than as a trainee or assistant). An application for a licence (or for registration) normally requires evidence to be submitted of the content and level of academic qualifications, which must satisfy certain criteria related to content and level of attainment. The criteria may be expressed in generic terms or may be linked directly to qualifications in that jurisdiction specifically. In many jurisdictions, candidates must also sit an examination which may test basic geological knowledge and adequacy of language skills. Professional practice examinations may also be required to test such things as: awareness of professionalism and understanding of ethical codes (also known as codes of conduct); knowledge of regulation and governance of their profession; and understanding of applicable law and legal concepts. In some jurisdictions, successful attainment of registration or a licence to practise in that jurisdiction allows geological or general geoscience practice in any geoscience sector, relying on adherence to a code of ethics to ensure that licensed or registered practitioners do not undertake work that falls outside their area of specialism and/or beyond their skill or qualification level. In others the practitioner is legally restricted to a specific area or areas of professional practice (e.g. hydrogeology, engineering geology, exploration geology etc.).

Having successfully attained registration or a licence to practise, a geoscientist must adhere to rules that are in force for maintaining that registration. These rules invariably include an absolute obligation to adhere to a code of ethics or code of conduct, as well as all regulations and laws applicable to professional geoscience practice in the jurisdiction concerned. Periodic submission of evidence of activities that contribute to continuing professional development (CPD) may also be required.

25.3.2 Self Regulation

Voluntary professional titles are awarded by professional geoscience organisations to those of their members who are judged by their peers to meet or exceed qualification and experience criteria. Professional titles held by geoscientists indicate to the public, employers or other professionals that the holder has achieved suitable academic training and a level of professional experience, skill and competence to perform tasks within their professional practice. It also confirms that the holder undertakes continuing education and training, demonstrating a personal commitment to stay up to date and informed within the sphere of their professional work as well as a personal commitment to adherence to a code of conduct or ethics.

The application requirements for voluntary professional titles in terms of providing information about qualifications may be similar (or identical) to those relating to registration and licensure, and there may be an examination (sometimes written but normally oral). Another common feature is a requirement to adhere to a code of conduct or ethics and agreement to be subject to disciplinary sanctions. However, they differ from compulsory registration or licensure in three important ways. First, the practitioner must demonstrate their experience and level of competence as a practitioner—these titles are not awarded immediately post graduation as a licence to practise or registration may be. Second, their voluntary nature underlines the personal professional commitment made by individuals who hold them. Third, central to these titles and associated regulation and disciplinary codes is assessment and being called to account by one's peers.

25.3.3 Professional Titles

Whether voluntary or related to mandatory registration or licensure, have become important in demonstrating the suitability of a professional to provide geological services. They are sometimes used as a 'threshold' level of experience and expertise in legislation or guidance. For example, all CRIRSCO (http://www.ccpg.ca/pgeoreg/index.php?lang=en&subpg=defprofgeo) codes and standards define geoscientists as 'Competent' or 'Qualified' persons for the reporting of mineral resources and reserves by reference to the amount and relevance of experience the person has and their membership of an approved professional geoscience organisation (at an appropriate level, usually expressed as holding a particular professional title). With some exceptions, the lists of approved professional geoscience organisations are common to the CRIRSCO codes (http://www.crirsco.com/background.asp) and standards and this allows recognition of the equivalent standing and meaning of professional qualifications (such as EurGeol, CGeol, PGeo, CPG etc.) internationally. Increasingly, professional geoscience organisations are exploring ways formally to recognise each others' professional qualifications so as to facilitate international mobility of geoscientists. This takes the form either of recognising absolute equivalence or, more usually, by 'mapping' the requirements of one qualification onto another so as to simplify application processes for holders of one qualification who wish to be recognised in another jurisdiction (e.g. taking validation of academic qualifications in one jurisdiction 'as read' in another).

25.4 The Practice of Professional Geoscience—Who Is Involved?

Geoscientists Canada has published on its website the following definition of professional geoscience practice (http://www.ccpg.ca/pgeoreg/index.php?lang=en&subpg=defprofgeo):

> the performing of any activity that requires application of the principles of the geological sciences, and that concerns the safeguarding of public welfare, life, health, property, or economic interests, including, but not limited to: investigations, interpretations, evaluations, consultations or management aimed at discovery or development of metallic or non-metallic minerals, rocks, nuclear or fossil fuels, precious stones and water resources; and investigations, interpretations, evaluations, consultations or management relating to geoscientific properties, conditions or processes that may affect the well being of the general public, including those pertaining to preservation of the natural environment.

In the authors' experience, those geoscientists who consider themselves to be in "professional geoscience practice" operate mainly in applied/industrial fields rather

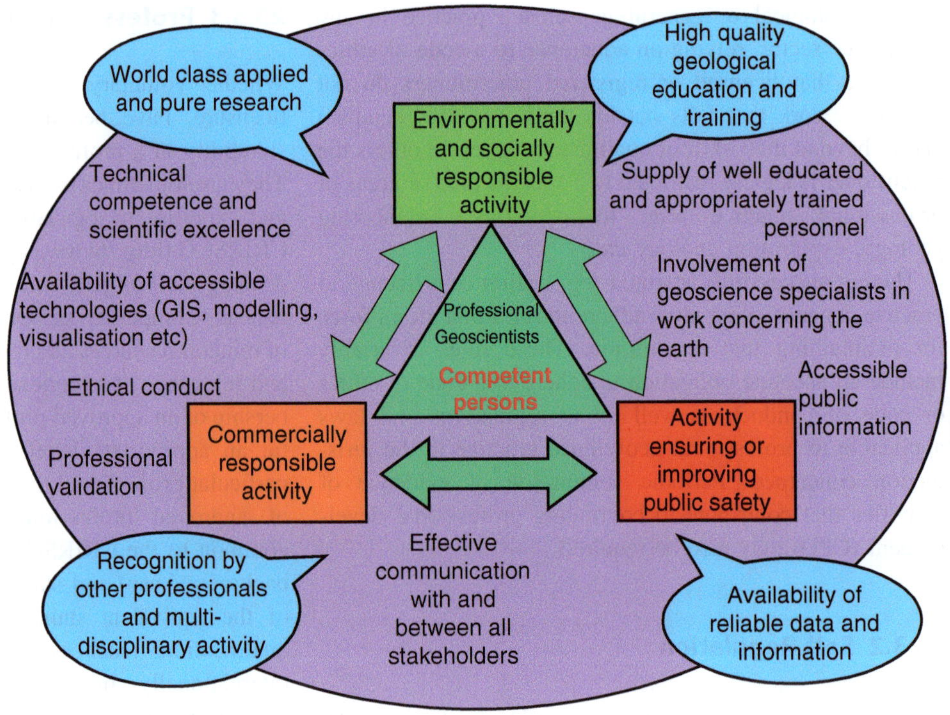

Fig. 25.1 Framework for considering how much more effectively geoscientists may serve and protect the public, maintain and develop their skills and push the boundaries of innovation and research based on a shared perception of professionalism

than in academia or government service. The membership of professional geoscience organisations (or those parts of scientific societies and associations that focus on professional matters) tends to be dominated by those for whom the link between the work that they do and service to and protection of the public is direct and obvious.

The diagram in Fig. 25.1 illustrates the authors' perception of the web of interfaces and inter-connections that contribute to effective professional geoscience practice. Contrary to widely held opinions, it is not just the professional practice of geoscience practitioners at the 'sharp end' of project delivery or advice or information to the public who have an impact on public safety, public information or sustainability. The centre of the diagram illustrates the balance between environmental/social responsibility, commercial responsibility, and ensuring or improving public safety that professional geoscientists must aim to achieve. The blue 'callouts' depict the excellent and relevant training and education, research, interdisciplinary cooperation, and data and information upon which professional geoscientists depend. The essential background and underpinning to these inter-linkages are the ethical, technological and political/procedural matters indicated in the purple area.

The following key questions summarise the framework illustrated in Fig. 25.1:

- Without understanding the skills and expertise needed by 'industry', how can educators in geoscience subjects prepare students for the workplace?
- Without understanding societal needs, how can researchers design research which is truly relevant to those needs?
- Without access to high quality graduates and excellent underpinning research, how can geoscientists in 'industry' develop and deliver their expertise effectively?

Natural Hazards, Disasters and Their Mitigation: An Overview with Reference to Indian Scenario

26

Vasudevan Desikachari

Abstract

Natural hazards are phenomena that are triggered by natural causes of geological, geophysical and hydro-meteorological origins. They occur in well-defined tectonic domains, like for instance the "Pacific Ring of Fire", stretching from New Zealand along the eastern edge of Asia, northwards across the Aleutians of Alaska and southward along the coasts of North and South America and which is the loci for most of the world's volcanoes and intense seismic activity. It is thus possible to say where these hazards like a volcano or an earthquake could be expected but as of today their forecasts and predictions remain difficult to achieve. Natural hazards are beyond human control and therefore when they strike a human habitation, they can instantly turn into disasters and wreak heavy loss of life and property, often pushing an unprepared Nation to virtual economic crisis. In recent years several major earthquakes have rocked megacities of the world causing contrasting damages in terms of fatalities and property. Haiti, a poor underdeveloped nation was nearly annihilated by a magnitude 7.0 earthquake in 2010 mainly because the country was unprepared to meet the impending danger, whereas well prepared nations in terms of disaster mitigation, management and crisis control withstood such hazard onslaughts with confidence for example, Chile in 2010 of a 8.0 magnitude earthquake and Tahoka of an earthquake-tsunami onslaught (9.0 magnitude temblor). This paper reviews briefly mitigation and disaster strategies that lead to maximum reduction in the devastating effects of hazards occurring in a given habitation, with special reference to India.

Keywords

Natural hazards • Disasters • Mitigation • India

26.1 Introduction

Natural hazards are phenomena caused by geological and hydro-meteorological processes that tend to shape the earth from the time of its inception till today and they operate totally independent of human control. These hazards when they strike a human habitation and infrastructures, transform into disaster. The severity of a disaster depends on several factors, such as the susceptibility of the area to a given hazard (volcano or even floods, cyclone etc.), its density of population and its economic status, infrastructures and their capability to sustain shocks and vibration given off by the hazard (say volcano, floods, cyclone etc.), its density of population and its economic status; its infrastructures and their capability to sustain shocks and vibrations given off by a hazard. Natural hazards seldom occur randomly as their occurrence is controlled by well-defined geologic/geographic horizons, for example, all great earthquakes and

Dr. Vasudevan Desikachari is a Retired Deputy Director General of Geological Survey of India.

V. Desikachari (✉)
Geological Survey of India, Kolkata, India
e-mail: dvchari44@gmail.com

G. Lollino et al. (eds.), *Engineering Geology for Society and Territory – Volume 7*,
DOI: 10.1007/978-3-319-09303-1_26,

volcanoes of the world occur in the "Pacific-Ring of Fire". In the modern world, therefore, a given inhabited locality anywhere in the world is aware of the natural hazards that it may have to encounter in a given period of time-like for example, seasonal floods or cyclones or if the locality is in an active tectonic domain, earthquakes. People of such localities are obviously expected to live in preparedness to face this situation with its available resourcefulness, failing which destruction would devour it. Two contrasting examples can be given here to exemplify this statement. Chile, a South American country was struck by a massive earthquake of magnitude 9.5 in 1960, the strongest tremor recorded in the twentieth century which brought loss of human lives and properties running into millions of U.S dollars. Chile, however, quickly reacted to the maladies of the disaster by formulating and adopting suitable measures of mitigation to safeguard its population in the event of such a hazard striking it in the future. Thus Chile adopted a uniform seismic code in all the major constructions of dwelling and major civil structures (hospitals, schools, roads, bridges, dams etc.) and adopted very strict designs and materials chosen for these constructions, showing great "Resilience" in mitigation. In 2010, the country was struck again by a major earthquake of magnitude 8.8 which although caused destruction of infrastructures, the damages were kept in check; fatalities and injuries were considerably less compared to the previous occasion. Compare this to a country which remained oblivious of its danger and thus succumbed to it almost totally. Haiti, an island country in the Caribbean plate was struck by a 7.1 magnitude earthquake in 2010 and being a very poor and under developed nation with poorly constructed dwellings and older buildings of poor seismic-sustenance, was devastated. There was heavy loss of lives in thousands and property in millions of dollars. The country's economy was in shambles and could not bear this loss. Yet another example of a country that did not anticipate a disaster due to hazard is India. Although 55 % of the land is declared prone to earthquake, effective measures were not adopted in the construction of human dwellings and the result was the tragedy in the form of a major earth-tremor that shook Bhuj in Gujarat, in January 26th 2001, causing a loss of 10,000 lives and several thousand injured besides total annihilation of thousands of structures over a wide area from the epicentre. The Rann of Kutch in Gujarat was affected by a devastating earthquake in the nineteenth century and therefore remained a seismically active area. The authorities, however, did not seem to take cognizance of this danger and faced the consequence. Contrast once again this with Japan, which, being aware of its living in active seismic domain in the world, took ample measures of "Preparedness and Resilience" in Mitigation measures with the result a mighty earthquake of magnitude 9 with accompanying tsunami which struck Tohoko in 2011, surprisingly resulted in a fatality of "only" 18,184 people and loss of property and structures attributable mainly to the tsunami, and, even this loss was due to an under-estimate of magnitude of "future earthquakes" (Tucker 2013) based on seismic studies, which did not predict quakes over magnitude 7.5 in the region.

This paper highlights the maladies of natural hazards and the mitigation measures to counter these with reference to India in particular.

26.2 Natural Hazards, Disasters and Their Mitigation

"Mitigation is the effort to reduce loss of life and property by lessening the impact of disasters" (FEMA 2007). In the field of disaster management, mitigation occupies the prime place. There are two aspects of mitigation, one, mitigation adopted to an area known to be prone to disasters caused by natural hazards and are prepared to meet the "Eventuality"; and the other the most important, is related to the mitigation in hazard-prone areas which are poor, under-developed and that choose to remain in oblivion to the impending dangers out of natural hazards, knowing full-well of living on hazard-prone zone. Accessibility to disaster-struck area becomes easier and clearer in the former enabling quick and prompt relief reaching affected areas; whereas in the latter areas, showing scant respect for "Preparedness" to meet disaster, mitigation becomes difficult solely because of inaccessibility to affected areas, caused by destroyed structures blocking the roads and street and difficulties in quick removal of debris to evacuate people from the affected areas. Example for well managed mitigation effort is afforded by relief measures made available to affected areas by Japan in the Tohoko earthquake of 2011 and Chile in 2010, whilst, Haiti, stood as a glaring example for badly managed or "nil-managed" mitigation effort by administrating authorities. FEMA (2007), has rightly pointed out that "every dollar spent towards mitigation programme helps the country to the extent of four dollars per person at the time of disaster". The hallmark of mitigation is to alleviate suffering and reduce the financial burden of the disaster-prone country and thereby make it nearly self-reliant to deal with disasters. Japan, New Zealand and Chile for example were self-reliant in managing the devastating effects of earthquakes in recent years thanks to their mitigation and disaster management agencies effective dealing of given situations.

Other mitigation measures that can be thought of include,

(a) Close-monitoring of active tectonic domains by a network of seismometers and ground GPS to evaluate and assess the seismicity of the area and to transmit the data collated to user-agencies on time for effective mitigation and disaster management.

(b) Adoption of suitable seismic code and choosing desired design and construction materials for building residential buildings of varied types as well as major civil constructions. Government agencies must keep a strict vigil in enforcing the construction of all types conforming to these adopted parameters.
(c) Vulnerable areas to be identified, for example densely populated areas, schools and other educational institutions, hospitals and if those buildings are old, retrofitting or repairs if any may be pursued.
(d) "Preparedness and Resilience" should be the watchwords for Mitigation management.
(e) In the management of construction, retrofitting and repairs of infrastructures, areas prone to landslides, floods and cyclones may also be kept in view.
(f) Monitoring drainage data of watersheds, weather forecast monitoring and their effective transmission to user agencies.
(g) macro and micro-level mapping of landslide prone areas; minimizing deforestation, stabilizing weak slopes of hills and providing suitable drainage to ease of water-logging and careful construction in areas noted for landslides and flooding.

Assessment of risks along vulnerable zones prone to disasters (Carcedo 2001) is yet another chapter in mitigation which involves collection of proper demographic data of human-cattle population and their distribution pattern and density. Status of infrastructures; condition of roads/bridges within the city and suburbs and the status of arterial roads are to be properly brought out and the data stored properly for use at the time of emergency (AUSGEO News 2008; Brown 1996; Laframboise 2012).

26.3 Indian Scenario

In the context of natural hazards and their occurrence world over and the disasters caused by them, India's position is very clear. According to AESGEO News (2008) in the Asia-Pacific region, India, along with its neighbouring countries of Pakistan, Nepal and China, stands as the most earthquake-prone countries in the world. As per a bulletin issued by NDMA (2011), 55 % of India faces threat from imminent earthquakes of destructible magnitude; 42 million hectares of land area in the country likewise is highly prone to disastrous riverine and coastal floods and 12 % to landslides. Earthquakes, floods and landslides form the major disaster-causing natural hazards in this highly populated country. Recent occurrences of earthquakes in the Himalayan area which is known for its tectonic instability and disastrous tremors that ravaged relatively "stable" part of the shield area of the country have clearly pointed out towards the need for a reassessment of "Natural hazard mitigation and disaster management policies" to come out with a suitable "blue print" to manage any adverse situations arising out of disasters effectively.

26.3.1 Earthquakes

Till 1967, India, excepting for its areas in the Himalayan territory was considered stable tectonically and therefore seismic activity of great magnitude was not expected in the areas other than the Himalayas. This scene changed dramatically in 1967, when Koyna area in Maharashtra, located in "relatively stable" part of the Indian Peninsular shield was struck by a magnitude 6.1 earthquake, forcing the scientific-earthquake engineering fraternity to revise the seismic zone map of the country by introducing Zone I, replacing the earlier Zone 0. The so-called stable part of the country was repeatedly shaken by earthquakes varying in magnitude between 6.0 and 7.7 and moderate to strong intensities over a period of 10 years between 1991 and 2001. These earthquakes are: 1991 Uttarkashi tremor of 6.6 magnitude; 1999 Chamoli tremor of 6.3 magnitude, the 1993 Killari tremor of 6.3 magnitude; 1997 Jabalpur earthquake of 6 magnitude and the 2001 Bhuj earthquake of 7.7 magnitude. This was followed by the tsunami of December, 2004 that originated off the coast of Sumatra. The historic but rather frequent occurrence of earthquakes of magnitude 7–8 recorded in the Himalayan region in the nineteenth century clearly established the great seismic activity underlying this area, which is a tectonic boundary marking the collision of the northern Himalayas against Eurasia. Wyss (2005) had carried out extensive study of the seismicity of the Himalayas and based on the parameters of historic earthquakes, their distribution and amount of slips accumulated by the contact plate motions since those earthquakes occurred, showed that the area is overdue for a very major tremor to occur. The prediction was proved by the occurrence of a major earthquake of 7.5 magnitudes that shook the Kashmir (Pakistan) and some parts of Kashmir and Punjab in India, in 2005 which caused a major loss of lives of 80,000 in Pakistan-occupied area. It is now clear that earthquakes in India occur in three geologic-set up, namely, the Stable-Continental zone of earthquakes; earthquakes associated with the Himalayan suture zone or collision zone and the Andaman-Sumatra subduction zone earthquakes (Kayal 2007). Seismic zone map of the country was revised in 2002, showing areas with probable intensities of earthquakes on the Modified-Mercali scale. It is clear that India needs to be more vigilant against earthquakes as a causative force of destruction in the 55 % of the country which has been declared as earthquake-prone. Measures of pre-disaster mitigation, such as providing earthquake-resistant structures to people; in the case of multi-storied building constructions every care must go into its construction so that the building suffers only minimum damage even in the event of a severe tremor affecting it. In the case of very large old structures, proper assessment should be made and decision taken accordingly either to go for suitable renovation involving retrofitting/and repairs or total demolition of the building.

Finally, whatever "blue-prints" are prepared by government agencies, as working models to counter natural hazards, the concerned authorities must enforce the "Do's" and "Do not's" so that mitigation envisaged works out successfully.

26.3.2 Floods and Landslides

Next to earthquakes, Floods and landslides are the most disastrous of natural hazards in India. The country is traversed by several major rivers, their tributaries and streams on the basis of which thrive major population, depending mostly on agriculture for its livelihood. India, being tropical, receives from heavy to very heavy rains, thunder-storms and cloud-bursts during monsoons; and when rain-water flowing into these rivers/streams exceed their capacity to contain, these rivers/and streams overflow, resulting in flooding of the adjoining areas of cultivation and population. The flooding originating thus wreaks havoc almost every year. The riverine floods of India especially in the Brahmaputra river basin and the Ganges basin, cause heavy damages of property, grains and loss of lives (human and livestock), in the states of Assam and Bihar respectively and there is as yet no clear tangible way found that could really control this flooding and thereby reduce the damage done. The government of India as well as the State government agencies are spending a lot of money in working out suitable measures of mitigation of flood and flood-related landslides and so far they have achieved very little towards their goal. Floods occur in the lower reaches of Himalayas too, where the problem is complicated by melting of glaciers in the upper reaches which augment the quantity of water flow into the major rivers and streams, causing floods. In June 2013, there was an unprecedented flooding of the pilgrim town of Kedarnath in the state of Uttaranchal, caused by torrential monsoon rain aided by melting of ice in the upper reaches. This caused landslides with such ferocity that an estimated 10,000 pilgrims who gathered in the temple town of Kedarnath lost their lives and a few villages in the neighborhood were totally destroyed. Although the flood and the accompanying landslides were caused by hydro-meteorological causes, it was accentuated by human contribution by way of indiscriminate deforestation, cutting of hill slopes randomly, mining and other engineering activity in the area of known fragile eco-system and unstable geology. Added to this was the reported deficiency in weather forecast and lack of implementation of remedial measures suggested to redeem malady by the authorities concerned. Buildings that were ravaged by flood and landslides were not sited on where they should have been and further, were weakly founded and poorly built without following any code of construction. Further the small area was over-populated.

Landslides, as such, forms yet another hazard of concern to India, affecting again greater part of the fragile Himalayas, north-east India and parts of Western Ghats towards south India. Flooding and heavy rains are again the chief causes that trigger those landslides, affecting hill-slope stability and where free movement of water is constrained by poor and choked drainage. Deforestation, cutting of trees and building activity further accentuate the process, causing instability of hill slopes, land and rock slip or both. Geological Survey of India is the nodal agency for carrying out landslide studies covering major areas of landslides.

26.4 Conclusion

Earthquake, Floods and Landslides are some of the most important natural hazards that have been causing destruction to society, since a very long time. The damage caused by way of fatality and property often reach staggering figure and affect in particular, economically backward and poorly prepared countries. It is here that Mitigation and Disaster management comes into play. The hallmark of mitigation is the effective implementation of precautions worked-out area specifically, such as, for instance, retrofitting and reconstruction of pre-existing structures affected during a major earthquake. Mitigation strategies are worked out as dictated by the area under study and they vary from place to place as dictated by parameters such as density of population, type of infrastructures and economic condition of the country and the social status of the people. Last but not the least in disaster management and mitigation is Educating the down-trodden in poor countries that are prone to disasters, as this goes a long way in alleviating sufferings due to hazards.

References

AUSGEO News (2008) AUSGEO News (Australian Government)

Brown WM (1996) Hazard maps help save lives and property. U.S.G.S. fact sheet, pp. 183–196

Carcedo A (2001) Natural disaster mitigation in the world and sustainable development—a risk analysis approach. Bull Geol Min 112(4):43–64

Dayton-Johnson J (2006) Natural disasters and vulnerability. OECD Development Center Policy Brief 29

FEMA (2007) Disaster management brochure

Kayal JR (2007) Recent large earthquakes in India. International Association of Gondwana research, Japan

Laframboise M (2012) Mitigating impacts, managing risks. IMF working paper

Li Q, Liu M, Yang Y (2002) The 01/26/2011 Bhuj India earthquake: intraplate or interplate. In Plate boundary zone. AGUGS Series, Mono 30:255–264

NDMA (2011) National disaster management information and communication systems. Government of India

Tucker B (2013) Reducing earthquake risk. Science 341:1070–1072

Wyss M (2005) Human losses expected in Himalayan earthquakes. Nat Hazards 34:305–314

Geohazards in the East Siang District of Arunachal Pradesh, India: Need for Geoethical Approach for Integrated Flood Mitigation Strategies

27

Swapna Acharjee

Abstract

The East Siang District represents one of the multi-hazard prone districts of Arunachal Pradesh, India covering an area of approximately 4,005 km^2 drained by the mighty trans-boundary river Siang. It is located in the foothills of the Himalaya and shows wide variation in lithology and topography. Pasighat census town, established in the year 1911, is the headquarter of the district which falls in zone III of the seismic hazard zonation map of North-Eastern region of India, and can experience peak ground acceleration of the order of 200–250 cm/s^2. Flash floods, landslide dam outburst floods (LDOFs) are the major geo-hazards in the district that causes heavy destruction of life and property. Other geohazards include landslides, hailstorm and forest fire spread across the year. The LDOF in Tibet on 11 June 2000 caused death of many people living in the downstream areas of the East Siang district. It also affected more than 10,000 people, submergence of villages and standing crops, and collapse of bridges in Arunachal Pradesh and Assam. The occurrences of flash floods have risen in the past decade causing partial damage to the township of Pasighat. A large numbers of rural link roads, culverts, suspension bridges and wooden bridges have been collapsed due to landslides, flash floods and LDOF time to time. In view of the fact that, Siang is trans-boundary river, temporal coverage in real time and at frequent intervals is 'thus' required for continuous monitoring of behavior of the river in relation to incessant rains, cloudbursts and on-going neotectonic activity. A geoethical approach is needed in establishment of integrated flood mitigation strategies by effectively using adequate technologies and sharing of scientific knowledge at trans-boundary levels.

Keywords

Geohazards • East Siang district • Siang river • Geoethics • Integrated flood mitigation strategies

27.1 Introduction

Disasters are as old as human history more particularly in the vulnerable areas of south-east Asia. It results from the combination of hazard, vulnerability and insufficient capacity or measures to reduce the potential chance of risk (Maskrey 1989). The East Siang District having an area of approximately 4,005 km^2 is one of the multi-hazard prone districts of Arunachal Pradesh, India. The distinctive physical setting of the region vis-a-vis tectonic activities in the Himalayas, has been significantly influencing the river

S. Acharjee (✉)
Department of Science and Technology, State Remote Sensing Application Centre, Arunachal Pradesh State Council for Science and Technology, ESS Sector, Itanagar, 791111 India
e-mail: swapna_a@yahoo.com

G. Lollino et al. (eds.), *Engineering Geology for Society and Territory – Volume 7*,
DOI: 10.1007/978-3-319-09303-1_27, © Springer International Publishing Switzerland 2014

Fig. 27.1 Damming of the Yigong River on May 4, 2000, the cause of LDOF on 11th June 2000 (http://earthobservatory.nasa.gov)

system. Palaeochannel reconstruction suggests gradual eastward migration of the Siang River in response to northeast ward and south eastward ground tilting due to neotectonic activity (Luirei and Bhakuni 2008). The geohazards in the district include flash flood, landslide dam outburst flood (LDOF), landslides, hailstorm and forest fire, spread throughout the year. Pasighat town falls in zone III as per seismic zonation and can experience peak ground acceleration of the order of 200 to 250 cm/s^2 (Mohan et al. 2008). Agriculture is the mainstay of the people of the study area. As the topography of the district is characterized by plains as well as hills, the agriculture pattern also varies from hill to plain areas. In the vast tract of alluvial and flood plains, the settled or permanent agriculture is predominantly practiced. The river Siang which is a transboundary river symbolizes extreme manifestation of nature's fury during heavy and prolonged rainfall. The mighty river flowing through the district has numerous tributaries which possess immense hydropower potential. The rainfall pattern, erosion, land-use pressure especially in the flood-plain belt intensify the flood hazard leading to disasters.

27.2 The Milieu

The LDOF on 11 June year 2000 was most devastating for the Pasighat town in the East Siang District. According to various reports it was about 300 million cubic meters of displaced debris, soil, and ice which dammed the Siang River in eastern Tibet. The dam was about 130 m thick, 1.5 km long and 2.6 km wide and was created in eight minutes (Fig. 27.1).

In April 2000, the discharge of the stream flow into the dam lake was about 100 cum/s with the water rising at the rate of 1 m/per day. This flood has affected over 10,000 people including submergence of villages and standing crops (Tewari 2004). Many bridges collapsed in Arunachal Pradesh and Assam including the rural link roads, culverts, suspension bridges and wooden bridges. Flash Floods have become a frequent hazard in the district. Flash floods in the month of June, 2013, damaged horticultural gardens, fish ponds and water pipelines besides a large area of crop fields in all circles of the district. Consequently it is observed that LDOF's and Flash floods deeply influence the East Siang district situated in the south eastern part of Arunachal Pradesh. We find some strata in the society are more prone to sufferings due to the disasters including people from weaker class, gender and age. In this perception it is felt that integrated mitigation strategies are very essential in the vulnerable catchments like that of river Siang. The satellite based observational networks have proved to be very effective in many parts of world. Satellite data captured in the near real time and adequate hydro-meteorological information always increases the integrated approach to assess risk, capacity building and time to respond. It also ensures disaster services using the available resources. The value of warning increases when people have a considerable amount of lead time. It is an era of information and communication technology therefore mitigating flood disaster is highly dependent on infrastructural capacity of the stakeholders. In a study based on remote sensing data under the DMIS programme (2004), some ground measures were suggested to alleviate the impact of flood hazard in the district, like

1. Strengthening of existing bunds/embankments.
2. Construction of rehabilitation shelters at safe levels.
3. Prevention of permanent structures in the severe and very high flood hazard zones and
4. Stabilization of slopes.

In the recent time the structural measures like construction of dams have raised conflict between the planners and the environmentalists therefore the principal non-structural option lies in use of appropriate technologies which will help the people living in the vulnerable catchments. A deep geo-environmental research using advanced technology for reducing the impact of the hazard in the catchment is apposite.

27.3 Importance of Geoethics in Flood Hazard Management and Mitigation

The trans-boundary catchments are very important focal point for geoethical intervention considering that the flood hazard in these catchments have enormous impact on several sections of the society covering one or more states and countries falling within the catchment. The Flood Mitigation plans for the vulnerable catchments are the matter of conscientious study for saving life and property of the people. Appropriate and adequate access ways for integrated flood mitigation strategies are the composite structure of proactive ethical practice for the improvement of the life of the people in the vulnerable catchments. For that reason a move toward conjugal planning by the stakeholders at appropriate level is the only way to live in synchronization with the catchment level disasters. Geoethical approaches in integrated mitigation strategies are very effective for environmental emergencies (Peppoloni and Di Capua 2012). It is succinctly felt by the global geoscientific community that geoethics is essential for judiciously assessing structural and non-structural mitigation strategies for flood hazard management, environmental security and sustainable development. However, the significant issue lies in striking balance equilibrium at policy level in the decision-making.

Acknowledments The author is grateful to Dr. Silvia Peppoloni, INGV, Italy for introducing the subject Geoethics. Further the author is thankful to the anonymous reviewer for suggestions that has incited improvement of the paper.

References

Luirei K, Bhakuni SS (2008) Geomorphic imprints of neotectonic activity along the frontal part of eastern Himalaya, Pasighat, East Siang district, Arunachal Pradesh. J Geol Soc India 71:502–512
Maskrey A (1989) Disaster mitigation: a community based approach. Oxfam, Oxford, UK. p 100. ISBN 0855981229
Mohan K, Joshi A, Patel SC (2008) The assessment of seismic hazard in two seismically active regions in Himalayas using deterministic approach. J Indian Geophys Union 12(33):97–107
Peppoloni S, Di Capua G (eds) (2012) Geoethics and geological culture. In: Reflections from the geoitalia conference 2011. Annals of geophysics, vol 55. p 3
Tewari P (2004) A study on soil erosion in Pasighat Town (Arunachal Pradesh) India. Nat Hazards 32:257–277

Part III

Resilience to Citizens and Citizens for Resilience. From Collaborative Risk Management to Knowledge Sharing of Natural Hazards

Convener: Prof. Marco Giardino—*Co-conveners*: John Clague, Elena Rapisardi, Massimo Lanfranco, Nick Roberts

As natural risks are part of our reality, the challenge is to focus on preparedness, an interdisciplinary issue for more effective disaster resilience. The "common and shared knowledge" approach, empowered by web 2.0/3.0 technologies, embodies the purpose of people-centered early warning systems: it could also improve efficacy and efficiency of rescue teams, thus enabling cooperation, knowledge sharing, and collaboration at all levels. This part encourages discussion on the applicability, opportunity, and constraints of the approaches, procedures and technologies to preparedness actions. (A) The "web 2.0 wave": threat or opportunity for disaster resilience? (B) Two-way emergency communications: empower or menace for governmental organizations. (C) ICT laws and regulations: dinosaurs in a glass store? (D) Is research ready for Open Data and Open Knowledge? (E) Cultural versus technological challenges in disaster resilience. (E) Web and mobile technologies: experiences and tools.

GeoMedia-web: Multimedia and Networks for Dissemination of Knowledge on Geoheritage and Natural Risk

28

Marco Giardino, Vincenzo Lombardo, Francesca Lozar, Alessandra Magagna and Luigi Perotti

Abstract

Geological knowledge is fundamental for the assessment and mitigation of natural risks, as well as for the successful exploration and management of georesources. The GeoMedia-web project aims to improve people awareness of geological issues by establishing a network of researchers, teachers, and scientific institutions sharing innovative multimedia solutions for the dissemination of Earth Science contents, focusing on natural hazards and geological heritage. Through the enhancement of awareness of geodiversity, we pursue to highlight evidences of past environmental changes and evolutionary processes, and also to make available information needed to accurately foresee future environmental changes, related risks and for providing adequate adaptation plans and/or prevention measures. The concept inspiring the GeoMedia-web project is based on principles of geoconservation and virtual reality applications. By means of collaborative sharing and creation of dissemination products, we want to fill the gap existing in Italy between these theoretical frameworks and their effective application in geodiversity action plans, and in the proactive management of natural risks.

Keywords

Multimedia • Geological heritage • Natural risk • Earth sciences • Education

28.1 Introduction

The project "GeoMedia-web: multimedia and networks for dissemination of knowledge on geoheritage and natural risk" pursues the establishment of a network, intended as a collaboration between a university team (*GeoMedia-team*) and a multiscale (local, regional, national, international) group of schools, governmental institutions, cultural associations, museums and enterprises (*GeoMedia-network*), working together to collect, test and design some innovative multimedia products for the dissemination of the geological knowledge, focusing on natural hazards and geological heritage.

This project is funded by the Italian Ministry of Education, University and Research (MIUR) under the Law 6/2000, for the promotion and execution of framework agreements with other government departments, public and private universities, relating to initiatives for the dissemination of scientific culture.

The GeoMedia-web project has been designed by a multidisciplinary team, integrating different skills, composed of the Department of Earth Sciences, the Centre for the Study of Media and Audiovisual (CIRMA, www.cirma.unito.it)

M. Giardino · F. Lozar (✉) · A. Magagna · L. Perotti
Dipartimento di Scienze della Terra - Università degli Studi di Torino, via Valperga Caluso 35, 10125 Turin, Italy
e-mail: francesca.lozar@unito.it

M. Giardino
e-mail: marco.giardino@unito.it

V. Lombardo
Dipartimento di Studi Umanistici and Dipartimento di Informatica - Università degli Studi di Torino, via S. Ottavio 20, 10124 Turin, Italy

G. Lollino et al. (eds.), *Engineering Geology for Society and Territory – Volume 7*,
DOI: 10.1007/978-3-319-09303-1_28, © Springer International Publishing Switzerland 2014

and the Centre for the Study of Natural Risks (NatRisk, www.natrisk.org) of the Turin University, based on the experience carried on with the Virtual Reality and Multi Media Park (VRMMP, www.vrmmp.it), a technological facility of Turin, dedicated to the development of multimedia with particular attention to the use of virtual reality applications.

Balancing the mixing of Earth Sciences knowledge, Information and Communication Technology (ICT) instruments, geomatics, geoethical principles and innovative didactic methodologies, the GeoMedia-web project aims to raise awareness, in schools of all levels and in the general public, about topics such as geological heritage and natural risk, fundamental for the protection and a better management of the territory (Gordon 2012; Gordon et al. 2012; Gray 2011; Henriques et al. 2011; Lucchesi and Giardino 2012; O'Halloran et al. 1994; Serrano and Ruiz-Flano 1995; Small 2005; Van Loon 2008; Wimbledon et al. 2000).

28.2 Timelines and Activities

The GeoMedia-web project is represented by two sets of activities, each one consisting of phases not necessarily consecutive, for an overall duration of 24 months starting from October 2013 (Fig. 28.1).

The first set of activities aims to the establishment of the network and the collection, storage and selection of existing multimedia products such as video, bi-and tri-dimensional animations, virtual tours, on geological heritage, geological processes, and natural risk. A database will be created with the products catalogued, referring to their technical features, their level of usability, the spatial and temporal scales represented (Fig. 28.2).

The second set of activities consists in testing some "case studies" in collaboration with educational and local institutions of the GeoMedia-network. Based on the feedback obtained, new multimedia products will be designed and developed. The database, the results of the project and the new multimedia products will be disseminated through the web and through a series of seminars, training courses, conferences, publications, both at the national and international level. The GeoMedia-team will monitor the network for the entire duration of the project, coordinating the functions for which the partners have joined the group, ensuring its dynamism and growth throughout the project.

28.3 Innovative Features

The GeoMedia-web project introduces some innovative actions in the dissemination of geological knowledge in Italy:

(1) the creation of a network of original collaborations, new for the variety of the territorial scales and the wide range of areas of interest, which is fundamental for a constructive sharing of the competences, aimed at the production of experimental materials and methods of high interdisciplinary value;

(2) the design and test of innovative interfaces including timelines and 3D graphics, with temporal and spatial interactivity in navigation and exploration of geological structures and processes (Fuchs et al. 2011; Jones et al. 2008; Tufte 1983; Tversky et al. 2002; Wood et al. 2005);

As a consequence, the GeoMedia-web project poses particular attention to:

- the use of innovative methodologies of environmental communication, implemented coupling qualitative and quantitative research (e.g. action research, focus group, interview, questionnaire, etc.), in which users play an active role (Balzaretti and Gargiulo 2009; Denzin and Lincoln 2005; Silvermann 2005; Elliott et al. 1993);
- the development of educational tools that integrate the Inquiry Based Science Education (IBSE) pedagogical approach, promoted by the European Commission, with the cooperative learning methods, the laboratory practice and interactive didactics (European Commission 2007, 2011);
- the clearing of the conceptual hurdles that are involved in the teaching/learning process of the Earth Sciences, with a particular reference to the visualization of geological forms and processes through the use of innovative multimedia tools that display realistic landscapes with high quality graphics (Hedberg and Shirley 1994;

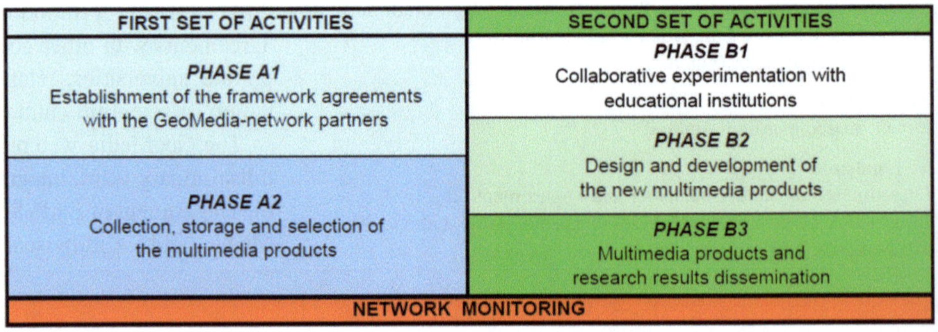

Fig. 28.1 Set of activities and main phases of the project. In the agreement, each partner chooses to collaborate on one or more phases

Spatial scales

CONTINENTAL	REGIONAL	LOCAL	PUNCTUAL	MINUTE	MICROSCOPIC
Thousands of km	from hundreds to tens of km	1-10 km	metric	<metre	invisible to the naked eye

Temporal scales

GEOLOGICAL	GEOMORPH-OLOGICAL	HYSTORICAL	HUMAN	SEASONAL	DAILY	CHRON-OMETRIC
millions of years	thousends of years	hundreds of years	years	months	days	hours, minutes, seconds

Fig. 28.2 Detail of the spatial and temporal scales used to catalogue the multimedia products

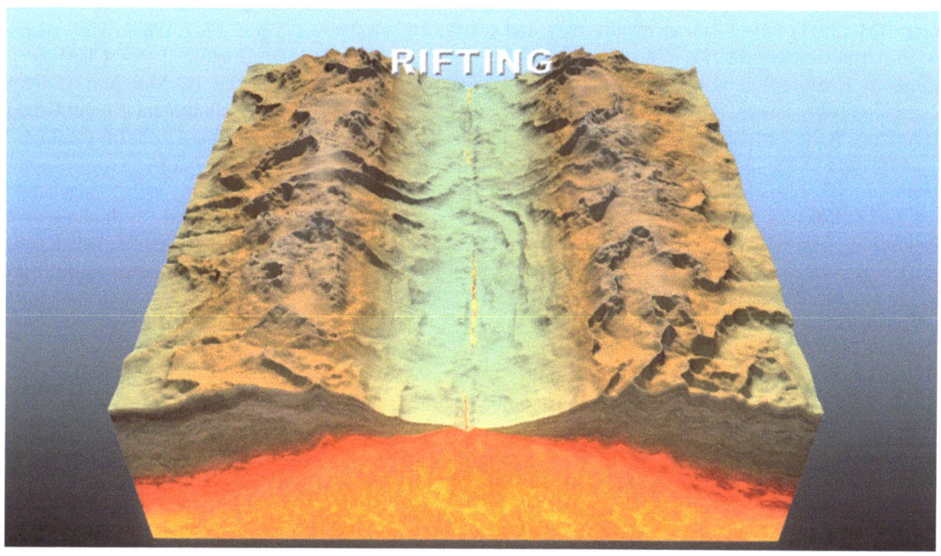

Fig. 28.3 Simulation of rifting in the video "ScopriAlpi", an introduction to the methodologies of Earth Science studies (Production Museo ScopriAlpi, Director Vincenzo Lombardo 2013)

Huntoon 2012; Mayer 2009; Qiu and Hubble 2002; Reynolds et al. 2002);

- the combination of virtual and real tours, promoting a widespread and rooted culture of natural risks based on the knowledge of the processes, of the dangers involved in the geological structure of the territory, of the responsible usage of the urbanized areas, of the perception of Earth dynamics (Fig. 28.3).

28.4 Foreseen Results

The GeoMedia-web project aims not only to establish a real network of universities, schools, museums, cultural associations, public and private organizations, but also to create a web-based network for the collaborative management of innovative multimedia products for the diffusion of the geological culture.

Innovative products will include models and animations of geological phenomena and natural risks, virtual and real field trips, and a system for sharing and disseminating new methodologies and didactical products.

Additional results of the project will provide a guiding network on the development of research and of methodologies that promote effective audio-visual teaching of Earth Sciences.

The final task will be the dissemination of the obtained results and products to a local and national scale audience, by promoting a deepened geological knowledge useful for the conscious management of local natural hazards.

References

Balzaretti E, Gargiulo B (2009) Comunicazione ambientale: sistemi, scenari e prospettive. Franco Angeli Editore

Denzin NK, Lincoln YS (2005) The sage handbook of qualitative research, 3th edn. SAGE, Thousand Oaks

Elliott J, Giordan A, Scurati C (1993) La ricerca-azione. Metodiche, strumenti e casi. Bollati Boringhieri

European Commission (2007) Science education NOW: a renewed pedagogy for the future of Europe. European Communities, Belgium

European Commission (2011) Horizon 2020: the framework programme for research and innovation. Communication from the commission to the European Parliament, the Council, the European economic and social committee and the committee of the regions, Brussels

Fuchs P, Moreau G, Guitton P (2011) Virtual reality: concepts and technologies. CRC Press, Boca Raton

Gordon JE (2012) Rediscovering a sense of wonder: Geoheritage, geotourism and cultural landscape experiences. Geoheritage 4:65–77

Gordon JE, Barron HF, Hansom JD, Thomas MF (2012) Engaging with geodiversity: why it matters. Proc Geol Assoc 123:1–6

Gray JM (2011) Other nature: geodiversity and geosystem services. Environ Conserv 38:271–274

Hedberg J, Shirley A (1994) Virtual reality in education: defining researchable issues. Educ Media Int 31(4):214–220

Henriques MH, Pena dos Reis R, Brilha J, Mota T (2011) Geoconservation as an emerging geoscience. Geoheritage 3:117–128

Huntoon JE (2012) Hidden, threats to science education. Eos 93(17):170–171

Jones RR, McCaffrey KJW, Clegg P, Wilson RW, Holliman NS, Holdsworth RE, Imber J, Waggott S (2008) Integration of regional to outcrop digital data: 3D visualisation of multi-scale geological models. Comput Geosci 35(1):4–18

Lucchesi S, Giardino M (2012) The role of geoscientists in human progress. Geoethics and geological culture: reflections from the Geoitalia conference 2011. Ann Geophys 55(3):355–359

Mayer RE (2009) Multimedia learning 2nd edn. Cambridge University Press, Cambridge

O'Halloran D, Green C, Harley M, KnillJ (eds) (1994) Geological and landscape conservation. Proceedings of the Malvern international conference on geological and landscape conservation 1993. The Geological Society of London

Qiu W, Hubble T (2002) The advantages and disadvantages of virtual field trips in geoscience education. China Papers 1:75–79. Available at http://science.uniserve.edu.au/pubs/china/vol1/weili.pdf. Accessed 01 Oct 2013

Reynolds SJ, Piburn MD, Leedy DE, McAuliffe CM, Birk JP, Johnson JK (2002) The hidden Earth: visualization of geologic features and their subsurface geometry. Available at http://reynolds.asu.edu/pubs/NARST_final.pdf. Accessed 01 Oct 2013

Serrano E, Ruiz-Flano P (1995) Geodiversity, a theoretical and applied concept. Geographia Elvetica 62:140–147

Silvermann D (2005) Doing qualitative research: a practical handbook. SAGE, Thousand Oaks

Small J (2005) Educational multimedia in the geosciences: can an adaptation to IMM technologies save geoscience from extinction? Multimedia applications in education. University of Southern Queensland

Tufte ER (1983) The visual display of quantitative information. Graphics Press, Cheshire

Tversky B, Morrison J, Betrancourt M (2002) Animation: can it facilitate? Int J Hum Comput Stud 57:247–262

Van Loon AJ (2008) Geological education of the future. Earth Sc Rev 86:247–254

Wimbledon WAP, Ishchenko AA, Gerasimenko NP, Karis LO, Suominen V, Johansson CE, Freden C (2000) GEOSITES an IUGS initiative: science supported by conservation. In: Barettino D, Wimbledon WAP, Gallego E (eds) Geological heritage: its conservation and management, pp 69–94

Wood J, Kirschenbauer S, Dollner J, Lopes A, Bodum L (2005) Using 3D in visualization. In: Dykes J, MacEachren AM, Kraak MJ (eds) Exploring geovisualization, Elsevier, Amsterdam, pp 295–312

Reason and Being of Engineering Geology

29

J. Javier Diez, Efren M. Veiga, Fernando Rodriguez and Rosa M. Paz

Abstract

This paper demonstrates the importance of a holistic comprehension of the Earth like a planet that is alive, not only in its Biosphere, looking at the atmosphere-ocean-crust-mantle interactions as its different sectorial expressions (climate, fluid-dynamics, morpho-dynamics, tectonics…) following the solar radiation and nuclear geothermal sources of energy. It considers the environmental incidence of different engineering activities to realize their underfeeding as the *raison*, and leads to that holistic formation as the *being* of the engineering geology.

Keywords

Thermal machine • Eustasy and isostasy • Subsidence • Fluid-dynamics • Morphodynamics

29.1 Introduction

Nobody can currently approach a territorial engineering problem against the planet living. Planet is living, though in a different way of the Biological processes, but its processes are such obvious and predictable/unpredictable as in the Biosphere. The same happens in the Fluid-sphere, where Climate may be considered like the living processes of the coupled answer of the Atmosphere/Ocean couple to the sun power. The work so relates coupling between general climate circulation and hydrological processes, establishing meteorological figures and river basin reaction to flood hazards and describing the whole (Thermo) Dynamical process of significant Flood events. We know now that waves and storm surges are coupled and that (Dynamic) Maritime Climate forms part of a more complex "Thermal Machine" including Hydrological cycle. The analysis of coastal floods could so facilitate the extension of that experience. All people accept that none maritime engineering work can be designed against the marine hydrodynamic processes (Per Bruun's Decalogue, Brunn 1962) and the same may be told about Geological processes and behavior. We consider fundamental that the same approach should be shown to all people in formation for any Engineering activity touching nature. This attention could be the beginning for an improvement of a holistic understanding of our planet.

J.J. Diez (✉) · F. Rodriguez
Universidad Politecnica, Madrid, Spain
e-mail: josejavier.diez@upm.es

F. Rodriguez
e-mail: fernando.rodriguezl@upm.es

E.M. Veiga
Universidad Politécnica, Madrid, Spain
e-mail: efrenmartin@gmail.com

R.M. Paz
Universidad Complutense Madrid, Madrid, Spain
e-mail: rosapaz@bio.ucm.es

29.2 The Thermal Machine

Climate can be considered the result of the functioning in the hydrosphere and atmosphere of complex thermodynamic "machine" whose temperature gradient is caused by the uneven heating via solar radiation of condensed surfaces (the earth's crust and oceans) on all areas of the planet, depending upon location and natural conditions. Subsequent radiation from these surfaces results in the uneven heating

G. Lollino et al. (eds.), *Engineering Geology for Society and Territory – Volume 7*,
DOI: 10.1007/978-3-319-09303-1_29, © Springer International Publishing Switzerland 2014

of the atmosphere above them, along with the generation of convective movements and horizontal gradients producing the winds, the displacement of air masses and the atmospheric circulation.

The atmosphere and oceans as a whole comprise a double system of interactive fluids in which great thermal and pressure gradients are established exclusively due to the energy provided by solar radiation. The fact that atmospheric warming is produced from the planet's concentrated surfaces leads to inversions and gradients that, conditioned solely by the Earth's rotation (Coriolis) and the distribution of the continents, results in their motion. These drive the movements of the oceans, which in turn are conditioned by their thermohaline gradients as well. As a consequence of this coupling of atmosphere and oceanic masses, and following the corresponding transitory period imposed by the planet's current rotation and topology, a complex double circulation, atmospheric and oceanic, resulted, which dictates the climate. The climate, then, is made intrinsically changeable by the intrinsically variable character of those interactive phenomena, but not by the nature of its own pattern.

29.3 Climate and Coastal Morph-Dynamics

Climate affects the pattern of oceanic winds, waves and currents, and the coastline is shifted inland or out to sea as a result of both sea level change and those ocean conditions (Mörner 1995), which are intertwined (climatic variability). The famous Bruun Rule (1962) determines the recession associated with a persistent increase in average sea level. The shoreline shifts account for other changes to the littoral barrier of which it forms a part, as appropriate, causing them to emigrate in response to these changes, but preserving them if there is no net erosion. An adverse situation takes place when the net erosion exceeds the net width of the barrier, though it could still be temporary if the winter profile is susceptible to reversibility. Dean (1977) extended Bruun's analysis to tides and other periodic variations which, since 1990, has also included the wave set-up effect. The directional variability of the swell also leads to natural lateral shifting of beach sands with seasonal variations of the beach shape which, together with the also natural seasonal variations in the profile, can cause temporary advances and retreats of the coastline by an order of magnitude comparable to those of the level variations once tide effects are isolated.

Climatic changes must have had a strong influence on the development of sailing routes along the history (let us consider the current Arctic Sea case). They must also have affected the access to land through their effect on beach slopes and on the configuration of bays, estuaries, lagoons and inlets, all of them linked to sheltered harbor areas. They could also be significant factors in landing and settling processes as well as other climatic and environmental considerations have affected life conditions, especially in low and wet lands.

On the other hand, the well-known influence of maritime climate on coastal morphology and morphodynamics does not appear to have been fully considered until recently; alongshore and onshore-offshore littoral transports, erosive and sedimentary processes and genesis and migration of barrier islands and other forms of sedimentary deposits have been thoroughly studied, though primarily in relation to current environmental impacts; however, other longer term morphological and dynamic pattern changes have not been so extensively considered (Mörner 1976; Emery and Aubrey 1991).

29.4 Hydrology and Climate

Climate is more than the addition of the spatial and temporal climatic phenomena and goes further than the temporal succession of the multiplicity of atmospheric and oceanic events identified as climatic. Climate is the consequence as a whole of the global dynamics (Thermal Machine) established on the Planet as driven by the thermal gradients generated by sun radiation on the planet surfaces. Therefore there is an aspect of locality composed by multiple local thermal machines varying on the time and across the mentioned surface, and a global aspect resulting from the whole local aspect but driving it also in an involving way.

The first climatic phenomenon appearing, due to the scarce air inertia, is the wind that displaces the atmospheric vaporised water and drugs ocean surface; it leads to a complex atmospheric mechanic-dynamic thermal circuit. Temperatures immediately generate evaporation and pressure gradients on the sea surfaces unleashing the also complex hydrological multiplicity of circuits. Wind drug forces finally generates (a) wind waves, (b) storm surges and setups and (c) currents, immediately affected by planet rotation (Coriolis) and thermo-salinity boundary conditions through density gradients, both of the affecting all power transportation through mass displacements (Broecker 2000). The last consequence of the increasing complexity and dimension of the ocean currents is the generation of the transoceanic currents, whose size and Thermal and kinetic inertia supposes a tremendous feedback on the much lighter atmospheric cycles. So the importance of the Gulf Stream (the greatest and warmest current on the planet) on the planetary climate, further than its inclusion in the Broecker's conveyor belt.

Following this reasoning no hydrologic global phenomenon may be wholly correctly understood out of that whole complex Thermal Machine. But the same analysis may be

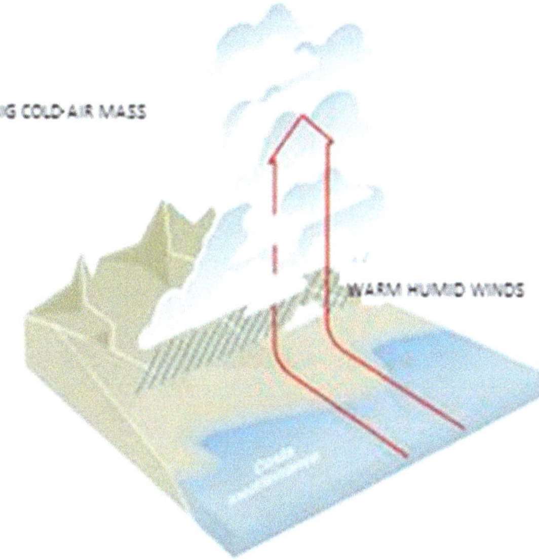

MAIN DYNAMIC FACTOR,

Wind. -It leads:

a) Through (Wave + Storm Surges):

to temporal **Sea level rise**

b) Through baric & convective circulation:

to inland **Rain** onto the wole basins

Fig. 29.1 Mechanism of the "cold drop"

driven in any local thermal machine affecting any hydrology determinant flood event (Fig. 29.1). The SMARTeST Valencia case study permits a right understanding of the local phenomenon known as "gota fría" ("cold drop/bubble") which is just the lighter reactivation on the still worm western Mediterranean basin of an extra-tropical cyclone having crossed above the Iberian Peninsula whose backside N-NE circulation has been drugging down since Polar Circle extremely cold air masses remaining over the whole Peninsula and more northern and southern.

The high cold/dry air mass, the hot sea surface layers, the drug winds and the coastal length sierras generate a convective air movement and a condensation process causing intense and extensive rains on great basin areas, and the sea level raises while wind drugs by wind and wave set ups plus low pressure storm surge. Flood (flash, pluvial or fluvial) may happen at any catchment of the basin but in coastal zones it can be accentuated by high sea level (Fig. 29.2). The situation is analogous in the case of extra tropical cyclones in the rest of Europe and other places of the planet, under hurricane/typhoon tropical conditions and even with Monsoon occasions. The duration of the high level corresponds to the duration of the cold drop event, and in small to moderate basins, as Valencia's and most Mediterranean, the coincidence of fluvial flood and high sea level is highly probable and affects mainly to Coastal Zones. In grater basins as Ebro, Rhone, Po, and most of North-Europe's that dramatic coincidence may happen or be delayed till a following low pressure episode, depending of the time of accumulation, snow melt conditions and basin regulations, but current sea level is always a main factor in most of floods, even further inland than mere coastal zones (SMARTesT FP7 Project 2013).

29.5 Geothermal Energy

The internal Geo-dynamics is a form of internal energy transfer of the Planet (Geothermal) to its crust. The very process of releasing that energy is able to bring about change is by means of vibrations in the tectonic plates or by volcanic eruptions. Tectonics and volcanism force local changes in relative sea level through the vertical movement of certain portions of the crust. Darwin noted these changes as early as during 1851 Beagle's on the island of Santa Maria in Chile. Changes along the edges of converging plates tend to be faster and more abrupt, whether positive or negative. Along diverging plates they tend to be slow and around 1 mm/year negative, which would be equivalent to between ½ and ¼ of the total change along the eastern coast of the United States. Tectonic phenomena are slow and difficult to detect, and are thus lumped under the category of subsidence.

Isostasy changes (Fairbridge 1983) are vertical movements in crust plates due to the Archimedean principle of buoyancy. Depending on their immediate causes, they can be glacio-isostatic, hydro-isostatic, erosive or sedimentary. Along with these, there are at least four other vertical movements which extend to the crust's surface and which, unable to be distinguished, are lumped in the subsidence category: (Diez 2000) (a) strict isostasy, positive or negative depending on whether the load on the tectonic plate is increased or lessened, in which sediments are the primary cause of the former and the melting of surface ice the cause of the latter; (b) indirect isostasy, caused by the bending, if present, of the plate affected locally by the preceding; (c) isostasy caused by the natural consolidation of sediments under their own weight; (d) isostasy caused by the anthropogenic compression of sediments, either by drainage or

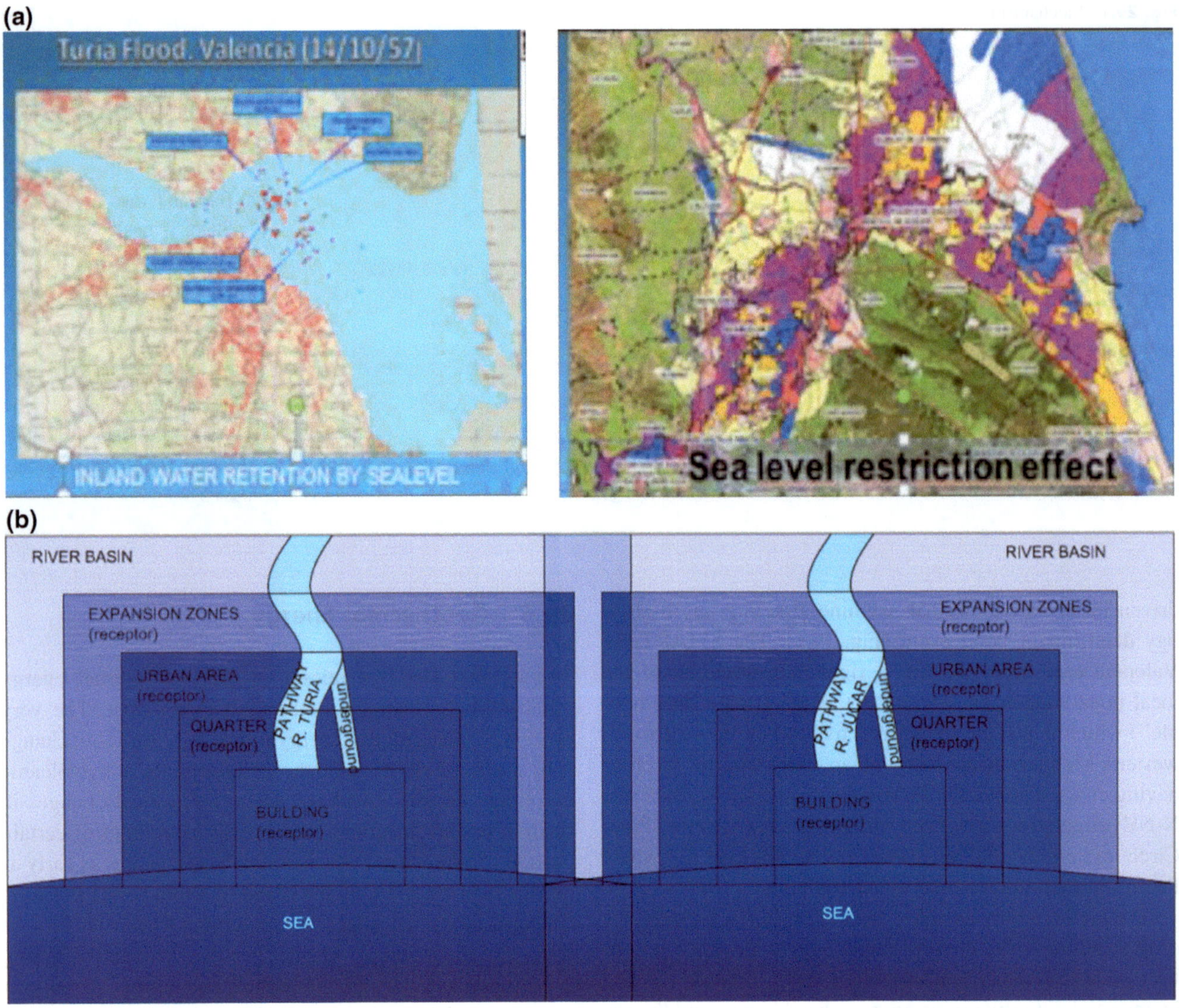

Fig. 29.2 **a** Flooding effect of the "cold drop" phenomena on Valencia Area: *left*, on Turia Basin; *right*, on Jucar basin. **b** Overlapping of Turia and Jucar floods on Valencia Area

under load. This last category is associated with the loss of interstitial fluids (water, gas or petroleum), as in Venice, and has been highlighted along the coasts and beaches of Denia, Alicante, Spain (Diez 1982).

29.6 Conclusion

The complex way which planet is "living" has been shown, as the interrelation of different kind of engineering affected processes.

That is the reason of the engineering geology, and it justifies a holistic formation on geology for all territorial engineering actions/actors.

The being is the comprehension of each natural problem in the right way and to see the real relation between them and between nature and anthropos.

That is the main ethic rule for all the actions related with territorial engineering. The considerations presented in the work so looks for a holistic approach to the Planet considering it as a real wholly living Planet.

References

Broecker WS (2000) Was a change in thermohaline circulation responsible for the Little Ice Age? Proc Natl Acad Sci U S A 97(4):1339–1342

Bruun P (1962) Sea level rise as a cause of shore erosion. Proc ASCE, J, Waterways Harbor Division 88:117–130

Dean RG (1977) Equilibrium beach profiles, U.S. Atlantic and Gulf Coast, Technical report no. 12, U. Delaware, Newark

Diez JJ (1982) Littoral processes in Denia coasts. Proceedings of SEME, pp 1.81–1.95. Brugge

Diez JJ (2000) A review of some concepts involved in the sea level rise problem. J Coastal Res 16(4), 1179–1184. (Palm Beach)

Emery KO, Aubrey DG (1991) Sea levels, land levels, and tide gauges. Springer, New York
Fairbridge R (1983) Isostasy and Eustasy. In: Smith DE et al (ed) Shorelines and isostasy. Institute of British Geographer, Academic Press, London
Mörner NA (1976) Eustasy and geoid changes. J Geol 84(2):123–151
Mörner NA (1995) Sea level and climate-the decadal-to-century signals. J Coastal Res Special (issue No 17: Holocene Cycles) 2:261–268. (Climate, sea levels, and sedimentation, Florida)
SMARTesT FP7 Project (2013). http://www.floodresilience.eu/. Accessed 8 Julie 2013

Human Factor in Resilience Systems

30

Human Role in the Environmental Risks Management

Iria Paz, Rosa M. Paz, Elvira Fernández and J. Javier Diez

Abstract

This paper shows the importance of a holistic comprehension of the Earth as a living planet, where man inhabits and is exposed to environmental incidences of different nature. The aim of the paper here summarized is a reflection on all these concepts and scientific considerations related to the important role of men in the handling of natural hazards. Our Planet is an unstable and dynamical system highly sensitive to initial conditions, as proposed by Chaos theory (González-Miranda 2004); it is a complex organic whole, which responds to minimal variations which can affect several natural phenomena such as plate tectonics, solar flares, fluid turbulences, landscape formation, forest fires, growth and migration of populations and biological evolution…. This is known as the "butterfly effect" (Lorenz 1972), which means that a small change of the system causes a chain of events leading to large-scale unpredictable consequences. The aim of this work is dwelling on the importance of the knowledge of these natural and catastrophic geological, biological and human systems so much sensible to equilibrium conditions, to prevent, avoid and mend their effects, and to face them in a resilient way.

Keywords

Natural/anthropic disasters • Post-traumatic stress • Human resilience • Capacity building • Risk management

30.1 Introduction

Populations have been driven by climatic and other environmental conditions, and have settled adapting to them and modifying them trying to avoid their damaging effects. Risks were increasing, however, as populations grew and expanded. Planning of recovery mechanisms after any event needs to incorporate the whole affection of involved populations, being the psychosomatic facet of this affection even more serious in the case of urban areas by the interactive synergies due to their human accumulations.

Among them, post-traumatic stress has a tremendous importance and it needs to be attended since the very beginning after the production of damages. Fast and planned prearranged reaction facing up them are so necessary as facing up killed or physical injured people. These situations lead to similarly serious consequences both in psychical and

I. Paz
URJC, Madrid, Spain
e-mail: iriapg@hotmail.com

R.M. Paz · E. Fernández
Universidad Complutense Madrid, Madrid, Spain
e-mail: rosapaz@bio.ucm.es

E. Fernández
e-mail: melviraf@bio.ucm.es

J.J. Diez (✉)
Universidad Politécnica Madrid, Madrid, Spain
e-mail: josejavier.diez@upm.es

G. Lollino et al. (eds.), *Engineering Geology for Society and Territory – Volume 7*,
DOI: 10.1007/978-3-319-09303-1_30, © Springer International Publishing Switzerland 2014

in physical answering of the affected population. Unlike other psychiatric disorders, post-traumatic stress disorder (PTSD) requires the exposition to an external and traumatic stress that, as defined in the Diagnostic and Statistical Manual of Mental Disorders (DSM-V, American Psychiatric Association 2013), is supposed to be an event implying vital risk, either dead or severe injure or threat, accompanying itself of intense feeling of fear, horror or impotence. The dramatic trace in humankind memory caused by the "Universal Flood" is inerasable, for instance, and was already narrated in the archaic literature of the Bible (Genesis) and the Gilgamesh Epic Poem.

Since the last decades, the answer to natural catastrophic hazards is based more on risk management (RM) than on risk mitigation systems, and in many European projects *Resilience* has acquired relevance versus *Resistance*, particularly related to Flood Risk (SMARTeST 2013). A main question of the Flood Resilient (FRe) Systems in urban environments is the "human factor", understood as "personal capacity" to face adversity. This is deeply affected under damage conditions, particularly when they repeat frequently in time; it is in fact a way of particular rheology of the individual personality. Formation of the human factor is rather than a problem of capacity building and reaches the philosophical field through the psychological introspection under sociological perspectives. In this sense, it is possible to speak of Human Resilience, a well-established concept today, which aims to approach the improvement of the human factor.

Thus, the resilience is currently appearing as a key factor to face natural geological hazards. Also, the specialized knowledge of the natural processes is essential, together with the accurate information, to prevent risks. That knowledge should go through accessible tools and policies to act properly. For that, it is crucial the study of the location, geological composition and climate context of regions, some of them highly vulnerable and with high human exposure. Therefore, the line of action and the analysis of the vulnerability should be oriented to risk management, by reinforcing the resilience of the social and economic system. The use of efficient tools of Engineering Geology, like geospatial models applied by means of geological and geographical information systems, probabilistic risk and vulnerability analysis platforms, technology of remote sensors in combination with digital models of different attributes and scenarios, are central to the implementation of prospective, corrective and reactive measures, though no more than the people capacity building and stress resilience. The analysis and integration of this multidisciplinary information, together with the people resilient psychological and technical capacity building, will allow optimize the risk management and the strength of the resilience.

Fig. 30.1 Human resilience allows the man to emerge from his own ashes, like "Phoenix Bird"

30.1.1 What Does Human Factor Mean?

The Human Resilience is the human capacity to recover from the suffered damage by different causes, although we are interested here on natural ones. It would be considered an acquired ability or a developed specific capacity, even, an art (the art to be o to make himself), that allows the man to emerge from his own ashes, like "Phoenix Bird" (See Fig. 30.1), after suffering a loss, a pain, a distress… damage.

Human Resilience is the capability to face the most adverse circumstances. We are born with this ability, which varies according to people, but it can also be trained, modified or increased throughout life or events (Santos 2013).

Resilience is as well an attitude, and it means that it can be modified not only outside-inside, but also inside-outside. We can learn how to be more adaptive, and show the other ones how to do it too. Another important factor is the copying style: optimism, creativity, and life sense, acceptance… increase our adaptation to the new situation and facilitate appropriate responses, while pessimism and negativism reduce that one (Paz et al. 2009, 2012).

30.1.2 Why Is Important to Increase Resilience?

Because it prepares person or organization to face changes, strengthens adaptation capacity and facilitates personal or collective growing in any circumstance (Santos 2013).

Resilience is a process with 4 consecutive phases (Santos 2013):

1. Adapt
2. Get over
3. Recuperate
4. Overcome.

30.2 The Human Role

The natural disasters are consubstantial to man life on the planet, unpredictable and insufficiently controllable in most of cases, as well as the range of their consequences. But, on the basis of experience and scientific advance of knowledge, the man can take advantage to them in order to reduce their damages in the future. Up to now, men were acting over its serious effects, and the efficacy of this actuation depends on the developmental level of the affected countries and societies.

Thereby, anticipating a natural disaster is crucial to minimize damages to improve the recovering and adaptation, thus, acting with operability.

Governments and societies must be aware that the "before" attenuates the "after". For that, it is essential the good land and urban planning and settlements with pertinent infrastructures, and the creation of national/regional/local disaster risk reduction systems. The adequate performance of the involved Institutions is required as well. But the human resilient factor, that makes possible dealing with shocks and responding to disasters, depends on the education of society and on the proper formation in responsibilities, being this a vital basis for the "before" and for the appropriate performance of "social and organizational resilience systems".

The management of disasters must contemplate the time situations "before", "during" and "after" the event (Salagnac et al. 2012). The "before" time requires proper urban planning and organization of societies: education of population, emergency plans, evacuation exercises, and, the most important, the psychological training to act "brain to brain". The "during" time consists of activation and implementation of emergency systems and international cooperation. The "after" time needs, as essential, the man recovery for the management of disaster.

Caplan (1985) sets prevention as a comprehensive action with 3 simultaneous intervention levels (Gómez-Acebo et al. 2013):

1. Primary prevention.
2. Secondary prevention.
3. Tertiary prevention.

In all of them we can forecast different strategies or programs to respond psychologically in a proper way.

30.3 Avoiding the Shipwreck

Traumas caused in the human being have tremendous consequences of panic and losses that have not been attended as others problems by human society, being the response very recent. Man is used to his biorhythm and ready to live his everyday under his plan for future, adapting his creative capability for each new venture. But he is not ever ready for unforeseeable and devastating situations, in spite of the flexible conditions of our species, to self-adaptation to the environment, that is, in spite of his capability for such vertiginous evolution. The latent adaptive mechanisms break out and active physiologically in the most critical moments, allowing our survival under unbelievable situations, as we can usually see in Media. But this also means great costs for the person due to the very serious psychical damage induced by the suffered trauma, what is already well known. For these reason is very important to act in advance.

30.4 How to Build Resilient Human Beings and Systems

In order to improve Human Resilience, we must consider that, despite being an individual capacity, it has social influences. In this way, resilient people often emphasize social support as a tool to begin their recover.

It is well known in Psychology that social network is essential to reduce the impact of traumatic events, so the interventions that facilitate the formation or strengthening of social nets are in the basis of the creation of resilient societies.

On the other hand we find the individual factor. In this sense, earlier positive experiences appear as a determining element in resilience. To provide positive experiences, we must count on the participation of educational systems, because they partially define our way of interpreting both our experiences and personality.

The construction of a well-organized personality is necessary to generate or improve resilience, and in that construction, not only school but parents are essential pieces of the puzzle. Programs like those of personal and familiar developing, psychosocial education, stress prevention, and similar are basic to help on this way.

Psychological factors involved in resilience may be: tolerance to frustration, self-esteem, copy strategies, self-consciousness, flexibility, humor, music education, emotional control, inner set of values and decision making.

Another and final critical factor is *THE WILL*: we must exercise our willing to be more resistant to stress, frustration, and unpredictable events on our lives.

30.5 Neurochemical and Genetic Basis

Recently, it has been demonstrated that the fear gene, *oprl1*, that produces the nociceptine receptor, is involved in post-traumatic stress. Laboratory assays prove that nociceptine and its agonist molecules, that stimulate this receptor, prevent *post-traumatic stress disease* (PTSD) in mice (Andero et al. 2013). It has been shown that genetic variants of the *oprl1* gene are associated in humans with higher risk of developing the disorder after exposure to trauma. The symptoms can include constant re-imagining of the traumatic event, an overall numbing to emotions, excess anxiety, and unpredictable bouts of anger. Thereby, neuroscience and genetics provide an essential tool to understand and prevent the posttraumatic stress response, strengthening the human resilience, *leitmotiv* of this work. Some researchers have created new memories by directly stimulating the mice *nucleus basalis*, which, as a consequence, released acetylcholine (ACh), a chemical involved in memory formation (Barad et al. 2006). The possibility of selectively creating some memories is so surprising as much as the capacity of deleting them. Recently, convergent evidence has implicated the amygdala in the extinction of fear. Other works inform that it is possible to selectively erase or disrupt unwanted memories. This could help to create a method for the interruption of unwanted remembrances, as some generated by addictions or traumatic events. "Our memories make us who we are, but some of these memories can make life very difficult" (Miller and Young 2013).

30.6 Conclusion

The brief and syncretic exposition above justifies the relevance of the human factor on the unleashing of any coordinated combination of actions, on its performance and on the result of any social aim; particularly on those pretending to make front successfully to risks derived from natural hazards. In within this context, we conclude as essential:

- The education of society in resilience to understand and develop it in adverse circumstances.
- Transforming traumatic experiences and stressor agents into stimuli to overcome them.
- Training to be able to live without fears to get over the limits of our mind thus building the resilience pillars.
- Understanding the neurobiological basis of trauma and resilience to prevent and to treat them.
- The reinforcement of the individual psyche to generate the synergy of collective resilient behavior.

References

American Psychiatric Association (2013) Diagnostic and statistical manual of mental disorders (DSM-V). A.P.A. USA

Andero R, Brothers SP, Jovanovic T, Chen YT, Salah-Uddin H, Cameron M, Bannister TD, Almli L, Stevens JS, Bradley B, Binder EB, Wahlestedt C, Ressler KJ (2013) Amygdala-dependent fear is regulated by Oprl1 in mice and humans with PTSD. Science Translational Medicine. Washington, DC

Barad M, Gean P-W, Lutz B (2006) The role of the Amygdala in the extinction of conditioned fear. Biol Psychiatry 60(4):322–328. (Elsevier. CA, USA)

Caplan G (1985) Principios de Psiquiatría Preventiva. Paidós, Barcelona

Gómez Acebo et al (2013) Resiliencia. Gestión del naufragio. LID, Madrid

González-Miranda JM (2004) Synchronization and control of chaos. An introduction for scientists and engineers. Imperial College Press, London

Lorenz E (1972) Predictability: does the flap of a butterfly's wings in Brazil set off a Tornado in Texas?. American Association for the Advancement of Science, Washington D.C.

Paz RM, Diez JJ, Paz I, Paz G (2009) Posttraumatic stress effects on population affected by catastrophic floods. Proceedings final COST C22 conference "road map towards a flood resilient urban environment", Paris

Paz RM, Díez JJ, Paz I, Fernández E (2012) Human factor in resilience systems. Implementing flood resilience. SMARTeST international conference, Athens. http://www.floodresilience.eu/

Miller, C. and Young, E. 2013. TSRI News Release. Florida

Salagnac JL, Diez JJ, Garvin S, Tourbier TJ (2012) Flood-management systems: paving the way to flood-resilient systems. SMARTeST international conference, Athens. http://www.floodresilience.eu/

Santos R (2013). Levantarse y Luchar. Cómo Superar la Adversidad con la Resiliencia. Conecta Editions, Random House Mandadori. Barcelona

SMARTeST (2013) FP7 project 2013. http://www.floodresilience.eu/

Recording the Onset of the Anthropocene

31

Martin Bohle

Abstract

The cumulative economic activity of humanity modifies planet Earth. Billions of humans are working hard to survive, to keep a good life or to get a better life. Awareness of global change is an issue for education and governance. The current time of global change requires its own name, e.g. "Anthropocene". Timing of its onset is a scientific endeavor. Comparing the stratigraphic records of sediments in seas and lakes should provide a method to determine the onset of the Anthropocene.

Keywords

Anthropocene • Sediments • Lakes • Oceans

31.1 Introduction

Research confirmed that mankind impacts on planetary ecosystems and global geochemical cycles (Zalasiewicz et al. 2011). Traces of human activities are found in various geochemical records throughout the Holocene—from the onset of agriculture to the onset of the industrial revolution. However, modern humanity is a driver of global change.

Since decades "The International Geosphere-Biosphere Programme" is bundling international research efforts to assess how mankind impacts, beyond climate change, on planet Earth. The research lead to the conclusion that a "profound transformation of Earth's environment is now apparent… The magnitude and rates of human-driven changes to the global environment are in many cases unprecedented for at least the last half-million years" (Steffen et al. 2004). Modern mankind's economic activity developed into an intrinsic part of planetary dynamics. This is noticed for climate change or complex geochemical cycles. Also, mankind moves more material as natural erosion and rivers, and it has increased global erosion rates by more than tenfold (Ball 2005). Deforestation, overgrazing of pastures, and crop cultivation increased weathering from not more than 30 Gt/year before the onset of agriculture to currently 50–80 Gt/year (Smil 2007).

The global impact of mankind's activities on planetary ecosystems and geochemical cycles is a recent phenomenon. This change should become part of the public debate (Biermann et al. 2012) because modern humans are "terra-forming" the planet in an involuntary and disorganized manner. Human population count several billion instead one billion a century ago. Economic activity increased by an order of magnitude in the last 50 years, even if many human beings are left at a low level of consumption of raw-material, energy and food. In that context, it seems appropriate to call the modern geological period "Anthropocene" and to establish its onset in present times and not in recent historical past. The simple implicit message of setting the onset of the "Anthropocene" at present times is: *Now we have to care about what we do to planet Earth.*

31.2 Name It to Drive a Paradigm-Shift

The human mind captures a "phenomenon" by naming it. Some, like Mark Pagel (2012), would say it needs a "meme" to integrate a "phenomenon" into thinking and

M. Bohle (✉)
Directorate General for Research and Innovation, European Commission, Brussels, Belgium
e-mail: martin.bohle@ec.europa.eu

G. Lollino et al. (eds.), *Engineering Geology for Society and Territory – Volume 7*,
DOI: 10.1007/978-3-319-09303-1_31,

culture. A "meme" is understood to be a way of thinking or behaving that spreads through writing, speech or other imitable phenomena from mind to mind. Notwithstanding how human mind-settings and cultural environments become common wisdom of societies, actions and values are based on the notions we use. Therefore mainstreaming a notion is a means to establish a hegemonic way of thinking and to favor a related culture.

Since eons humans tackled the philosophical question: What is "our place" in the world? Myths and minds are full of related "memes" responding to that question. It is a bold philosophical insight that mankind impacts involuntarily on planetary ecosystems and global geochemical cycles. This insight breaks with eon-old paradigms; namely that we are surrounded by "eternal nature" to that we have to succumb, that is indifferent towards us, or that mankind will have to master. Naming the recent geological times "Anthropocene" would mint a different paradigm involving values such as stewardship of the Earth (Tickell 2011).

Support for a paradigm-shift has to be based on reliable science. One option is to refer to the classical techniques that are used to describe geological periods, thus to look for a change in the stratigraphic record. I like to discuss this approach for the sake of its simplicity and the conservatism of the underpinning idea. The associated scientific techniques may well be complicated, but they may be less rule-shifting than counting the mine-shafts or density of road-network as it has been proposed.

31.3 What Is Taken on Record?

Since some years, earth scientists look to the stratigraphic record to pin down the "Anthropocene"; for example the working group on Anthropocene within the activities of the International Union of Geological Sciences. Many findings are reported: Pollen records in peat show vegetation changes by agriculture. Ice cores from Antarctic, Arctic (Greenland), or mountain glaciers contain lead from fuel additives. Nuclear fall-out from atomic-bombs or nuclear accidents serve as reference-tracer, so called "golden spikes". The steady increase of carbon-dioxide concentrations indicates humanity's growing technological and economic power.

Each tracer is valuable; however the impact of humanity on the globe should be measured by a comprehensive range of parameters, which show the composite stage of global geochemical cycles and global ecosystems. Where to find these parameters? I would like to argue that contemporaneous marine and lacustrine sediments provide a global composite picture of the onset of the "Anthropocene". What is taken on record in these sediments?

Different varying isotope ratios are a first feature to consider. Water molecules react slightly differently in the hydrological cycle depending on the oxygen isotope; oxygen isotopes are fractionated in a temperature depending manner; the burning of coal and oil influences the ratio of carbon isotopes in the environment, and carbon isotopes are fractionated in biological processes (Langmuir and Broecker 2012). Likewise, the fixing of atmospheric nitrogen on industrial scales, by the Haber-Bosch process, led to a change of the global nitrogen isotope composition (Holtgrieve et al. 2011). Global pollution can be measured: lead from fuel-additives, mercury from industrial processes, or radionuclides from nuclear tests, nuclear accidents and processing of nuclear waste. The decay products of radionuclides provide for a clock to date a sediment layer (Herman et al. 1995).

31.3.1 Example: "Lacustrine Sediments"

Lacustrine sediments incorporate isotopes and pollutants (Heyvaert et al. 2000). Detailed stratigraphic records develop in lakes with seasonal sedimentation patterns and little perturbation of the sediment layers. Seasonality of sedimentation patterns can be caused by seasonal throw-flow or seasonal cycles of plankton communities.

Sirocko (2012) shows that the lakes in the German Eifel-region provide a very detailed stratigraphic record. These lakes ("Maar"), which are formed by single volcanic eruptions when groundwater is coming into contact with hot lava or magma, have limited through-flow, small catchment area and low surface/depth ratio. The deeper water layers and the bottom of these lakes are sheltered. The lakes also show strong seasonal biological and hydrological cycles.

Layered sediments of German Maar lakes (Sirocko 2012) could be studied on a sub-annual time scales providing a detailed record of regional climate and local human activity from paleolithic to most recent times. A combination of techniques such as freezing in situ sediment samples before withdrawing them and dating layers using Cesium isotope (^{137}CS) revealed events of the last decades including the radioactive fall-out from the Chernobyl accident in 1986.

31.3.2 Example "Marine Sediments"

Marine sediments incorporated isotope and pollutants (Fukuea et al. 1999). In addition, some other processes are going-on in seas only, which reflect human activity. First to mention is littering. Litter is present in all oceans and is a global problem. The concentration of litter, such as plastic and glass, is particular high in shelf-sea regions and coastal zones but litter is found also in the deep sea (Galgani 2000). Second, the hydraulic regime of many rivers is modified by man and their sediment load discharged into the sea is modified, and the sedimentation rates in coastal plains shift (Stanley 1996). Third, bottom trawling in shelf-seas is reworking the sea bottom to a depth of several centimeters. Massive bottom trawling, also in deeper shelf waters, happens since some decades up to the point that the sea bottom gets smoothed (Puig et al. 2012). Bottom trawling is a more powerful means to modify the sea bottom than bottom currents. Bottom trawling mixes the top sediment-layer and destroys the benthic communities that rework sediments. Bottom trawling modifies the physical and biological features of the top sediment layer. Summarizing, marine sediment layers that are formed currently should have different physical, chemical and biological characteristics as in the past. The sediment layers are tagged with isotope-signals, radionuclides and litter-fossils. In particular, the sediment layers in coastal seas and shelf seas should currently change most prominently.

31.4 Conclusion

The onset of the Anthropocene is been recorded by the sediments forming currently in lakes and seas. Marine sediments should change more prominently than lacustrine sediments because of the combination of physical, chemical and biological process and the abundant litter. However, the layering of lacustrine sediments should be finer and less perturbed than the layering of marine sediments easing precise dating by radionuclides from fall-out of nuclear incidences.

Comparative studies of lakes and shelf-seas in different geographical settings and climate zones should be undertaken to describe the global composite picture of the on-set of the "Anthropocene", defined as the accumulated impact of mankind's economic activity on the globe. Consequently, public debate should reflect the paradigm shift of humanity's place on the globe, e.g. making *Global Stewardship* the operational meme of the "Anthropocene."

Disclaimer The views expressed in this paper solemnly engage the author.

References

Ball P (2005) The earth moves most for humans, agriculture and excavations shape the landscape more than rivers and glaciers. Nature. (doi:10.1038/news050307-2), http://www.nature.com/news/2005/050307/full/news050307-2.html

Biermann F, Abbott K, Andresen S, Bäckstrand K, Bernstein S, Betsill MM, Bulkeley H, Cashore B, Clapp J, Folke C, Gupta A, Gupta J, Haas PM, Jordan A, Kanie N, Kulvánková-Oravská T, Lebel L, Liverman D, Meadowcroft J, Mitchel RB, Newell P, Oberthür S, Olsson L, Pattberg P, Sánchez-Rodríguez R, Schroeder H, Underdal A, Camargo Vieira S, Vogel C, Young OR, Brock A, Zondervan R (2012) Navigating the Anthropocene: improving earth system governance. Science 335:1306–1307

Fukuea M, Nakamuraa T, Katoa Y, Yamasakib S (1999) Degree of pollution for marine sediments. Eng Geol 53(2):131–137

Galgani F (2000) Litter on the sea floor along European coasts. Mar Pollut Bull 40(6):516–527

Herman J, Kershawb PJ, Bailly du Boisc P, Gueguenіatc P (1995) The distribution of artificial radionuclides in the English channel, southern North Sea, Skagerrak and Kattegat, 1990–1993. J Mar Syst 6(5–6):427–456

Heyvaert AC, Reuter JE, Slotton DG, Goldman ChR (2000) Paleolimnological reconstruction of historical atmospheric lead and mercury deposition at Lake Tahoe, California-Nevada. Environ Sci Technol 34(17):3588–3597

Holtgrieve GW, Schindler DE, Hobbs W-O, Leavitt PR, Ward EJ, Bunting L, Chen G, Finney BP, Gregory-Eaves I, Holmgren S, Lisac MJ, Lisi PJ, Nydick K, Rogers LA, Saros JE, Selbie DT, Shapley MD, Walsh PB, Wolfe AP (2011) A coherent signature of Anthropogenic nitrogen deposition to remote watersheds of the Northern Hemisphere. Science 334(6062):1545–1548

Langmuir CH, Broecker W (2012) How to build a habitable Planet. Princton University Press, Princeton, p 718

Pagel M (2012) Wired for culture. Origins of the human social mind. W.W. Norton & Company, New York, p 423

Puig P, Canals M, Company JB, Martín J, Amblas D, Lastras G, Palanques A, Calafat AM (2012) Ploughing the deep sea floor. Nature 489:286–289

Sirocko F (ed) (2012) Wetter, Klima, Menschheitsentwicklung, (3rd ed). Theiss, p 208

Smil V (2007) Global material cycles. In: Cleveland CJ (ed) Encyclopedia of Earth. Environmental Information Coalition National Council for Science and the Environment, Washington, D.C. http://www.eoearth.org/article/Global_material_cycles

Stanley DJ (1996) Nile delta: extreme case of sediment entrapment on a delta plain and consequent coastal land loss. Mar Geol 129(3–4):189–195

Steffen W, Sandersoon A, Tyson PD, Jäger J, Matson PA, Moore III B, Oldfield F, Richrason K, Schellnhuber HJ, Turner II BL, Wasson RJ (2004) Global change and the Earth system: a planet under pressure, executive summary. IGBP Secretariat Stockholm, p 42

Tickell C (2011) Societal responses to the Anthropocene. Phil Trans R Soc 369(1938):926–932

Zalasiewicz J, Williams M, Haywood A, Ellis M (2011) The Anthropocene: a new epoch of geological time? Phil Trans R Soc 369(1938):835–841

Searching for Connections to the Raison d'Être of Engineering Geology in Codes of Ethics

32

Robert E. Tepel

Abstract

Engineering geology has no generally acknowledged and formally stated Raison d'Être. This lack could be because engineering geology practice has long been so closely integrated into civil engineering practice that its Raison d'Être appeared to be nothing more than to serve the needs of civil engineers. Today, our scope of services has expanded beyond direct and immediate application to engineered projects. A 21st century Raison d'Être for the profession should match our 21st century practice; it should be derived from a description of the benefits the profession provides to humankind. This study seeks insight into how engineering geology benefits humankind by examining a subset of thirteen of forty-nine codes of ethics under which engineering geologists might practice. All thirteen contain an ethical requirement that stands out as supportive of a Raison d'Être for the profession based on its societal value. This special clause imposes a mandatory duty on the professional to hold paramount the public health, safety, and welfare in all professional work. The connection between this ethical requirement and our 21st century Raison d'Être is revealed by asking the question, *how* does our practice meet the ethical requirement to protect the public health, safety, and welfare? The proposed answer (Raison d'Être) is that engineering geologists protect the public health, safety, and welfare by finding, defining, and communicating actionable information about, geologically-sourced risks and their associated hazards that can affect the health, safety, and welfare of the public, as they live in and use their natural and built environments.

Keywords

Professional ethics • Public recognition • Raison d'Être

32.1 Introduction

Engineering geology has no generally acknowledged and formally stated Raison d'Être. This is not from lack of effort, but from difficulty in deciding what engineering geology is really all about (Tepel 2004b, 2012a, c). Engineering geology today has evolved and expanded from the original scope and workplace setting of the practice. Today, many engineering geologists do not work in an engineering office but instead practice directly before the public and provide engineering geology services that do not have immediate and direct input into engineered projects. This type of practice opens the door to expanding our thinking about the societal value of engineering geology practice; it requires a Raison d'Être for our profession that recognizes its independent, stand-alone, nature but also recognizes our traditional close link to engineering practice. The value of adopting a formal Raison d'Être statement for engineering geology is manifold. It will provide a sense of unity and direction to the members of the profession, provide focus

R.E. Tepel (✉)
Consultant, 767 Lemonwood Ct., San Jose, CA, USA
e-mail: rtep99cml@earthlink.net

G. Lollino et al. (eds.), *Engineering Geology for Society and Territory – Volume 7*,
DOI: 10.1007/978-3-319-09303-1_32,

for promoting the profession and its services, and inform the public in plain terms how it can benefit from our services.

Engineering geologists have engaged in introspective pondering about the underlying philosophical rationale of their practice (why it exists and its value to society) almost from the beginning of the discipline. The focus of 20th century authors included the utility of geology to the engineer, the role of the geologist in the engineering organization, the workplace relationships between geologists and engineers, how the profession of engineering geology could promote its application, and defining engineering geology and its relationship to other geo-professions. These efforts were not satisfying. The sense of dissatisfaction at the time over the lack of resolution about the nature of engineering geology was summed up by Müller-Salzburg (1976) as: "However, it appears to me that the development of the science in recent years does not seem to head in a direction that one could be satisfied with. Again and again we seem to divert from the right path and one could say that many of us do not even seem to know the ultimate goal. This explains why one gets such different answers to the basic question what Engineering Geology really is."

Of importance to the goal of this paper, in the late twentieth and early twenty-first centuries, engineering geologists began to explore the basic nature of their unique work products and methods, and to discuss the concept of risk management through engineering geology. Pioneering authors defined the fundamental character of the work products and methods of engineering geology practice in a way that led to a new understanding of the societal value of the work of the engineering geologist. The closely related work process/product concepts of the *site geological model*, the *total geological history model*, and the *site characterization* were developed. Contributors included Fookes (1997), Fookes et al. (2000), Hatheway (2000, 2002), Davison et al. (2003), Baynes and Hatheway (2003), Knill (2003), Hempen and Hatheway (2004), Joint European Working Group (2004), Task Force (2004), Baynes and Rosenbaum (2004), and Sullivan (2010). Knill (2003) in particular linked engineering geology practice to risk management and risk reduction, and this leads to recognizing that our practice supports the societal goals of public safety and risk reduction. Inspired by these efforts, I explored concepts about the nature of engineering geology and a suitable Raison d'Être for engineering geology (Tepel 2004a, b, 2008, 2009, 2010, 2011, and 2012a, b, c). While the works cited here provide a substantial base for my analysis, my current work did not delve into ethical codes or guidelines in languages other than English, and is largely limited to western European-Anglo-North American settings.

32.2 Thesis and Method

My thesis is that we can and should develop a Raison d'Être for engineering geology that is based on the societal value of our practice: how it benefits humankind (or "society"). This Raison d'Être should be worded such that it accommodates both the traditional benefits of our practice to the engineering profession and the societal benefits that flow from our modern expanded scope of activity—work that does not provide direct and immediate input into engineered projects. As one method to research how our practice benefits society, I reviewed codes of ethics applicable, or potentially applicable, to engineering geology practice (Tepel 2012b). These codes fall into two groups: (1) statutory codes of ethics imposed on the profession by jurisdictions that have adopted licensure for geologists (including engineering geologists), and (2) professional organizations, including those dedicated to geology or engineering geology and those professional engineering organizations of which engineering geologists might be members. The premise as to government-imposed codes of ethics is: if we want to know how our practice benefits society (government here is a proxy for society), why not see what society says those benefits are by seeking them in government-imposed codes of ethics? The premise as to professional organization codes of ethics is: these codes are presumably developed with some level of altruism that is directed to the societal benefits of professional practice and those benefits should be stated or implied in the codes.

32.3 Results

Almost all of the forty-nine codes reviewed in Tepel (2012b) contain a "primacy clause" that specifies the highest (primary or paramount) ethical level of duty the professional must meet. Of the forty-nine codes, nine regulatory codes and four professional organization codes contain similarly worded primacy clauses requiring, in mandatory phrasing, that the professional shall place the public health, safety, and welfare paramount. The thirteen codes and their primacy clauses are listed in Tables 32.1 and 32.2 (derived from Tepel 2012b).

32.4 Discussion and Conclusion

One American state licensure board, eight Canadian provincial licensure boards, (Table 32.1), and four professional organizations (Table 32.2) have similar strong and unequivocal clauses that make paramount the duty of the licensed geologist (or member) to protect the public health,

Table 32.1 Jurisdictional codes of ethics potentially applicable to engineering geology practice with primacy clauses that place the public interest paramount

Jurisdiction	Primacy clause excerpts (Analytical commentary by author is in italics)
Alberta	Professional engineers, geologists and geophysicists shall, in their areas of practice, hold paramount the health, safety and welfare of the public and have regard for the environment. *Straightforward, strong, and unequivocal*
British Columbia	Members and licensees shall...hold paramount the safety, health and welfare of the public. *Straightforward, strong, and unequivocal*
Manitoba	Each practitioner shall regard the physical, economic and environmental well-being of the public as the prime responsibility in all aspects of professional engineering and professional geoscientific work. *Strong and unequivocal*
New Brunswick	Engineers, geoscientists, and members-in-training shall...hold paramount the safety, health and welfare of the public.... *Straightforward, strong, and unequivocal*
New Hampshire	The licensed professional geologist shall hold paramount the safety, health and welfare of the public. *Strong, Straightforward, and unequivocal*
Newfoundland and Labrador	...professional engineers and geoscientists shall...hold paramount the safety, health and welfare of the public... *Straightforward, strong, and unequivocal*
Nova Scotia	Professional Geoscientists shall.... Hold paramount the safety, health and welfare of the public and the protection of the environment. *Straightforward, strong, and unequivocal*
Ontario	A Professional Geoscientist shall...regard his or her duty to public safety and welfare as paramount.... *Straightforward and strong, bit does not reference the public health*
Saskatchewan	...members and licensees shall: hold paramount the safety, health and welfare of the public and the protection of the environment .... *Straightforward, strong, and unequivocal*

safety, and welfare. (The word "welfare" is taken to include the economic value of property and other wealth (assets) of individuals, families, businesses, and societal assets, as well as the societal benefit in avoiding or reducing losses from geologic hazards.) The Canadian boards listed in Table 32.1 all license both engineers and geologists and apply the same code of ethics to both professions. The ethical requirement imposed by governmental licensure boards can be viewed as society, organized as government, telling the profession what it thinks are the most important societal benefits of its practice—benefits so important to society that they must, in event of conflict, be given priority over the benefits of the professional's practice to clients or employers, or to the professional as a person. Table 32.2 lists four professional organizations of which engineering geologists might be members that place strong and unequivocal ethical requirements on their members to place the protection of the public health, safety, and welfare paramount among their members' ethical duties. These four organizations, representing the best interests of their members, support the statutory requirements noted in Table 32.1.

Our Raison d'Être, however, is not simply to protect the public health, safety, and welfare; all the design professions do that. The connection between the ethical requirement to hold paramount the protection of the public health, safety, and welfare and our 21st century Raison d'Être can be discerned by asking the question, *how* does our practice meet the paramount ethical requirement to protect the public health, safety, and welfare? The suggested answer is that engineering geologists protect (and enhance and safeguard) the public's health, safety, welfare by finding, defining, and communicating actionable information about, geologically-sourced risks and their associated hazards that can affect humans, their health, safety, and well-being, and their institutions and property, as they live in and use their natural and built environments.

To state this proposed Raison d'Être in capsule form, engineering geologists help people and their institutions recognize and manage geologically-sourced risks and their associated hazards.

A formal and complete statement of the proposed Raison d'Être of engineering geology might be phrased thusly:

Engineering geology protects, enhances, and safeguards the public's health, safety, and welfare by finding, defining, and communicating actionable information about, geologically-sourced risks and hazards that can affect humans, their health, safety, well-being, and their institutions, and property, as they live in and use their natural and built environments.

It is here proposed for discussion that it is this set of benefits that engineering geologists provide to humankind that justifies the existence of the profession: it is our Raison d'Être.

32.5 Application

The proposed broad Raison d'Être of engineering geology fits well with the broadened scope of 21st century engineering geology practice: out work supports societal goals of sustainability, livability, protection of cultural heritage, and

Table 32.2 Organizational codes of ethics potentially applicable to engineering geology practice with primacy clauses that place the public interest paramount

Organization	Primacy clause excerpts (Analytical commentary by the author is in italics)
American society of civil engineers	Engineers shall hold paramount the safety, health, and welfare of the public. *Strong, straightforward, and unequivocal*
Institution of civil engineers	Members of the ICE should always be aware of their overriding responsibility to the public good. A member's obligations to the client can never override this.... *Strong and unequivocal*
National society of professional engineers	Engineers shall hold paramount the safety, health, and welfare of the public. *Strong, straightforward, and unequivocal*
Society for mining, metallurgy, and exploration	The first responsibility and the highest duty of members shall at all times be the welfare, health and safety of the community. *Strong, but only if community is a synonym for public*

environmental protection. The ethical standard that we as practitioners must hold paramount the public health, safety, and well-being is well suited to guide our 21st century practice in the framework of the societal goals of environmental protection, sustainability, and risk reduction. It is also applicable to our traditional; practice that offers direct and immediate input into engineered projects. Formal adoption of a modern Raison d'Être for engineering geology by our professional associations and by IAEG is encouraged.

References

Baynes FJ, Hatheway A (2003) Responsibilities of engineering geologists in practice. Perspect No 55 AEG News 46(2):13–15. (Assoc Eng Geologists)

Baynes FJ, Rosenbaum MR (2004) Discussion arising from the 1st Hans Cloos lecture, by John Knill. Bull Eng Geol Environ 63(1):89–90

Davison L, Fookes PG, Baynes FJ (2003) Total geological history: a web-based modelling approach to the anticipation, observation and understanding of site conditions. In: Turner K, Rosenbaum M (eds) New paradigms in subsurface prediction, Springer Berlin Heidelberg

Fookes PG (1997) The first glossop lecture. Geology for engineers: the geological model, prediction and performance. Q J Eng Geol 30:293–431

Fookes PG, Baynes FJ, Hutchinson JN (2000) Total geological history: a model approach to the anticipation, observation and understanding of site conditions. Invited paper, Geoeng 2000 conference, Melbourne Australia, published in Ground Engineering vol 34, no 3, pp 42–47

Hatheway AW (2000) Richard H. Jahns distinguished lecture

Hatheway AW (2002) Site characterization, *Raison D'être* of engineering geology. In: Tepel RE (ed) Visioning the future of engineering geology, sustainability and stewardship, association of environmental and engineering geologist, special publication 19, vol. 1. Association of Environmental and Engineering Geologists, Denver, Colorado. CD-ROM

Hempen GL, Hatheway AW (2004) Introduction to the symposium. In: Tepel RE (ed) Visioning the future of engineering geology: site characterization, association of environmental and engineering geologists, special publication 19, vol. 3. Association of Environmental and Engineering Geologists, Denver, Colorado. CD-ROM

Joint European Working Group (2004) Professional tasks, responsibilities, and co-operation in ground engineering. Final report to the societies, Helmut Bock, Chairman, transmitted to Dr. Niek Rengers, President of the IAEG, 07 June 2004, p 20

Knill J (2003) Core values: the first Hans Cloos lecture. Bull Eng Geol Environ 62(1):1–34

Müller-Salzburg L (1976) Geology and engineering geology. Reflections on the occasion of the 25th anniversary of the death of Hans Cloos. Bull IAEG 13:35–36

Sullivan TD (2010) The geological model. In: Williams AL, Pinches GM, Chin CY, McMorran TJ, Massey CY (eds) Geologically active; proceedings of the IAEG congress, Auckland, New Zealand, 5–10 Sept 2010

Task Force (2004) Core values of engineering geology and the commission's work, a working plan produced for discussion. IAEG News 32(1):1–4

Tepel RE (2004a) The engineering geology site characterization as the core function of engineering geology practice. In: Hatheway AW, Hempen GL (eds) Symposium proceedings, visioning the future of engineering geology: site characterization, special publication 19. Association of Engineering Geologists, Denver, Colorado. CD-ROM

Tepel RE (2004b) Toward a philosophy of engineering geology. In: Tepel RE (ed) Symposium proceedings, visioning the future of engineering geology: understanding conflict and cooperation in the practice of engineering geology, Association of Engineering Geologists Annual Meeting, Dearborn, Michigan, special publication 19. Association of Engineering Geologists, Denver, Colorado. CD-ROM

Tepel RE (2008) What is the societal value of the science and practice of geology? AEG News 51(4):28–31

Tepel RE (2009) The core attributes of engineering geology: a USA perspective. 10th international congress of the international association for engineering geology and the environment, Nottingham, United Kingdom. In: Culshaw MG, Reeves HJ, Jefferson I, Spink TW (eds) Engineering geology for tomorrow's cities, geological society, London, engineering geology special publication 22, pp 273–276

Tepel RE (2010) Risk management as the essence of licensed engineering geology practice. AEG News 53(2):32–34

Tepel RE (2011) Who is your client, really? AEG News 43(3):34–35

Tepel RE (2012a) (abstract) Renaissance and rediscovery in engineering geology: the search for Its 21st century *Raison d'Être.* 4th international professional geology conference: earth science-global practice, Vancouver, British Columbia. Conveners: European Federation of Geologists, Australian Institute of Geoscientists, and American Institute of Professional Geologists, hosted by Association of Professional Engineers and Geoscientists of British Columbia

Tepel RE (2012b) The "primacy clause" in codes of ethics: a window into licensure board and professional association values. AEG News 44(1):32–37

Tepel RE (2012c) Response: engineering geology for the 21st century: it isn't only about engineering. Short version published in AEG News 55(4):15–16. Full version posted at http://www.aegweb.org/docs/about/tepel.pdf?sfvrsn=0

33 Web Participatory Framework for Disaster Resilience: Coping with Information Deluge

Elena Rapisardi, Sabina Di Franco and Marco Giardino

Abstract

A clear and correct information on hazards and in Disaster Resilience (DR) is crucial. But the "web era" changed the information models of production and distribution, allowing citizens to take part in the overall process; and this change is having a strong impact also in DR Information. Validated scientific information on natural hazards, far from being a common heritage, requires both a high degree of cooperation and a wide interdisciplinary approach. It involves experts, scientists, practitioners, civil servants, but also citizens, volunteers and media representatives. This complexity should be taken into account, without raising barriers, but avoiding the risk of misleading or «unsafe» information spreading, fostering a true information preparedness, through collaborative-knowledge building approach. DR Information is a key element all along the disaster cycle: from preparedness to disaster reduction, and the challenge is to organize data and information adopting a user-centered approach. We envisage an open approach to structured knowledge so to benefit from web tools (including Social Media) and to build precision and common understanding. In this perspective, a joint team UNITO-NatRisk and CNR IIA conceived a web project "Natural Hazards Wikisaurus" (NHW), whose aim is to sustain and support a common understanding through the implementation of an augmented «terminology tool»: a collaborative digital source of validated information and knowledge on Natural Hazards and Civil Protection.

Keywords

Disaster resilience • Web 2.0 • Terminology • Hazards • Information

33.1 The Collaboration: A 2.0 Dilemma?

The post-Gutenberg revolution is an ongoing process. The web as a platform—web 2.0—is not only an intermediary, but becomes a mediator (van Dijck 2013) reshaping every-day life, but also, in the long term, socio-economic systems, cultural model, and mental structures. The web 2.0, since its early beginning, has been seen as enabling factor for equality, participation, democracy, collaboration: a new cyber-territory populated by open and democratic communities. From a cultural perspective, this revolution could be seen as a new utopia, but as a matter of fact the web 2.0 has in its DNA the potential to reshape human society as a whole, particularly the information society. The production model of content (either information or knowledge, products and services) is "reformed": the lowering of means of production costs generate a new social-production player, the "produsage" (Bruns 2007). User becomes creator, producer and distributors of any kind of content; as in "citizens sensors", "citizens science", or "citizens participation", the suffix citizen stands for a change where individuals become a key

E. Rapisardi (✉) · M. Giardino
NatRisk, Interdepartmental Centre for Natural Risks, University of Turin, Torino, Italy
e-mail: e.rapisardi@gmail.com

S. Di Franco
CNR-IIA National Research Council – IIA, Rome, Italy

G. Lollino et al. (eds.), *Engineering Geology for Society and Territory – Volume 7*,
DOI: 10.1007/978-3-319-09303-1_33,

player of the information and of civic society. The web era frees the information, and we as individuals are getting accustomed to search on the web to make any kind of decision or to express our opinion. Social media and crowd sourcing have considerable promise for supporting collaborative and innovative ways that reshape the information production and distribution; however now the debate is facing an important concern related to true-false issues, focusing on validation, and liability (Bohannon 2013). The internet could be also seen as an "information deluge", a continuous and rather chaotic flow, or an endless library with no filtering function, a true and false endless continuum. This brings to the idea of "encyclopedia" (and of Wikipedia) as a tool to organize, manage and filter knowledge, to allow communication, knowledge transfer, education, and sense-making. "Multi-way information exchange systems. These increase the capacity of communities to engage in coordinated actions by making available and sharing timely accurate information about risk" (Comfort et al. 1999).

In this perspective the importance of a correct information and communication, is a key factor to be adopted and included in disaster information management policies in order to contribute to reduce societal vulnerability, and therefore risk, through a common knowledge and information (Simpson and Katirai 2006). Without any doubt the massive use of Social Media during recent major and minor disasters highlighted a huge need of clear, correct, free and trustworthy information. In this framework, web and mobile technologies are a key element to innovate information processes and management in DR, disregarding that technology is not a "neutral" element, and that it is not by itself able to ensure validity and reliability. Technology is extraordinary, but an extraordinarily useful tool; consider what Duncan Edwards, from the Institute of Development Studies, says on open data: "A common narrative in many "open" development projects goes along the lines of "provide access to data/information → some magic occurs → we see positive change." (Edwards 2013). Hence it would be an advantage to discuss and define shared models and cultural approaches, in order to overcome both web rhetoric and cyber-utopism.

33.2 The Four Pillars

«Information as open commons» represents a great opportunity in Disaster Resilience, where validated scientific information on natural hazards, far from being a common heritage, requires both a high degree of cooperation and a wide interdisciplinary approach (United Nations 2005). It involves and should commit not only experts, scientists, practitioners, civil servants, but also citizens, volunteers and media representatives. It is based on knowledge, information, skills and competencies (abilities) pertaining to various fields such as: scientific, legal, logistics, historical, organizational, psychological, sociological, cultural, health.

As a matter of fact, nowadays academics, experts and institutions are called to take the responsibility for a more effective knowledge transfer and exchange (technical, operational, historical, social), abandoning "Elites Knowledge", in favor of a new commitment with the «open age», so to allow the switch from «protection/passivity» (focused on disaster event) to resilience, provided that knowledge "users" are able to make use of knowledge, data, information. Building a resilient knowledge and practicing a resilient information approach, requires a high level of collaboration and interaction across disciplines, and stakeholders (scientists, practitioners, governments, citizens and media).

How to implement collaboration in the disaster resilience domain? How to use web 2.0 potential to implement a wide collaboration? Four pillars are here proposed.

The *first pillar* is focused on a structured content availability: data and information sharing approach, in order to guarantee easy and transparent access usage of specific and validated knowledge and data. Key issues: terminology tools, open geodata and standards, knowledge based framework. For example, in DR this could be accomplished supporting the wide spreading of the open data initiative, and making available (share and explained) specific scientific knowledge on hazards and risks, such as scientific definitions and related concepts.

The *second pillar* is focused on comprehension: implementation of tools and means to allow a better understanding. The web is not one, there are different ways (tools and languages) to represent and display data and information: interactive hazard maps, timelines on historical disaster. For example during flood events, a tool to increase understanding of the hazard processes and evolution, could be an interactive map, showing hazards layers with clear explanation of hazard degree and references, the flooding occurrence in that specific area (historical data), and sensors showing real time information about waters levels measurements and the probability of flooding spreading areas based on collected historical data.

The *third pillar* is focused on linked information and data: tools to set up relations between the topics of knowledge and information.

We envisage the development of a true and concrete Linked Data for DR to build a common language and understanding so to avoid misleading information, and to increase preparedness. From this perspective metadata and ontologies are the first building blocks to build a semantic ecosystem for DR and par conseqúence for civil protection and emergency management.

The *fourth pillar* is collaboration, intended as a praxis (act) driven by participatory approach, responsibility and awareness, through knowledge, competence and multidisciplinary

(Goodchild 2007). Collaboration does not mean that "anybody can do any job", hence collaboration should consider and be aware of the limits of individual knowledge and trustworthiness. For example, Citizens Science in disaster resilience should include training programme for citizens, so to include them as contributor in observations of phenomena (natural processes and events).

33.3 NHDR Semantics[1]: An Approach to Reframe Knowledge and Information

What can we do in face of the huge amount of information needed and offered on the web? How to make a selection, how to improve the searchability and access to information (easy to find and usable)? How to organize individuals and society learning process? How to build a "common language"?

The web information "flood" seems to be «a treasure without a map». Democratization of the «*quest*», freedom of access and usage of information is the challenge of our cultural and societal development, as web is definitely changing how we produce, consume and interconnect information. In this perspective a *context-aware solution* (Hendler and Berners-Lee 2010) could be a model also for knowledge awareness in Natural Hazards Disaster Resilience (NHDR), with the aim to make the scientific content more explicit. In this perspective, the need for a correct and precise vocabulary in DR is an easily understandable issue, whilst not easy to implement (Denaux et al. 2011).

Terminology could be a source of ambiguity and terminology associated with identifying and communicating risk is a relatively new science (Christensen et al. 2003). The confusion and misunderstanding on natural disaster terms is often underestimated in its consequences.[2] It is needed to start back from «words», from their meaning and relations between concepts and terms, as precise comprehension could improve a more resilient behavior.[3] One example: let's consider the difference between natural phenomenon (event, process, factors, related effects…), natural hazard and risk. A natural phenomena that can be harmful to human life and societies, it is related to the probability of its occurrence; whilst risk is defined by probability (hazard), vulnerability and exposure. It is not uncommon that hazard is used as synonym of risk, and viceversa, losing the concept of probability related to a specific hazard, and at the same time the notion of the calculated impact on the area concerned. Whilst, a clear distinction between phenomenon, hazard and risk highlights the approach to natural phenomena, the role and responsibility of human beings in managing the territory and the action to be taken to reduce the impact of disasters. Although interrelated, phenomenon, hazard and risk identify also different activities and knowledge to be considered when talking about natural phenomena: the study and the monitoring, the preparedness and the prevention, the emergency management and reconstruction (EU-Project MONITOR 2007).

On the other hand during emergencies is crucial to understand each other on the base of a common terminology, where terms and associated concepts should be both comprehensive and unambiguous. Ultimately, terminological tools[4] could represent the first step to build a *koinè* and sharing a common specific language.

33.4 The NH WikiSaurus Framework: A Work in Progress

By considering these facts a joint team UNITO-NatRisk and CNR IIA conceived a web project (in the framework of an Earth Sciences Ph.D program) "Natural Hazards Wikisaurus" (NHW) to combine two previous experiences: "HyperIspro"—a civil protection wiki—and "Earth Thesaurus" of environmental terms. The project aims to implement an augmented «terminology tool» conceived as a collaborative virtual source of validated information and knowledge on Natural Hazards and Civil Protection, to sustain and support a common understanding. Thesaurus represents the building block of knowledge, enabling the "know-how" on NH, whilst the Wiki selects and organizes the praxis of natural hazards management (Stuckman and Purtilo 2011). The NHW project[5] is structured as follows:

Glossary

- structured categories to organize wiki entries are taken on the basis of Earth thesaurus;

[1] In this context "Semantic" is referred to the term as it is used in semiotics, and so related to concepts.

[2] There is a lack of field research on public understanding and comprehension of natural hazards characteristics and dynamics, on emergency management, and also on the efficacy of communication campaigns and tools. Far from being a merely research object, findings from this type of researches could also give a deeper insight so to improve and ameliorate the communication policies of central and local Governments.

[3] It is clear that Science is neither static nor immune from debate and controversies amongst scientist of the same domain field, and among scientists from different domains. Moreover the public should become aware that science has not solved all the questions and issues, but it is an ongoing and evolving process.

[4] The most suitable tools for DR terminology domain are: Lexicons, lists of terms relating to a particular subject; Glossaries, alphabetical lists of terms peculiar to a field of knowledge with definitions and explanations; Thesauri, structured controlled vocabularies, covering the terminology of a specific knowledge domain.

[5] SW specifications: wordpress cms, plugins.

- wiki term contains the definition and an explanation, each term is attributed to a specific category,
- web editor can create and edit links to external and internal sources using the Contextly[6] function so to add manually selected related links; the Contextly links are showed below the body text of the terms;
- in the text body the wiki and glossary terms are highlighted as active links
- in case of term ambiguity, either because it is a term used in plain language or pertaining also to other disciplines or fields, a warning box is showed containing some explanation on the wrong usage of the term
- each term has a set of automatically generated metadata (Dublin core[7] compliant) and metadata fields to be filled

Function to be implemented an interactive hierarchical map shows the equivalent thesaurus term in the sidebar so make explicit the relation of the terms with the others (broader, narrower, related terms, synonyms, antonyms)

Collaboration policy The NHW is configured to allow selected subscribers to collaborate, once their request is approved by the members of the scientific committee, on the basis of their acknowledged competences. Comments are allowed to whom subscribe to the platform. This project is based on the commitment and the engagement of researcher, academics and practitioners, called to collaborate and actively contribute to the project as a competent user collaborative-knowledge building. Active members will be entitled to publish new wiki entries and to broaden the published ones. Discussions on entries will be taken into account, and included in the text.

NHW is designed as a "matrix" model to be used in "practice": a structured and collaborative web platform based on validated information on geosciences to support a common understanding.

The tool here proposed could be seen as a mere technical project, however it is much more a culture based approach to knowledge, collaboration and sharing, an enabler to support a better understanding and comprehension of natural hazards and disaster resilience content. Hence, this tool if "embedded" in any DR website could represent a knowledge base to improve Information Quality (Wilson 2002), and at the same time contribute to a more precise understanding of DR content. In this perspective NHW could become a "CMS function" that enables to parse text for defined terms, by adding links to the NHW post or glossary page that contains the definition of the term used.

[6] http://contextly.com/. Contextly's editorial tools, includes a related links service that can be managed also manually to improve the precision of related content. Designed by digital journalist Ryan Singel for digital journalists.

[7] Dublin Core Metadata Initiative—http://dublincore.org/.

33.5 Towards Semantic Web for Disaster Resilience

The effort in the field of Semantic Web and Linked Data could be to draw the «map» and enhance the ability to create infinite roads and tracks in the World Wide Web. Either we call it Semantic Web, Linked Data, Web of Things or Web 3.0, the overall aim of this research field is to improve the sense of direction in the current chaotic mass of contents, with the help of social machines. The Web Semantic approach underlines that ontologies define a common vocabulary, with machine-interpretable definitions of basic concepts in specific domain, and giving relations amongst terms. So ontologies are a way to share and reuse information, knowledge, and data in a given domain, among people or software agents (Musen 1992; Gruber 1993). If DR Web sites would refer to the same ontology of terms, then computer agents could extract and aggregate information from all these different sites. The agents can use this aggregated information for user queries or as input data to other applications.

Ontologies describe and represent areas of concepts, allow the integration of data and are a way to represent knowledge in a domain; metadata allow the semantic interoperability and evaluation of data sources in terms of reliability and trustworthiness (Nativi et al. 2012). We envisage the development of a true and concrete Linked Data for DR to build a common language and a common understanding of this domain so to avoid misleading and inappropriate information, and increase preparedness. From this perspective the next step of the project is to move towards metadata and ontologies as building block to build a semantic ecosystem for DR and for civil protection and emergency management.

References

Bruns A (2007) Produsage: towards a broader framework for user-led content creation. In: Proceedings Creativity & Cognition 6. Washington, DC

Bohannon J (2013) Who's afraid of peer review? Science 342(6154):60–65. doi:10.1126/science.342.6154.60, http://goo.gl/YM699D

Christensen JH, Hewitson B, Busuioc A, Chen A, Gao X, Held I, Jones R, Kolli RK, Kwon W-T, Laprise R, Magaña Rueda V, Mearns L, Menéndez CG, Räisänen J, Rinke A, Sarr A, Whetton P (2007) Regional climate projections. In: Climate change 2007: the physical science basis. Contribution of Working Group I to the Fourth Assessment Report of the Intergovernmental Panel on Climate Change [Solomon S, Qin D, Manning M, Chen Z, Marquis M, Averyt KB, Tignor M, Miller HL (eds)]. Cambridge University Press. New York, NY, USA

Comfort L, Wisnerb B, Cutterc S, Pulwartyd R, Hewitte K, Oliver-Smithf A, Wienerg J, Fordhamh M, Peacocki W, Krimgoldj F (1999) Reframing disaster policy: the global evolution of vulnerable communities. doi:10.3763/ehaz.1999.0105

Denaux R, Dolebear C, Hart G, Ditrova V, Cohn AG (2011) Supporting domain experts to construct conceptual ontologies: a holistic approach. Web Semant Sci Serv Agents World Wide Web 9:113–127

Edwards D (2013) The revolution will NOT be in open data, guest post on open knowledge foundation blog, 21, Oct 2013, http://goo.gl/vslJpi

EU-Project MONITOR (2007) Hazard monitoring for risk assessment and risk communication. Risk ontology, an ontological basis for risk management

Goodchild MF (2007) Editorial: citizens as voluntary sensors: spatial data infrastructure in the world of web 2. 0. Int J 2:24–32

Gruber TR (1993) Toward principles for the design of ontologies used for knowledge sharing. International workshop on formal ontology, March, 1993, Padova, Italy

Musen MA (1992) Dimensions of knowledge sharing and reuse. Comput Biomed Res 25:435–467

Hendler J, Berners-Lee T (2010) From the semantic web to social machines. Artif Intell 174(2):156–161

Hester AJ (2011) A comparative analysis of the usage and infusion of wiki and non-wiki-based knowledge management systems. Inf Technol Manage 12:335–355

Lukashevich NV (2011) Concepts in formal and linguistic ontologies. Autom Documentation Math Linguist 45(4):155–162

Mugridge R (2007) The thesaurus: review, renaissance, and revision. Libr Resour Tech Serv 51(3):341–342

Nativi S, Craglia M, Pearlman J (2012) The brokering approach for multidisciplinary interoperability. Int J Spat Data Infrastructures Res 7:1–15

Simpson DM, Katirai M (2006) Indicator issues and proposed framework for disaster preparedness index (Pi), Center for hazards research and policy development, working paper 06-03

Stuckman J, Purtilo J (2011) Analyzing the Wikisphere: methodology and data to support quantitative Wiki research. J Am Soc Inf Sci Technol 62(8):1564–1576

United Nations (2005) International strategy for DR Hyogo framework for action 2005–2015: building the resilience of nations

van Dijck J (2013) The culture of connectivity: a critical history of social media. Oxford University Press, Oxford

Wilson P (2002) How to find the good and avoid the bad or ugly: a short guide to tools for rating quality of health information on the internet. BMJ 324(7337):598–602. doi:10.1136/bmj.324.7337.598

Italian Civil Protection 2.0: Towards a Resilient Information Management Framework

34

Elena Rapisardi, Sabina Di Franco and Marco Giardino

Abstract

The web 2.0 revolution is a twofold reality: technological and cultural. The widespread of web 2.0 technologies is changing Disaster Resilience. Information in science communication is the ability and the capacity to transfer scientific knowledge to enable the understanding of communication content. Particularly a clear and correct information on hazards and emergency matters is crucial, either for practitioners and population, to cope with disaster and to allow collaboration, and to take the best decision. Hence, web 2.0, allowing a better information sharing and widespread, could represent a leverage to foster an Information Preparedness approach, so to better cope with disasters. However a Web Information Preparedness in Disaster Resilience requires an overall strategy that can have an impact both on the cultural side (knowledge sharing and collaboration) and on the organizational one. How civil protection bodies can benefit of this global changes to better inform citizens? How to implement and manage a new Resilient Information Preparedness approach enabled by web technologies?

Keywords

Web 2.0 • VTCs • Information management • Disaster resilience

34.1 Web 2.0: The Challenge for Institutionalized Mediation Models

Timely and accurate information is recognized as integral to emergency response. Amongst the principles of Information Management proposed by Office for the Coordination of Humanitarian Affairs (OCHA) in 2002, Accountability and Verifiability are key to evaluate the reliability and credibility of data and information, highlighting the need for an information that is "accurate, consistent and based on sound methodologies, validated by external sources, and analyzed within the proper contextual framework" (OCHA 2002). Accountability and Verifiability of information becomes crucial, particularly in the emergency/disaster phase when information flows on the internet at a very fast speed, and information demand reaches its peak (Limbu 2012). As shown in Google Trend search trends for the words, fault and liquefaction, from 2009 (Abruzzo Earthquake) to 2012 (Emilia Romagna earthquake), two search peaks occur exactly when the two earthquake stroke. Presumably internet users were looking for the words fault and liquefaction, that belong to the seismic discipline, to better understand the earthquake as a natural phenomenon (event, process, factors, related effects…). The problem is related to search findings: how to discern content accountability and

E. Rapisardi (✉) · M. Giardino (✉)
NatRisk, Interdepartmental Centre for Natural Risks, University of Turin, Turin, Italy
e-mail: rapisardi@gmail.com

M. Giardino
e-mail: marco.giardino@unito.it

S. Di Franco (✉)
National Research Council IIA, Rome, Italy
e-mail: difranco@iia.cnr.it

G. Lollino et al. (eds.), *Engineering Geology for Society and Territory – Volume 7*,
DOI: 10.1007/978-3-319-09303-1_34, © Springer International Publishing Switzerland 2014

verifiability, and also how institutions respond to the request for information, how they are using web 2.0 tools to build web presence and delivering information? Though Italian Legal Framework on Civil Protection and Public Administration regulations[1] outline duties and define web guidelines, the increased demand of information from citizens in emergency, and the state of the art of the institution in the web sphere, shows a critical the gap between institutional communication, usually by press release or through traditional media, and web information stream produced by the citizens. However a pervasive and fast propagation of citizens generated information, although exposed to the risk of inaccuracy or of a low level of liability, is filling the vacuum created by the dependency from old mediation models, exacerbated by a great heterogeneity and dispersive of the presences of institutions on the web, and by a usage of social media more as a channel for traditional communication rather than as channel to interact, and to give information on public utilities (e.g. emergency, weather alerts).[2] In case of emergency, the acceleration of the speed of information flow calls institutions not to react with a time-to-web strategy during the event, but to build an overall web strategy of information preparedness. In this perspective, the "peace time" can be understood as the time to build and consolidate a web presence or web reputation, aiming at helping people to understand the phenomena and the model of intervention of Civil Protection. Be prepared should not be limited to self protection behavior, but requires an effort to make explicit both the scientific knowledge on the natural hazards, and the actions to undertake to reduce the potential risk. But this could be accomplished only by planning an information management strategy, carried out by teams able to manage the overall information process.

34.2 Civil Protection Plan as Information Management Preparedness

Information management covers 'the various stages of information processing from production to storage and retrieval to dissemination towards the better working of an organization; information can be from internal and external sources and in any format.'[3] Data collection, processing, analysis and dissemination are the different phases of the information management cycle. Hence the information management cycle should be included in each phase of the overall disaster cycle; and for each phase of the information management cycle tools, strategies, resources, competences will have to be defined. The new web technologies allow the implementation of information systems, tools, that can support a more efficient Information Management. In the Italian Civil Protection framework, the actor closest geographically and responsible according to the law is the mayor, who has also the duty of informing population. The Law 100/2012 recognize the higher-level of Civil Protection Plan with other planning tools and therefore the Civil Protection Plan is the main tool to start gradually to reduce the exposed elements at risk, and prevent new and inconsiderate use of soil increase the danger of the territories. Hence the Plan is not only operational, but constitute an increasing source of communication content. The Plan is therefore both Preparedness in terms of response and recovery planning and a tool for "data and information" preparedness. In this perspective a local Civil Protection website should be developed translating the civil protection plan into web content, setting up a first building block to organize knowledge and information focused on the territory, and with a user-centered perspective. But in order to give effectiveness and efficiency to such a tool, some operational principles should be identified, to guide information management and exchange activities. The website should be developed according to acknowledged web design principles and standards: : Information Architecture, Visual Design and Visual Representation, Cascading Style Sheets (CSS), Usability, Accessibility, Search engine optimization (SEO), Content Management System (CMS) (preferable OpenSource -OS- and Free, e.g. Drupal, Wordpress), Analytics to monitor website user and performances. The content structure should enhance operational principles for information exchange activities in emergencies defined by OCHA (Office for the Coordination of Humanitarian Affairs) in 2002 (Accessibility, Inclusiveness, Inter-operability, Accountability, Verifiability, Relevance, Objectivity, Humanity, Timeliness, Sustainability).

Figure 34.1 shows an example of the information architecture blueprints of a site of a local civil protection: the conceptual framework of the web content is based on the organization of knowledge in the domain: scientific, risk and emergency communication. The innovative elements of this architecture are: the definition of a unique taxonomy (thesaurus), SEO and metadata (e.g. dublicore compliant); integration with external terminological tools (Glossary, Wikis, Thesaurus) and interaction with other web platforms, to improve understanding and the explanation of the text; responsive for navigation via mobile phones; feed (e.g. Xml,

[1] "Linee guida siti web delle PA 2011", "Vademecum Pubblica Amministrazione e Social Media", (e.g. L. n. 225/92 e L. 100/2012), D.lgs. 33/2013.

[2] OPERA, Unità di Ricerca del Centro di Ricerca GIUnO, Comuni 2.0, Utilizzo dei social network nei comuni italiani di medie e grandi dimensioni, Reggio Emilia, Università degli studi di Modena e Reggio Emilia; #FacebookPA (2013), A cura di Giovanni Arata, Nexa Center for Internet and Society, Politecnico di Torino.

[3] Association for Information Management 2005 (See http://www.aslib.co.uk).

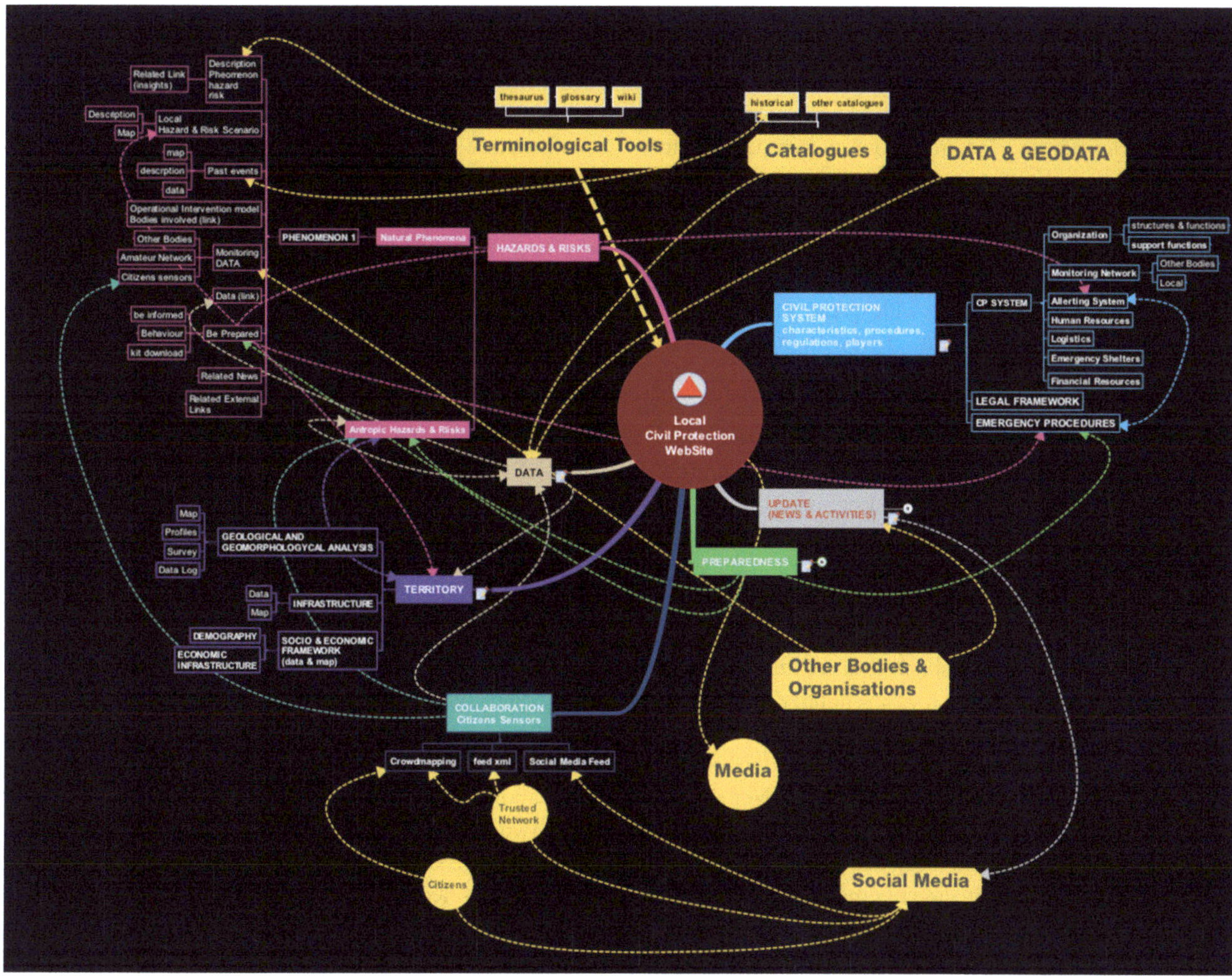

Fig. 34.1 IA—blueprint local civil protection web site—larger view http://goo.gl/VZhB1z

GeoRSS, API),[4] crowdmapping and Social Media; outgoing feed towards social media platform, categorized RSS.

The website is divided into seven sections interrelated.

Civil Protection System: Who, What, How and When of the Civil Protection Organization, including legal framework and operation model of intervention.

Hazards and Risk: this section is organized by "phenomena" labelled in plain language (e.g.. Floods instead of Meteo-hydrogeological). Starting from the "phenomenon" is a way to make clear that a phenomenon is a natural process that can be an hazard and then a risk depending on the probability, exposure and vulnerability of the event. Each phenomenon is briefly illustrated in general terms (with link to web sites that explain in details the issue), and its characteristics in the local area are explained in terms of risk scenario; the visual representation can vary, but interactive maps, video and commented illustration/photo are suggested. Other contents: past events in the area with brief description, how to be prepared; operational model of intervention, the related data and geodata (also feed from sources e.g.: institutional, amateur, and citizen).

Territory: overview of the area from geological, geomorphological, and climate perspective (map, profiles, survey, data log); infrastructure (data and map); socioeconomic context. An interactive map could visualize all this information at glance.

Preparedness: self-protection behavior and information sources by risk.

Data[5]: list of OpenData and Open Geodata (metadata information) organized by hierarchical Category, link to open data websites; data section can support transparency

---

[4] Extensible Markup Language (XML), RSS (Really Simple Syndication) standard web feed format, GeoRSS standard for encoding location as part of a Web feed; application programming interface (API).

[5] The state of the art on open data at national level is heterogeneous, so is the adoption of EU INSPIRE (Directive 2007/2/EC). See also The Humanitarian Exchange Language (HXL) initiative for a "community agreement on data standards", and to develop "a technical infrastructure for automating simple peer-to-peer data exchange".

through validated data, and feed the situational awareness scenario.

News/Articles and Alert pushing via email-subscription

Collaboration/VGI: besides social media platforms, several other tools can be used for crowdsourcing or participatory activities, both in peace time and in emergency (e.g. crowdmap or mobile field data collection tools); an accurate planning is needed in order to carry out this activity, considering also the available resources and skills.

Visualization: MAPs: "The widespread engagement of large numbers of private citizens, often with little in the way of formal qualifications, in the creation of geographic information, a function that for centuries has been reserved to official agencies. [...] But, collectively, they represent a dramatic innovation that will certainly have profound impacts on geographic information systems (GIS) and more generally on the discipline of geography and its relationship to the general public." (Goodchild 2007). The "democratization of GIS" empowered by web 2.0 technologies, contributed to the growing popularity and usage of location based information: mapping and mashups, to share information and to collaborate. On the basis of a vast scholarly literature and wide debate on this topic, we can assume the idea that maps could represent an "information service" to deliver information, related to hazard, risk and emergency. Moreover, in Disaster Resilience, web maps could ameliorate information sharing, support response activities, decision making and last but least to give guidance in case of emergency. Visualizing spatial information on a map, could potentially (as it is not data evidence) increase comprehension and understanding, support awareness raising, hence improving information to the public. Assuming that the geodata are available in a standard formats, there are several web mapping tools allowing map visualization. A web interactive map shows a representation of several characteristics of the territory either related to geology and geomorphology, or hazards/risks, or emergency management. In Fig. 34.2 an example of an interactive map on spatial seismic information - geology, hazard, risk, shelters- [6] that could be embedded into a web page and displayed on a mobile device.

Such web model,[7] tailored for civil protection purposes, has the potential to answer to several information needs both internal and external. A clear focus on phenomena, hazards and risks (provided that are expressed in explicit language and supported by terminological tools), both in a general perspective and in a local one, is a way to explain and ameliorate the comprehension and the understanding of when and how a phenomenon becomes an hazard, a risk or an emergency. In this perspective earth sciences and scenario planning competences are key to implement and develop a scientifically coherent content structure. Information should become a true user experience increasing citizen knowledge both of the territory and of the phenomena.

34.3 Who Manages Information Management?

Crowdsourcing enabled by web technology (mobile, internet) allows individuals to become observers and witnesses. A concrete and important example is the experience of Ushahidi set up in 2007 as a non-profit software company that develops free and OS software for information collection, visualization, and interactive mapping: the Crowdmap platform Ushahidi is so emblematic not only as experience of "citizens as sensors", but also as enabler of a structured and organized practice of crowdmapping. Undoubtedly, Volunteered Geographic Information (VGI) (Goodchild 2007), and other type of user-generated-content, either spontaneous or from trusted networks, can effectively contribute to increase situational awareness, through their observations from the field. However this new practice raises two issues: the validity, precision and accuracy of information (risk of information) and the skills necessary to produce or select valuable observations. The Ushahidi Crowdmap has been designed envisaging validation and approval as key tasks to be performed so to make citizens reports accountable. Furthermore this seemingly simple validation process can be "distributed" and run by a subset of the "crowd", or by groups of skilled volunteers placed in different places, which are named Volunteer Technical Community, (VTC) as the Ushahidi Stand By Task Force (SBTF) launched in 2010 (Capelo et al. 2013). VTCs give support performing tasks such as media tracking, validation, geolocation, mapping, data cleaning, translation, and social networks monitoring; they add value to crowdsourced data and information, playing a crucial role in the information management process, also by giving support to organizations, particularly local organizations, who have neither the means nor skills to manage information/data with new technologies. The ability to create flexible and open team, where people with different skills are called to work together remotely, is reshaping the relationship between responders, scientists, citizens, experts, volunteers and last but not least, institutions. VTCs, SBTF experiences highlight that digital spontaneous volunteering could be at risk in disseminating false or misleading information (Robson 2012), or their action becomes less effective when there is a lack of coordination with the institutional bodies and responders. These weak points suggest the definition of new

[6] Google Crisis Map, an OS tool to create and visualize maps for crisis, humanitarian, and non-profit purposes. Format: KML, GeoRSS, WMS, Fusion Tables, Google Maps Engine.

[7] In the next future, the web model of information management here proposed could be potentiated by linked data technologies.

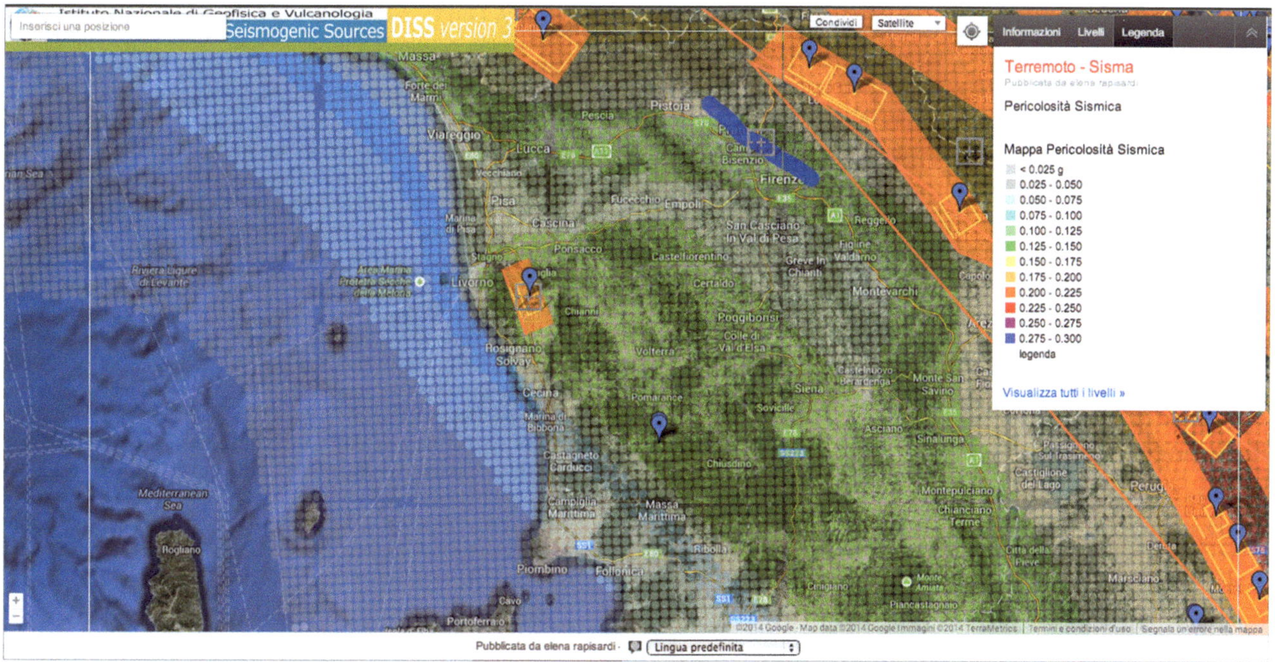

Fig. 34.2 Google crisis map civil protection—live maps http://goo.gl/2E0cgm and http://goo.gl/5LfUyl

models of collaboration between the stakeholders (institutions, responders, media, volunteers and citizens) envisaging the organization of a new interconnected volunteering, where rules of responsibility (rights and duties) are explicit and that could interact with spontaneous VTCs or bounded networks. (Rapisardi et al. 2013). Without any doubt if Italian Civil Protection would commit with a new model of Information Management, transparent, collaborative, and interoperable, should modify its information management process and acquire new competences and skills. Therefore, the issue to be discussed is not only the sustainability of the technological tools (e.g. proprietary vs OpenSource), or how to cope with the reduction of budgets, but "who" will be in charge to manage a new Information Management, how this new task will be included in the operational model, and how to better engage citizens spontaneous contributions. The information process is changing quickly calling institutions and public administration to a change. The rigidity of the institutional information flow should accept this new challenge, where the Web Information Management Process, understood as the flow of information between senders and receivers (internal and external, crowdsourcing, participatory), should become an essential, recognized and organized structure within civil protection system. In the Italian Civil protection framework this could be accomplished innovating and updating the communication function [F3],[8] as defined by the Method Augustus.[9]

Provided that the civil protection volunteering is well defined and protected by the law, an acknowledged and independent National Task Force (NTF) could be set up, on the basis of the experience and the best practices of SBTF or of the Digital Humanitarian Network (DHN),[10] but in the framework of national laws on volunteers associations. Including VTCs approach into the Civil Protection System should be done on condition that the Italian NTF is ruled under the same conditions of the other civil protection volunteers organization. During an emergency, the NTF, a *super partes* task force, could be integrated and supported by a local and spontaneous SBTF, but both VTCs should comply with a multi-skilled and multi-competent team schema: individuals from various disciplines and professions (e.g. geologists, engineers, GIS specialists, emergency managers, communication experts, web developers, web copywriters, law experts), whose skills and competencies are verifiable and tested. Such formalized, but flat, distributed and flexible team should also engage local and spontaneous teams, ensuring a full support to local bodies in managing information: structuring information and improving easy access, monitoring the web sphere and

[8] See also http://www.slideshare.net/BBarsanti/crisiscamp2

[9] The method Augustus is a planning tool in the field of emergency adopted by the Department of Civil Protection of the Italian Republic.

[10] In 2012 SBTF, GISCorps, MapAction, Humanitarian OpenStreetMap Team (HOT), ESRI, Geeks Without Bounds, Translators Without Borders, Statistics Without Borders, Humanity Road and UN Online Volunteering Service, joined the Digital Humanitarian Network (DHN), launched by A. Verity, OCHA and P. Meier, iRevolution.

developing solutions and tools on the fly in order to respond to information demand from the public, to ameliorate situational awareness and to give support to spontaneous citizens participation (either VTC, or in the field). In this perspective, Information Preparedness, is a fundamental leverage to foster such a radical change in information management and processes, and should be embedded in the disaster cycle, by conceiving an overall web information strategy, supported by a clear and open organizational structure, based on collaboration, citizens engagement, participation, open knowledge and transparency.

References

Capelo L, Chang N, Verity A (2013) Guidance for collaborating with volunteer & technical communities in sudden onset disasters, digital humanitarian network

#FacebookPA (2013) Quanti sono e cosa fanno gli enti locali italiani su Facebook, A cura di Giovanni Arata, Nexa Center for Internet and Society, Politecnico di Torino

Goodchild MF (2007) Editorial: citizens as voluntary sensors: spatial data infrastructure in the world of Web 2.0

Limbu MK (2012) Integration of crowdsourced information with traditional crisis and disaster management information using linked data. Master's thesis. Masters Program in Geospatial Technologies. University of Münster, February 28, 2012, Münster, Germany

OPERA, GIUnO, Comuni 2.0, Utilizzo dei social network nei comuni italiani di medie e grandi dimensioni, Università degli Studi di Modena e Reggio Emilia

Rapisardi E, Giardino M, Cordola M, Monticone M., Pelosini R, Poncino S (2013) Web information emergency management. The role of information in the framework of the Italian Emergency Management, GeoItalia 2013

Robson ES (2012) Responding to liability: evaluating and reducing tort liability for digital volunteers. Woodrow Wilson International Center for Scholars, Washington, D.C.

United Nations Office for the Coordination of Humanitarian Affairs (2002) Symposium on best practices in humanitarian information exchange, Geneva, Switzerland, 5–8 Feb 2002

Geomathics Tools and Web-Based Open-Source GIScience Public Participation Field Approach for Field Data Collection: Supports for a Civil Protection Exercise in a Hydrological Risk Scenario

35

Bacenetti Marco, Bertotto Stefania, Ghiraldi Luca, Giardino Marco, Magagna Alessandra, Palomba Mauro and Perotti Luigi

Abstract

The improvements in the Information and Communication Technologies, coupled with the observed increase of frequency and intensity of extreme meteorological events, pushed many software houses to realize specific applications for the management of the communications between institutions and citizens during the emergencies. In this context, work methodology and specific instruments have been developed in order to analyze web based tools for field data collection, by mobile devices, and visualization, within a specific application customized for supporting civil protection activities during the emergencies. Methods and tools have been recently tested during a civil protection exercise managed by Piemonte Region, achieving good results both in the field data collection and in the web data management and visualization phases.

Keywords

Web mapping • Hydrological risk • Mobile devices • Geo-database

35.1 Introduction

Last decades have been characterized by two important aspects: (1) the technological improvements in the Information and Communication Technologies (ICTs) and the raise of GIScience (Goodchill 2010), with the resulting change in the information management and accessibility and (2) the climate changes increasing the frequency and the intensity of extreme meteorological events (Brunetti et al. 2001, 2004), which many times caused emergencies situations including economic and social damages, whose importance was determined by the local vulnerability.

B. Marco (✉) · B. Stefania · G. Marco · M. Alessandra · P. Mauro · P. Luigi
Earth Sciences Department – GeoSITLab, Torino University, Turin, Italy
e-mail: marco.giardino@unito.it

G. Luca
Regione Piemonte - Museo Regionale di Scienze Naturali di Torino, Via Giolitti 36, 10123 Turin, Italy

In the framework of this technological and climatic "trends" many software houses and developers communities have been realizing applications whose aim is to improve the management of the communication flows between emergency managers and citizens during the different phases of disaster cycle, especially in those of the preparedness and mitigation ones. Consequently, many scientific institutions, international organizations and public administrations started to test and to use these applications on mobile devices and Web-platforms. Surprisingly only few application tailored to set-up for field collection during the response phase, probably due to the fact that collection of information during the emergency requires specific questionnaires based both on the type of natural risk and the information that need to be collected.

35.2 Motivation and Goals

During an emergency it is extremely important to act in a very short time. Therefore, it is essential to get situational awareness. The near "real time" submission, analysis and

G. Lollino et al. (eds.), *Engineering Geology for Society and Territory – Volume 7*,
DOI: 10.1007/978-3-319-09303-1_35, © Springer International Publishing Switzerland 2014

presentation of data, could be very useful for decision makers and for managing communication flows towards media and citizens. On the basis of these considerations work methodology and specific tools have been developed in order to achieve the following tasks:

- Collection, homogenization and geo-referentiation of historical information and thematic maps within a geo-database;
- Creation of questionnaires for field data collection;
- Development of the questionnaires into mobile devices (Tablets and Smartphones);
- Training of the staff involved in the field data collection;
- Web-based data management and data visualization.

The methodology and the tools have been tested in preparation for and during a civil protection exercise managed by the Piemonte Region and the Verbano Cusio Ossola (VCO) Province and planned within the Interreg project STRADA: Adaptation strategies to climate change for natural risk management in the transboundary Italy-Switzerland territory (Progetto STRADA 2013).

35.3 The Verbania Civil Protection Exercise

On 22th April 2013 in the municipality of Verbania, in the northern sector of Piemonte Region, a civil protection exercise has taken place. Many human and material resources were implicated, with the aim to test different aspects related to collaborative management of a hydro-meteorological risk scenario. Some technical meetings between the civil protection bodies of Piemonte Region and VCO Province and the GeoSITLab laboratory of Torino University were carried out to analyze the Verbania area and to provide the more useful technical solutions. The drill focused on a hydrogeological risk scenario consisting of a progressive increase of the Maggiore Lake water level, with a meteorological context of severe precipitations, similar to the conditions affecting the municipality in the autumn 2000. The main issues covered were:

- extensive flooding in the areas next to the lake;
- roads interruption;
- landslide phenomena interesting the National Road 34 connecting Italy and Switzerland;
- monitoring activities on the basis of the hazard maps realized by the Geological Survey of the Verbania Province.

In order to cover the whole area interested by the drill, twenty peoples were trained and then divided into five groups located in the most critical sectors previously identified.

35.4 Methodology and System Architecture

In order to achieve the determined goals, different actions and tools have been implemented and developed.

35.4.1 Geo-Database

In order to have a detailed knowledge of the natural hazards and the potential vulnerability characterizing the area, the following data and information have been collected and organized within a geo-database:

- Geological and geomorphological outline of the area;
- Climate analysis based on the correlation between precipitations data and the increasing water level of the Maggiore Lake;
- Landslides inventory based on the national IFFI project data available for the Piemonte region (Colombo and Ramasco 2007);
- Census data based on the national inventory;
- Historical data (hydrometric and pluviometric) registered in the past events;
- Spatial data concerning the critical infrastructures and buildings (public and religious) located in areas particularly vulnerable;
- Risk and vulnerability maps realized by the geological service of the Verbania province (Progetto STRADA 2013);
- VCO province civil protection emergency maps.

35.4.2 Questionnaire for Field Data Collection

Data collection was carried out using the Open Source software Open Data Kit (ODK), developed at University of Washington—Department of Computer Science and Engineering (Open Data Kit 2013). ODK consists of a suite of modular applications allowing people for:

- Building customized forms for field data collection using the XForm standard.
- Collecting information using an application installed on mobile device running on android operating system.
- Managing and organizing data into a Web server repository (Fig. 35.1).

In order to make easy the compiling and to reduce the errors, the questionnaires have been designed taking into account the following characteristics: (a) Appropriate number of fields for carrying out fast and accurate surveying; (b) Usable also by people lacking of specific technological knowledge; (c) Assisted compilation by means of select menu and skip logic.

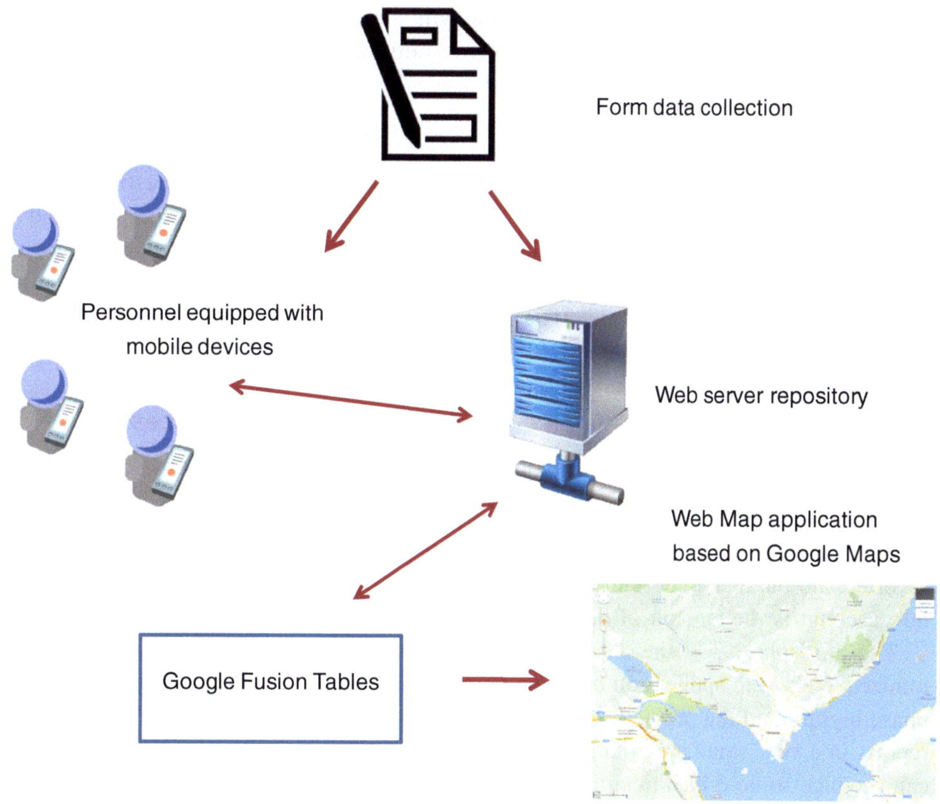

Fig. 35.1 The system employed for managing the data flow through the ODK software

The questionnaire consists of three main sections and several subsections that could be made visible and editable only in response to previous questions. The first section, always visible and mandatory, aims to identify the operator and the time of the survey, while the second one identifies the type of activity carried out. On the basis of the response, specific subsections could became visible and editable. Finally, the third section deals with the opportunity to acquire multimedia files through the tools provided by the mobile device (Fig. 35.2).

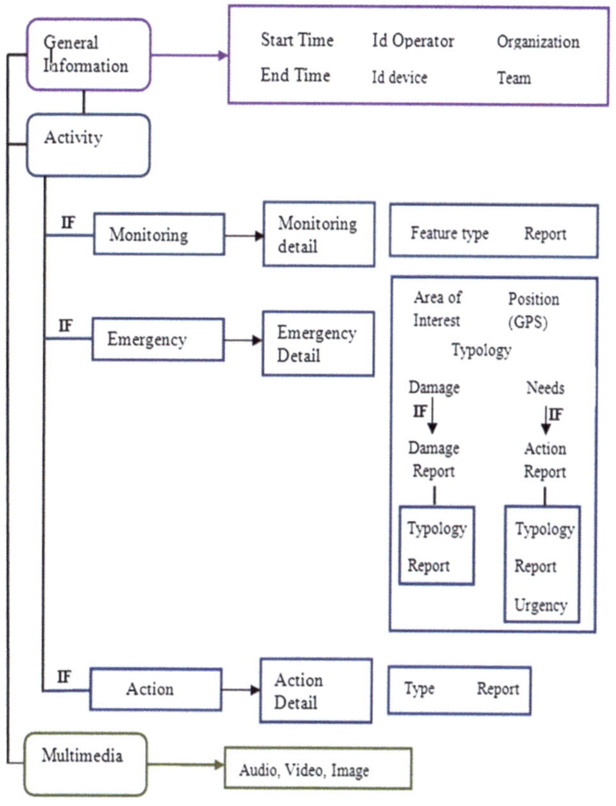

Fig. 35.2 The flow chart describing the structure of the form build using the ODK software

35.4.3 Web-Based Tool for Data Management and Visualization

Data stored in the repository were automatically published in the Google Fusion Tables where they were analyzed and evaluated. In order to share information through the Web, a Web-based mapping application was developed using a customized Google Maps interface which would fulfill the following characteristics:

- Optimization for visualization on mobile devices;
- Power to add thematic maps;
- Power to filter data on Google Fusion Tables in order to have different products: the first for internal Civil Protection use and the second for public information;

- Power to visualize data with customized symbologies;
- Power to add geographic and alphanumeric information by operators.

35.5 Discussion and Conclusion

The first phase concerned the field data collection. It didn't show any particular technical problems, aided by the good network coverage all over the area. The personnel submitted more than 40 reports in less than 3 h; the average time to complete the questionnaire was approximately 7 min, while the time to send data to the server has been almost 25 s. The publication on the Fusion Tables has recorded a shift of about 3 min.

The second phase concerned the data management on the server. Also in this case any particular problems have been reported, although an important factor which has to be taken into account is the amount of data we plan to manage.

The third phase was the data visualization using a customized Web-Map application based on Google Maps. This phase highlighted the problem of having a large number of reports in a very small area that, despite the different icons, can produce some difficulties in discerning the most recent report. To overcome the problem a priori identification of monitoring station and critical points (e.g. hydrometers, bridges) has been performed. This arrangement resulted in a considerable reduction of the icons used to categorize data on the map because the information recorded in the database were updated manually and inserted in a temporal sequence in the attributes tables associated with the geographic feature. This allowed a significantly reduction of the time used to fill out the questionnaire. Furthermore, the system allowed having a complete and up-to-date database usable by the persons responsible for the management of the information flows.

References

Brunetti M, Colacino M, Maugeri M, Nanni T (2001) Trends in the daily intensity of precipitation in Italy from 1951 to 1996. Int J Climatol 21:299–316

Brunetti M, Maugeri M, Monti F, Nanni T (2004) Changes in daily precipitation frequency and distribution in Italy over the last 120 years. J Geophys Res Atmos 109:D05. doi:10.1029/2003JD004296

Colombo A, Ramasco M (2007) Analisi del dissesto da frana in Piemonte. In: APAT, Rapporto sulle frane in Italia. Il Progetto IFFI—Metodologia, risultati e rapporti regionali. Rapporti 78/2007, p 81–120

Goodchill MF (2010) Twenty years of progress: GIScience in 2010. J Spat Inf Sci 1:3–20

Open Data Kit (2013) http://opendatakit.org/. Accessed 15 Sept 2013

Progetto STRADA (2013) http://www.progettostrada.net/it_IT/home. Accessed 15 Sept 2013

A Missed Window of Opportunity

36

Massimo Lanfranco, Elena Rapisardi
and Marco Giardino

Abstract

In October 2010 a rainstorm hit some municipalities in Liguria (NW Italy) resulting in flash floods and debris floods. Rescue operations were carried out without situational awareness due to phone and cellular line failure and civil protection plan deficiency. Relief operations were carried out to recover to previous state with no lesson learned analysis nor legislative upgrade. A window of opportunity opened to fulfill the disaster management cycle theory. On October/November 2011, when a wider event struck both Tuscany and Liguria regions causing 12 fatalities, the situation was, more or less, the same as the past year: the window of opportunity closed without a reduction in risk exposure. After this last event Ligurian Region administration started to re-evaluate hazard mapping and risk assessment methodologies, since debris floods were not considered in official hazard maps and urban planning regulations. But a month later, the temporary building restrictions were reformed to low strict level. Different to Northwest Italy extensive floods of 1994, that led Italy to a new approach to hazard planning and risk evaluation, Liguria 2010/2011 events did not to seemingly involve, bureaucracy and civil society on the new resilience concept and on the use of the emerging technologies and social media.

Keywords

Mediterranean rainstorm • Crisis management • Disaster risk reduction • Resilience • Web 2.0

36.1 Introduction

Since the publication, in 2010, of the UN/World Bank booklet "Natural Hazards, UnNatural Disasters", an international agreement on fundamental contribution of intentional human behaviour on natural disaster has been achieved.

The "act of god" line of thought (Dynes and Drabek 1994), so linked to the "reward and punishment" Catholic's view of historical events, was finally overcome, with a better integration of engineering and social science approaches to disasters studies (U.S. report "Disaster Resilience: A National Imperative" in Showstack 2012) focusing on resilience and communities engagement, and

M. Lanfranco (✉)
European Commission, Joint Research Centre (JRC), Institute for the Protection and Security of Citizen (IPSC) Global Security and Crisis Management Unit, Via Fermi 2749, TP267, 21027 Ispra, Italy
e-mail: massimo.lanfranco@unito.it; massimo.lanfranco@jrc.ec.europa.eu

E. Rapisardi · M. Giardino
NatRisk, Interdepartmental Centre on Natural Hazards in Mountain and Hilly Environment, University of Torino, Torino, Italy

E. Rapisardi · M. Giardino
Department of Earth Sciences, University of Torino, Via Valperga Caluso 35, 10125 Torino, Italy

G. Lollino et al. (eds.), *Engineering Geology for Society and Territory – Volume 7*,
DOI: 10.1007/978-3-319-09303-1_36,

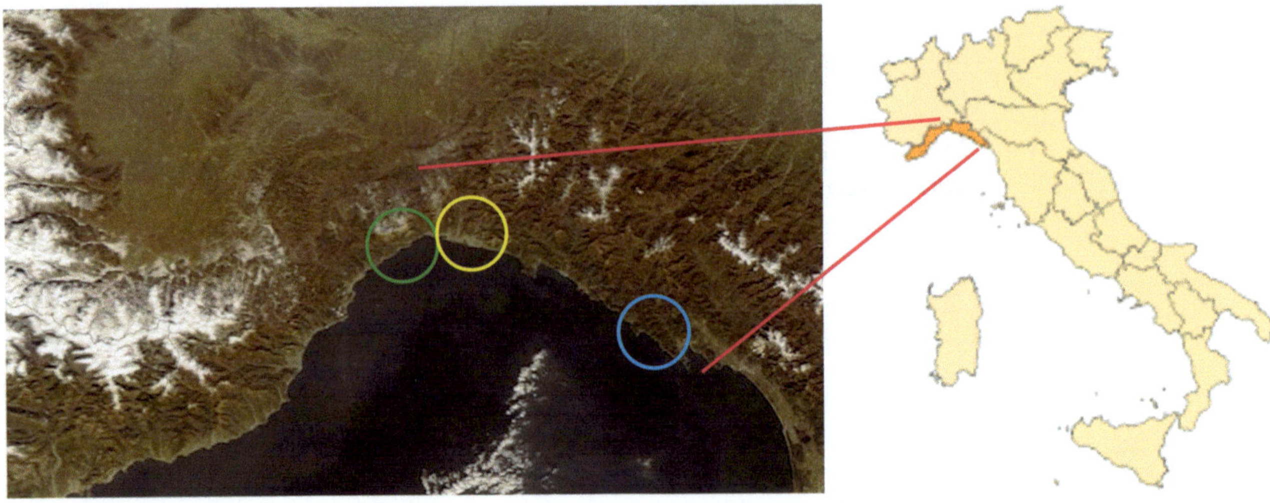

Fig. 36.1 Liguria region: location and physiographic setting; flood event locations: in *green* Varazze (04/10/2010), in *blue* Cinque Terre (18/10/2011), in *yellow* Genoa (04/11/2011)

pointing out that disaster risk reduction strategy should include social factors and enhanced risk communication.

But at the same time an Italian Civil Protection Department book stated: "The term "natural disaster" means an event due exclusively or mainly, to the forces of nature that cause, in an average short time, a significant damage" (ISAT 2006).

The wider debate on disasters risk reduction (DRR) led by the United Nations underlined a mandatory culture switch from "war" against hazards (and pure relief action *after* disasters) to "preparedness", in order to decrease the vulnerability of our societies (Wisner et al. 2004). In this perspective the key word "resilience" fosters a wider cultural change that should drive the risk management and emergency response towards a participatory dimension of all the key players: scientific communities' experts, civil protection bodies, media, citizens, volunteers, civil society. The aim is to integrate all the actors in a network that will increase the knowledge in prevention, preparedness and response.

The 2010 flood event that affected the towns of Varazze and Sestri Ponente (Liguria—Northern Italy) was the perfect example of anticipatory catastrophe that hit the following year the Cinque Terre and the City of Genoa, causing 10 deaths (Fig. 36.1).

Lessons highlighted from the 2010 event, either on extreme danger posed by debris flow/debris flood, imprecise hazard mapping, absence of vulnerability assessment and response enhancer web 2.0 tools, were not discussed in scientific papers or institutional congresses, leading to a substantial failure in life saving mitigation strategies (Berkhout et al. 2006).

36.2 The Hazard

Hydrometeorological events can trigger hydraulic or geological processes (floods or landslides) depending on slope gradient. In Liguria, since the hilly environment is so close to the sea shore, these processes are a mixed occurrence of water and sediments that range between sediment flood to debris flow. This involves continue upgrading of torrent beds and strong difficulties in modelling extreme events since the available of sediment volumes is still partially unknown.

A worst characteristic of the Ligurian floods regards the meteorological associated parameters: strong rain events are always linked to southern wind that forces big swells and raises sea level along the shore, both disturbing the river discharge.

36.2.1 Rain Rates and Sea Level Rise

Tyrrhenian Liguria, a region in north-west Italy, is characterised by complex orography that leads to strong storms and heavy rainfall due to orographic lift. The recorded rain rates are among the highest in Europe: annual rainfalls of up to 2,000 mm are typically recorded for the 200 km length of coast between Savona and La Spezia. Heaviest rain occurs mainly in early autumn, when tropical-storm-type convective cells develop due to the high sea surface temperatures experienced offshore of this region. The resulting storms discharge usually up to 500–600 mm of rain in 24 h, at rates of up to 100 mm/h. In the last 5 years a continuous series of local events fostered weather forecast accuracy and radar

Fig. 36.2 Varazze 2010 flood: debris flow along a culvert (100 m long)

based quasi-in-time web precipitation location (i.e. http://93.62.155.214/~omirl/WEB/). Across this region as much as 948 mm of rain have been recorded in 24 h, like during the 7–8 October 1970 event (the highest rainfall in Italy and the second in Europe).

The 2010 (Lanfranco and Rapisardi 2011) and 2011 events showed high precipitation rates especially in short time (≤6 h); the 2011 Genova 1 h rain rate (at Vicomorasso rain gauge) was the highest recorded in Italy (181 mm).

The flash floods hydrometeorology is complex and in Liguria it is complicated by tyrrhenian catchment fragmentation. Uncertainty, even given meteorological RADAR near-time forecast, is large than the timeline of early warning systems.

Another level of uncertainty that affects current hazard mapping and risk scenarios is generated by future sea-level rise. The IPCC AR5 (International Panel on Climate Change—Fifth Assessment Report, accessible at http://www.ipcc.ch/) contained several statements with a global average increase between 0.2 and 0.8 m in last 30 years. However, several studies suggest that future local amplitude could exceed 2 m based on paleo-climatic evidence (EEA 2012).

As afore mentioned, the sea level and storm surge strongly affected the watershed discharge and turned steady flow water profile to unsteady one.

36.2.2 Geological and Geomorphological Landscape

The geological setting of the Liguria region is highly complex due to the contact between the Alps and Apennines chains. Strong and pervasive metamorphic pattern complicates the lithological complexity driving each catchment to totally different attributes. The geomorphological heterogeneity is increased by all the morphogenetic agents that range from maritime degradation (0 m a.s.l.) to glacier environment (Punta Argentera 3,297 m a.s.l.). The watershed divide, height of approximately 1,000 m a.s.l., is located 5–10 km from the coast, defining short and steep tyrrhenian basins with a poorly hierarchical catchment pattern. The Po Plan side presents a gentle morphology and a more articulated watershed pattern.

The alluvial deposits, that formed all the tyrrhenian side urban terrain, are subjected to a progressive compacting also related to fresh-water pumping out.

Wildfires are the main accelerators to erosion and runoff (Parise and Cannon 2012). Wildfires are recurrent due to hot summers and windy early autumns. Yearly burned woods maps are mandatory and generate direct restriction to building. However, they are linked neither to Basin Plans nor to Civil Protection Plans.

Flash flooding, especially when associate to suspended-sediment movement and bed aggregation (that are not modelled in actual hazard mapping), coupled with very short concentration time, is the most severe hazard in the region (Fig. 36.2).

36.3 Social and Technical Vulnerability

Vulnerability describes the extent to which a system is susceptible to be harmed by hazards or threats (from natural, man-made or terrorism events). In recent years, the concept of resilience has been introduced, starting from ecological systems and moving into climate change issues and disasters in general (see www.floodsite.net and the definitions in Report: T32-04-01 Language of Risk).

Definitions of risk are commonly probabilistic, relating the probability of occurrence of a hazard with a consideration of the likely consequences (damage). This is one of the easiest factor to be measured by post-event analysis.

Unfortunately, on this point of view, the historical records focused on geophysical events, and related structural costs, and not on the social dimension of the associated emergencies and disasters (Adger and Kelly 1999, Haque 2005).

36.3.1 Human Development and Urban Planning

Liguria has been inhabited since the Roman Age and this strongly contributes in the social construction of modern disasters. Roman settlements and their Middle Age enlargements were mainly located on one side of final stretch of tyrrhenian catchments. This location guarantees

Fig. 36.3 Genoa 2011 flood: the Bisagno Torrent full bank flow (note cars still crossing the bridge), at the same time a little catchment (Rio Fareggiano) flooded the city streets killing four woman and two children

Fig. 36.4 Varazze municipality in 1700 A.D. (I.G.M.I.) and in 2013 (Google Earth)

fresh water reservoirs, easy sanitation and rapid access to maritime transports. On the other side ancient landscape shaping constrain to post World War II urban growth. Main road (Via Aurelia) and railway locations, near to the seaside, strongly increase flood risk, since the earthworks contribute to "bowl effect" in gentle deep terminal plains (Mileti 1999) (Fig. 36.3).

Starting from the 20th century, the seaside settlements have seen a low lands urban growth and a parallel urban sprawl on the surrounding hills. Pressures for economic development have rose in the last years, with house value burst up to 20.000 €/m for near shore or best sea view flats (Fig. 36.4).

36.3.2 Building Codes and Levees

Building codes had been established by Italian national law in 1941, but distance limitation to the rivers were set by Royal Decree 523/1904 and hydrogeological limitation in hilly areas were fixed by Royal Decree 3267/1923. However, every construction is possible if it can be demonstrated that is not at risk, this evaluation was (and indeed is) carried out by applying statistical exercise to rainfall rates. The problem is that the historical records are limited to the last 50 years and long time occurrences (100–200 years) are not consistent. Climate change is therefore enhancing foresight missing (http://www.floodsite.net).

Ligurian Basin Plans never consider the complex human dimension of risk planning nor integrate individual conditions in vulnerability analysis. However, Basin Plans pure passive defence approach leads to the loss of social memory. The presence of levees drove residents and local officials into a false sense of security (Colten and Sumpter 2009). Moreover economic resources were totally deflected toward the "hard" flood protection system, leaving the "soft" part (preparedness, evacuation plans and emergency response) without funding. For example in 2011 (the year after the worst flood event in the last 50 years) the Varazze municipality budget for the Civil Protection office was only 2.000 € while 7 M€ flood protection works were made available and 10 M€ more were claimed to build new levees where the 2010 flood hit the urban area (private damage of the 2010 flood was up to 80 M€). No additional monetary funds were foreseen to reduce flash floods and debris floods up in the hills (replanting burned areas, stopping large woody debris, improving real-time observations). Levees and associated urban growth drove to a shift from a more resilient community to a less resilient social system, where people are decoupled from natural events and unable to cope with them.

36.4 Web 2.0 Enhancer

The recent disasters all around the world demonstrate that web and crowdsourcing are powerful tool for crisis management (Rapisardi et al. 2011).

But while in the 2010 Liguria flood events, the information sharing and "parallel" (unofficial) communication channels were not ready to lead early warning and relief effort, in 2011 there were tentative from private citizens and local newspapers. Unfortunately the call for a better integration that came from scholars remained unheard:

> without an improvement in alert and warning systems, with which Web 2.0 tools could help, the next severe storm in the region could have worse consequences (Massimo Lanfranco cited in Showstack 2011).

The web 2.0 revolution is twofold: technological and cultural. The former has led the first steps while latter has still to come in disaster risk reduction strategies and tactics. The new < information/share/collaborate > era will enable:

- wider citizens engagement in preparedness, planning, early warning, relief, restore;
- faster relief with improved situational awareness;
- better communication strategy both bottom/up and top/down;
- vulnerability reduction and resilience enhancement with local storytelling.

36.5 Conclusions

Disasters (for the etymology of the word itself) are infrequent in the life of most urban areas where preparedness and mitigation action can reduce significantly economical and human lives losses. Crisis and emergency management, incorporating the social response, constitutes the link between the hazard mapping and the risk assessment. How people will react to unforeseen events is part of crisis management. And disaster is just this: a foreseen adverse event not avoided with prevention or mitigated with preparedness (Montz and Gruntfest 2002).

Natural hazard studies had abandoned pure "technical" approach (geological, geomorphological and geophysical) to focus on a comprehensive approach to vulnerability. Here, vulnerability is not seen more as an amount of victims or a percentage of buildings or infrastructures destroyed, but as communities could and success to cope with natural disasters. The coping capacity is expressed by the concept of resilience, tied to social factors that vary over time and in space.

Our society is increasingly vulnerable (i.e. less resilient) as it is linked to individual and collective property, lifelines and critical infrastructure that can be easily disrupted by natural disasters else without serious loss of life.

Based on the existing data (IPCC 2013; Lung et al., 2013) the increase of dangerous floods in Ligurian basins can not be related to the growing trend of rainfall (although this possibility does exist) but can certainly be linked to multiplier factors as wild fire, ocean level rise and urban growth.

Building resilience is not only building levees; exposure to flash flooding and, therefore vulnerability to losses, will continue to increase until we will move to measures to reduce social vulnerability, rather limiting and reducing exposure by relocating buildings and activities at risk and improving early warning and relief, in order to significantly reduce the risk to human life and faster recovery.

The key is to improve crisis management with better and faster information flow between officials and scientist and to foster risk communication to the public, either scientific advice or warnings.

> It is that solutions are not to be found primarily in new technologies or better use of existing ones. The difficulties note stem from social factors. Social problem can only be dealt with socially; technological improvements can only address technological problems (cited from Quarantelli 1991)

But this sentence not addresses web 2.0 revolution, which has a technological side but deals on "social" side of disasters: community's engagement, knowledge sharing, storytelling, proactive response. On this point of view the web 2.0 is the only step forward that we can do to deal with future disasters in time of scarcity of economic resources.

References

Adger WN, Kelly PM (1999) Social vulnerability to climate change and the architecture of entitlements. Mitig Adapt Strategies Global Change 4:253–266

Berkhout F, Hertin J, Gann DM (2006) Learning to adapt: organisational adaptation to climate change impacts. Clim Change 78:135–156

Colten CE, Sumpter AR (2009) Social memory and resilience in New Orleans. Nat Hazards 48:355–364

Dynes RR, Drabek TE (1994) The structure of disaster research: its policy and disciplinary implications. Int J Mass Emerg Disasters 12(1):5–23

EEA (2012) Climate change, impacts and vulnerability in Europe 2012. EEA Report No 12/2012. http://www.eea.europa.eu/publications/climate-impacts-and-vul1nerability-2012

Haque CE (ed) (2005) Mitigation of natural hazards and disasters. Springer, The Netherlands

IPCC (2013) Climate change 2013: the physical science basis. In: Stocker TF, Qin D, Plattner GK, Tignor M, Allen SK, Boschung J, Nauels A, Xia Y, Bex V and Midgley PM (eds) Contribution of Working Group I to the Fifth Assessment Report of the Intergovernmental Panel on Climate Change. Cambridge University Press, Cambridge, United Kingdom and New York, USA, p. 1535

ISAT - Istituto per le Scelte Ambientali e Tecnologiche (2006) Disastri naturali. Conoscere per prevenire. Officine Grafiche Garzanti. Varese (in italian)

Lanfranco M, Rapisardi E (2011) Urban resilience to severe storms. An Italian case-history of local community response to natural disasters. Geophys Res Abstr 13:EGU2011-6770-1

Lung T, Lavalle C, Hiederer R, Dosio A, Bouwer LM (2013) A multi-hazard regional level impact assessment for Europe combining indicators of climatic and non-climatic change. Global Environ Chang 23:522–536

Mileti D (1999) Disaster by design: a reassessment of natural hazards in the United States. Joseph Henry, Washington

Montz BE, Gruntfest E (2002) Flash flood mitigation: recommendations for research and applications. Environ Hazards 4:15–22

Parise M, Cannon SH (2012) Wildfire impacts on the processes that generate debris flows in burned watersheds. Nat Hazards 61:217–227

Quaratelli EL (1991) More and worse disasters in the future: social factors involved. University of Delaware, Disaster Research Center, preliminary paper 173

Rapisardi E, Lanfranco M, Di Lolli A, Lombardo D (2011) Web and mobile emergencies network to real-time information and geodata management. doi:10.2788/73045

Showstack R (2011) Using web 2.0 technologies to help during flood events. Eos, vol 92, No. 18

Showstack R (2012) Report focuses on improving resilience to disasters. Eos, vol 93, No. 33

Wisner B, Blaikie P, Cannon T, Davis I (2004) At risk. Natural hazards, people's vulnerability and disasters. Routledge, London, p. 471

Pyramids, Toxic Wastes and Nuclear Reactors Containments: A Lesson Drawn from History with a Risk Manager Perspective

37

Franco Oboni and Cesar Oboni

Abstract

Risk Assessments for "perpetuity" (geo-engineering) projects, i.e. projects that should last "forever" and/or receive "perpetual care", are raising in number and criticality. These project are oftentimes linked to the storage of wastes containing toxic, not easy to neutralize, not necessarily radioactive, compounds. No prior Human generation had to tackle this problem because: (a) produced volumes were insignificant or (b) there was no real understanding of "perpetuity". This paper compares the "historic" world-wide rate of major accidents of Tailings Dams and Nuclear Reactors to previously published acceptability criteria and codes. The paper shows how a generic modern "excellent quality" dam probability of failure can be estimated, how the initial probability of failure will evolve during the dam life, as care and monitoring are released in the post production phase and under different hazards. The paper then explores selected Human geo-structures survivability experience and finally suggests a model for long term risk evolution of Tailings Dams, with particular emphasis on post production/closure.

Keywords

Risk-assessments • Perpetuity • Tailings-dams • Nuclear reactors • Acceptability-tolerability

37.1 Introduction

Humanity is confronted for the first time with Risk Assessments for "at perpetuity" projects, that is projects that should last "forever" and/or receive "perpetual care". These project are linked to the storage of wastes containing toxic difficult to neutralize, but not necessarily radioactive, compounds. We will first compare the "historic" rate of failure (major accidents only) of Tailings dams and Nuclear Reactors world-wide to public acceptability criteria and codes, then define the long term evolution of the risks (major accidents) generated through the life of a Tailings Dams (TD), focussing on the long lasting post-production phases/closure.

Models and data used in this paper were published in 2013 (Oboni and Oboni 2013).

37.2 Tailings Dams (TDs) and Nuclear Reactors' Failures

Rates of TDs failures and major nuclear reactors' accidents to date were empirically estimated and both are compared to societal and technical acceptability thresholds to understand if present and foreseeable performances are aligned with expectations (Oboni and Oboni 2013). Risk is defined as the

F. Oboni · C. Oboni (✉)
Oboni Riskope Associates Inc., Vancouver, Canada
e-mail: coboni@riskope.com
URL: http://www.riskope.com

F. Oboni
e-mail: foboni@riskope.com

G. Lollino et al. (eds.), *Engineering Geology for Society and Territory – Volume 7*,
DOI: 10.1007/978-3-319-09303-1_37, © Springer International Publishing Switzerland 2014

product of the probability of failure by the related consequences expressed in casualties, leaving aside all environmental and physical direct or indirect consequences, for space reasons the sake of simplification. Casualties for TDs are estimated to a minimum of nil, a maximum of ~500, and an expected value of ~ 80 casualties. Casualties for Nuclear 5+ accidents are estimated to a minimum of nil and a maximum of 3,500 casualties, with a "best estimate" at 890 casualties (Oboni and Oboni 2013).

37.2.1 TDs Rate of Failure and Long Term Behavior Forecast

As reportedly slope instability is the highest cause for TDs failures, this paper focusses on that particular failure mode. The proposed methodology could, however, easily be expanded to cover other failure modes (Table 37.1).

If we consider the hydro dams failures in the decades 1989 and 1999, based on an "average number of dams" of 30,000, we get $p_f = 3 \times 10^{-6}$ to 10^{-5}. The statement above is in good agreement with the common understanding and empirical knowledge that TDs are generally of "lesser quality" than hydro dams. Interestingly many different industries around the world consider values below 10^{-6} to 10^{-5} as the boundary of what is humanly credible (meaning that below that range of probability an event is generally considered "incredible").

The Silva et al. 2008 methodology is used to estimate the p_f of "excellent" TDs at 10^{-5} to 10^{-6}. "Excellent" means top quality structure, well engineered, undergoing serious QA/QC, with a minimal Factor of Safety (FoS) of 1.4–1.5. Should inspections become occasional and measurements/monitoring ceased the probability will raise to 6×10^{-5}, a value also obtained using the Silva et al. methodology.

It is also possible to "simulate" TDs long term complete abandonment. For example, if we look at the case of phased release from operational life down to long term closure, the probability of failure will increase each time the standard of care is released. For initial FoS in the 1.3–1.5 range, the difference between operational, we are assuming that the dam under examination is initially an "excellent" structure, and abandoned varies respectively between 1.5 and 2 orders of magnitude: for FoS = 1.5, $p_f = 10^{-6}$, with possible increase to 10^{-4} and higher if the same structure falls in total neglect.

37.2.2 Major Nuclear Accidents

As of Feb. 2, 2012, 435 nuclear power plant units with an installed electric net capacity of about 368 GW were in operation in 31 countries and 63 plants with an installed capacity of 61 GW were under construction in 15 countries. The cumulative nuclear reactor operating experience amounted to 14,745 years.

Table 37.1 Summary of historic rate of failure of Tailings dams around the world

Where	When (decade)	p_f	Approx p_f
World-wide	Around '79	44/(3,500 × 10)	10^{-3}
World-wide	Around '99	7/35,000	2×10^{-4}
US	Around '79 and Around '99	7 or 8/ (1,000 × 10)	7 or 8×10^{-4}

Table 37.2 Worldwide nuclear accident of Scale 5+

Level 5	Level 6	Level 7
Accident with wider consequences	Serious accident	Major accident
First Chalk River (1952) Windscale (1957), Lucens (1969), Three Miles Island (1979)	Kyshtym (1957)	Chernobyl (1986) Fukushima (2011)

To date the world has seen the occurrence of a number of major nuclear reactors accidents (rated 5 and above on the International Nuclear Event Scale by the International Atomic Energy Agency). For Fukushima we consider one accident, although more than one reactor was involved, to ensure the list is made of "independent" accidents. Assuming seven accidents, the "historic" world average rate of Scale 5+ accidents is: 4.75×10^{-4} Scale 5+ accident/annum (Table 37.2).

This value is rather unexpected as is falls well within the realm of credibility and compares to the range of TDs. The surprise is even higher when considering the high level of regulation of one industry compared to the other.

37.3 Comparing Risks of Two Industries

Figure 37.1 displays tolerability criteria developed independently by various authors over more than thirty years (Morgan and Lave 1990; Whitman 1984; ANCOLD 2003). ANCOLD (2003) acceptability criteria (Fig. 37.1) are compatible with Comar (1987), Wilson and Crouch (1982) and later criteria published in the field of chemical industry, such those from Renshaw (1990) who defined societal risk acceptability as fatality of one individual per year of risk exposure. Many reputable publications point at a probability (of a casualty per annum) of 10^{-4} (similar to ANCOLD lower bound) as being the limit of "safe", however with a lower limit of 10^{-6} for unwillingly exposed public.

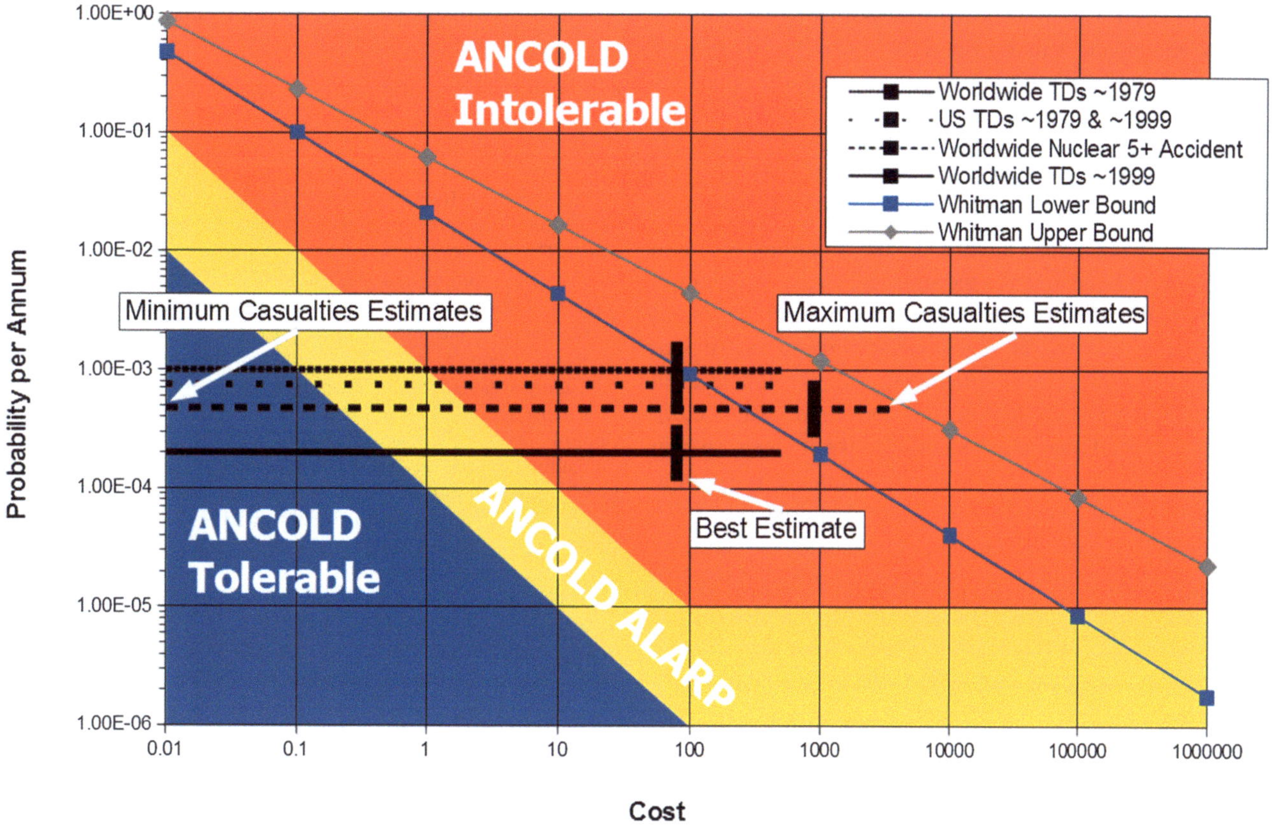

Fig. 37.1 Different acceptability criteria, cost expressed in casualties

"Simplified" metrics are difficult to apply to nuclear accidents, because of the long delays to health effects. However, if we look at Whitman tolerability thresholds the Nuclear 5+ accidents fall between lower and upper tolerability thresholds. The TD 79–99 failures fall within societal acceptability lower bound, unlike the more ancient period or US failures. Should a TD be abandoned and undergo on the long run a number of natural hazards hits, the risks will become socially intolerable even if there are less than ten casualties.

37.4 Human Geo-Structures Past Experience

When looking at ancient Man-made structures to attempt to draw some conclusions on long term survivability, we can cite the following: Europe Tumulus of Bougon 4800BCE; Africa Pyramid of Djoser 2667–2648 BCE Earliest large-scale cut stone construction; America Sechin Bajo 3500 BCE the oldest known building in the Americas and also, however way more recent, Cahokia. Apart from the Tumulus of Bougon, these structures are very similar, both in shape, size and they are numerous. Aside these pyramids we only know a few dozens of older excavated structures. Pyramids represent the largest family of long-term easily-visible surviving structures around the world and are a feature of many civilizations.

We are not implying that the builders understood what shape would "hold" through millenia . We are only noticing long term survivors belong to the same type of structure, independently of their purpose. Thus we can see a similarity with our modern need to indicate, forever, that a given site contains large and toxic hazardous matters. We should use these lessons to define the design parameters of "sign posts" we will leave for future generations, to trigger their attention to areas that have to be taken care at perpetuity.

37.5 Conclusions

Unless a TD is designed, built and cared-after like a hydro-dam, which means "at perpetuity" high level monitoring and care (TD cannot be breached, unlike hydro dams) its probability will raise very significantly. No residents should be allowed downstream of the said structure, within reach of possible run-out from a breach, to ensure ANCOLD compliance. Risk assessments have to be sophisticated enough to allow p_f estimates compatible with the ANCOLD tolerability thresholds. Standard practice matrix-based risk assessments Oboni et al. (2013), (Oboni and Oboni 2013) cannot be used as they lack the necessary finesse and resolution and could actually severely mislead TD owners/operators to the point of exposing them to severe liabilities

In order to ensure "visibility" of the area for future generations we cannot use living creatures, not even very long lasting trees as they are too vulnerable, but use carefully selected rock, if possible without any mortar or filled joints (physic-chemical alteration, air contaminants, acid rain, etc.). We have to make the sign big, so that even future deposit of alluvium, soil etc. will not easily cover it; make it massive, no openings, or sealed ones; wide so that it will accommodate differential settlements, will not topple even if parts are removed (vandalism, terrorism, wars, etc.). Carefully select foundation/location. Make it steep, so rain water will 'flash-wash' the faces and eliminate vegetation, grasses, etc. Make it flush (the Great Pyramid was flush until recently, with sharp angles, so it cannot be confused with a natural feature even if it gets partially eroded or if it is "buried" under wind blown sediments etc. Create myths and a specific clergy, caretakers forever, make sure legends will convey a sense of danger and mystery, so that future generations will respect the "symbol" and take care of its surroundings.

References

ANCOLD (2003) International commission on large dams, Australian National Committee. Guidelineson risk assessment, 2003

Comar C (1987) Risk: a pragmatic de minimis approach. In: Whipple C (ed) De minimis risk. Plenum Press, New York, pp xiii–xiv

Morgan G, Lave L (1990) Ethical considerations in risk communication practice and research, Risk Anal 10:355–358

Oboni F, Oboni C (2013) Factual and foreseeable reliability of Tailings dams and nuclear reactors: a societal acceptability perspective. Tailings & Mine Waste, Banff

Oboni F, Oboni C, Zabolotniuk S (2013) Can we stop misrepresenting reality to the public? CIM 2013, Toronto

Renshaw FM (1990) A major accident prevention program. Plant/Operations Progress 9, no. 3 (July), pp 194–197

Silva F, Lambe TW, Marr WA (2008) Probability and risk of slopes failures. JGGE, ASCE

Whitman RV (1984) Evaluating calculated risk in geotechnical engineering. J Geotech Eng 110:2

Wilson AC, Crouch E (1982) Risk/benefit analysis. Ballinger Publishing Company, Cambridge, pp 92–93

Part IV

Standards, Guidelines and Best Practices for Engineering Geology

Convener: Dr. Peter Bobrowsky—*Co-conveners*: Keith Turner, Ariane Locat

Professionals in all disciplines frequently rely on specialized documentation that ranges from compilations of the best practice examples to recommended technical procedures in more formal guidelines to rigid regulatory or prescriptive standards which are legally binding. Such documentation is regularly produced for a variety of topics relevant to the engineering geology community for a suite of topics including aggregate studies, landslide risk management, subsurface mining, infrastructure construction, groundwater extraction, and so on. The aim of this part is to encourage individuals to communicate/share their personal, national, or specialized experiences, lessons learned, successes, and failures with fellow professionals. Our aim is to internationally exchange knowledge gained about such documents. They provide much needed guidance and structure to practicing engineering geologists and they underlie our professional obligations to ensure the health, safety, and well-being of society.

Working with Engineers, as an Engineering Geological Consultant

38

Robert Goldsmith

Abstract

An outline of the purpose of engineering geology includes the application of earth sciences to the practical requirements of engineering. Whether it's a small site investigation or a major project the engineering geologist must provide a clear appreciation of what is required for a particular project. The success of the relationship between geologist and engineer depends upon good communication, maintaining a team approach to problem solving and engaging with the people concerned about particular issues. Working with engineers requires geologists to develop skills in investigation, analysis, communication, negotiation and application. The engineering geologist must have expectations that covers a diverse range of issues from soil erosion to rock mechanics. First and foremost is the need for careful observation, be prepared to conduct extensive fieldwork and to collate a large quantity of historical and current data. Through all this is carried the inherent understanding of earth processes that contribute to the site conditions under study.

Keywords

Engineers • Purpose • Skills • Communication • Teamwork

38.1 Introduction

An outline of the purpose of engineering geology includes the application of earth sciences to the practical requirements of engineering. Whether it's a small site investigation or a major project the engineering geologist must provide a clear appreciation of what is required, a focused set of data and recommendations that are specific to the particular project. The success of the relationship between geologist and engineer depends upon good communication, maintaining a team approach to problem solving and engaging with the people concerned about particular issues. As quoted by Dr H Golder in about 1960 "the scientist is interested in the right answer, the engineer is interested in the best answer, now".

To appreciate how the geologist and engineer relationship has evolved a brief history of the development of engineering geology is necessary and some examples of major international projects where the relationship evolved are also provided. The necessary skills for the maintenance of a good working relationship are discussed and finally selected working examples from personal experience are presented.

38.2 Development of Engineering Geology

Many early engineers were also active geologists such as William Smith who worked on canal projects through England in the nineteenth century and was able to assess geological conditions to work through his engineering

R. Goldsmith (✉)
SMEC International, Melbourne, Australia
e-mail: Robert.goldsmith@smec.com

G. Lollino et al. (eds.), *Engineering Geology for Society and Territory – Volume 7*,
DOI: 10.1007/978-3-319-09303-1_38, © Springer International Publishing Switzerland 2014

challenges. Thus, the link between geology and engineering has long been established. Legget (1962) stated that engineering geology is concerned with the application of scientific study to the art and practice of engineering. The findings of a pure science are applied to the specific problems of the engineer. Variability for the earth sciences in nature must be considered while preparing the precision required for designs of structures. Although characteristics may vary the fundamental geological principles do not.

In order to discuss the role of the engineering geologist's relationship with the engineer we have reached today it is of value to look at some landmark projects and evaluations of our profession. The important developmental writings by Legget and others in the 1960s and 1970s provided a basis for how we have evolved in the subsequent five decades. The iconic projects such as the Snowy Mountains Hydro-electric Scheme in Australia, the Glen Shira Hydro-electric Project in Scotland and the Roseires Dam in Sudan have demonstrated the critical importance of the engineering geologist. Each of these projects in its own way has contributed to the development of engineering geology.

The Snowy Mountains Scheme, comprising numerous dams and power stations connected by many kilometres of tunnels, saw key developments in engineering geology in the 1950s and 1960s, including setting definitions for rock weathering and strength, developing standards for core drilling and water testing and working with engineers to solve problems of rock support in tunnels and caverns (Moye 1955). It is interesting to note however that the engineering team was part of the Scientific Services division for the construction project, it was not part of the design team.

The Glen Shira project, completed in Scotland in 1957, and located in the Dalradian metamorphic complex presented many baffling geological problems (Paton 1956). By combining the geological knowledge, application of rock mass assessment of variable weathering, strength and permeability three dams and interconnecting tunnels, including an inclined shaft could be sited taking advantage of the variable terrain to develop this unique pumped storage hydropower scheme.

The Roseires Dam on the Blue Nile in Sudan was completed in 1964 as a combined concrete barrage flanked by earth embankments. The foundations for the concrete section of the dam in complex high grade metamorphics intruded by dykes had to be approved for construction. Knill and Jones (1965) report on the development of a rock mass classification system based on rock type, weathering and structures that provided the engineers with a simple guide for acceptance or otherwise for foundation preparation. During dam raising by 10 m, completed in 2012, this author developed a rigorous rock classification based on currently accepted systems to aid for the analysis of both stability of the raised dam and for the new rock foundations beneath the web extensions of the concrete buttresses.

38.3 Role of the Disciplines

The combined roles of the geologist and the engineer need to include cooperation and partnership that in some ways is a union of opposites. Legget (1962) again provides us with a simple set of guidelines. His view was that the geologist analyses conditions as they are found; whereas the engineer considers how to change existing conditions so that they suit the plans. The geologist cites problems that exist and suggests the possible consequences; whereas the engineer has to find and implement a method to mitigate the situation or prediction. In the end the engineer requires a clear picture of the geological conditions related to a specific project, presented as concisely as possible with practical utilization in view. For the geologist to be able to satisfy this requirement there should be a list of specific questions from the engineer in a usable format and with some quantifiable set of design criteria. This last item is often the most difficult as the natural variability of a site must be assessed, so that a value or range of values can be taken and used with some confidence in the precise design.

38.4 Necessary Skills in the Consulting World

Working with engineers requires the engineering geologist to develop skills in investigation, analysis, communication, negotiation and application. This specialist must have the resources to address a diverse range of issues from soil erosion to rock mechanics. First and foremost is the need for careful observation, to be prepared to conduct extensive fieldwork and also to be able to collate a large quantity of historical and current data. Along with this is the inherent understanding of the earth processes that contribute to the site conditions being studied.

38.5 Personal Experiences

The following examples indicate the differences between the geological and engineering disciplines and the ways these have been approached.

An example of a significant difference between the thought processes of the geologist compared with the engineer was seen while investigating the Slaty Creek brick clay pit in central Victoria, Australia. This pit has been worked for 50 years and so has good exposures of basement

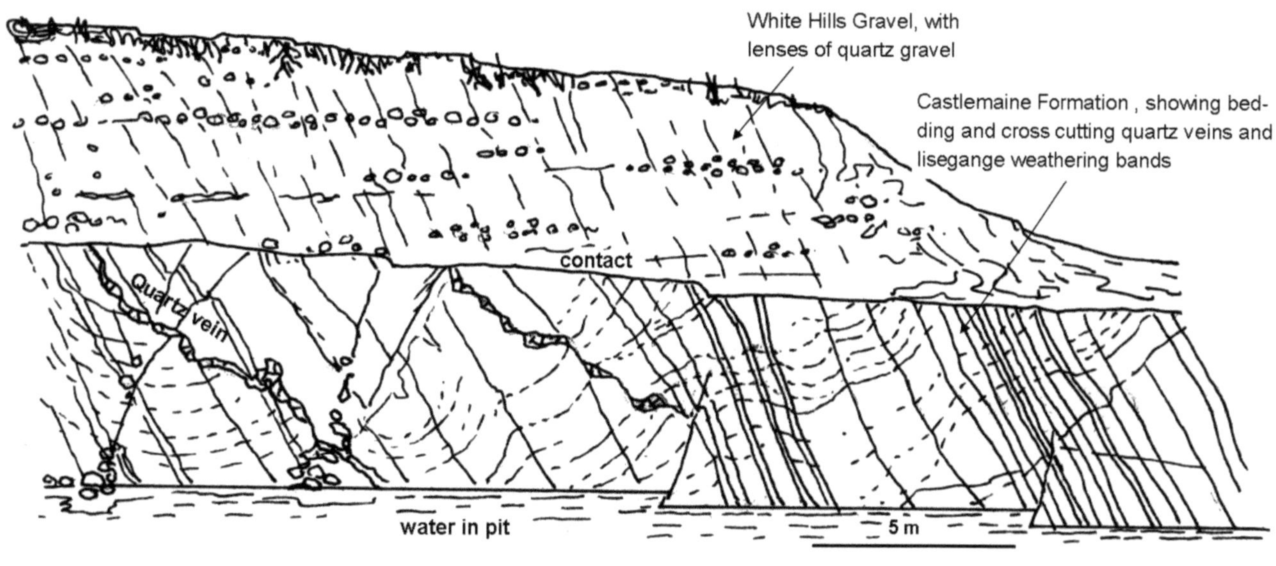

Fig. 38.1 Sketch of profile at Slay Creek brick pit, Victoria, Australia

rocks beneath a gravelly soil sequence (Fig. 38.1). The geological conditions consist of a basement of Ordovician Age Castlemaine Formation consisting of interbedded claystone and sandstone that is deeply weathered. The exposure in the pit walls shows the bedding is steeply dipping, with cross cutting quartz veins, in addition to distinctive liesegang weathering rings intersecting the steeper structures in a wavy pattern. Overlying this is a layered sequence of the much younger White Hills Gravel, of Tertiary Age, with numerous alluvial beds of white, rounded quartz gravel derived from the gold bearing quartz veins that are seen in the basement rocks below. The immense age difference between the two units of about 400 million years is appreciated.

What the engineer sees is a mass of soil with some linear patterns and areas of the milky white quartz. Consideration is given to how to extract the materials for a nearby dam extension project. The clay and the gravel will need to be kept separate. By cooperation the plan for material separation along the stratigraphic contact and selecting the best clay by following along strike in the steeply dipping basement sequence can be worked out together on site.

A second example has been the use of presentation style for better communication in deciding on the importance of geological structure on foundation stability. An existing dam, completed in 1969, recently required upgrading to meet increased spillway flow designs. A key issue has been the recognition of significant shear zones within the interlayered schist and gneiss foundations. These features are persistent, commonly clay-filled and have low shear strengths. Intersections of such structures appear to form wedges with potentially unfavourable orientations. As the engineering geologist assigned to this project it has been necessary to demonstrate to the engineers how the planned spillway extension works relate to the geology. It was found that a block diagram was more helpful in providing this three dimensional model than using plans and sections. Whereas it could be shown by stereonet analysis that the intersection line of Seams A and C does not potentially form an unstable wedge the diagram in Fig. 38.2 can be readily visualized by a non-geologist to see how various planes intersect a rock mass.

For the third example an existing concrete arch dam has required a better understanding of shallow dipping structures in bedded quartzite on the right abutment to withstand high spillway discharges (Fig. 38.3).

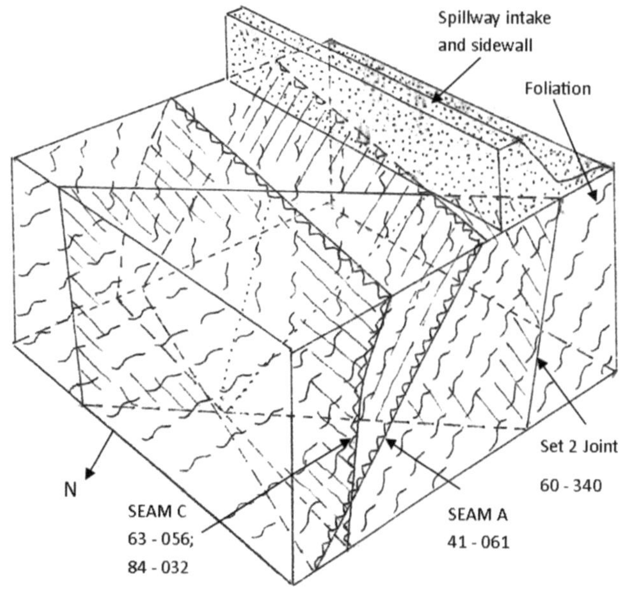

Fig. 38.2 Block diagram of geological structures in relation to spillway wall

Fig. 38.3 Right abutment of arch dam with variably bedded quartzite mass that required stability assessment

The bedding in the quartzite is marked by silty to clayey beds, with some breccia beds of sedimentary origin. The issue for this project was whether to reinforce this rock mass or not. The stability of the rock mass was dependent on the shear strength of the weaker bedding surfaces. Site measurements revealed high variability in both dip and dip direction, plus a waviness factor of individual beds to consider. Despite such variation the design engineer wanted a set of parameters in order to design the reinforcement. The engineering geologist was able to work with the engineer to use the natural variability and select an appropriate range of shear strengths. Either by engineering judgment or by more rigorous statistical and probability analyses combined with risk assessment the decision for the appropriate design parameters can be made. This is a two-way process requiring cooperation between the geologist and the engineer.

In addition to these examples the geologist and other project team members must often work within strict terms of reference for such clients as international aid organizations, government agencies, military organizations and large private companies. The team must conduct evaluations in terms of the nature of the client, the site conditions, logistical requirements, communications issues, expected outputs and actual outcomes.

38.6 Conclusions

In engineering projects the union of and collaboration between geologist and engineer has been well established for some time. The inherent differences in their respective approaches to project issues can be used to advantage to reach conclusions and facilitate decisions. The geologist must adapt by understanding the engineer, be innovative in ways of communicating and look to the solution instead of at the problem.

References

Knill JI, Jones KS (1965) The recording and interpretation of geological conditions in the Foundations of Roseires, Kariba and Latiyan Dams. Geotechnique 15(1):94–124

Legget RF (1962) Geology and engineering, 2nd edn. McGraw-Hill, New York

Moye DG (1955) Engineering geology for the Snowy Mountains Scheme. J Inst Eng Aust 27:287–298

Paton J (1956) The Glen Shira hydro-electric project. Proc. Inst. Civil. Eng. (London) 1(5):593–632

A Case Study of Intelligent Compaction Used in Road Upgrades

39

Robert Leyland

Abstract

Compaction is one of the more important processes in roadway construction. It is needed to achieve high quality and uniformity of materials and ensure design performance. Current procedures using conventional compaction machines and limited Quality Control/Quality Assurance procedures may result in inadequate and/or non-uniform material densities, which can be one of the major factors that result in premature pavement failure. One of the most recent advances in related technology is that of intelligent compaction (IC) systems. As a whole such systems are said to provide numerous advantages including increased productivity, proactive compaction process adjustment, reduced spatial variations in compaction and greater data coverage compared to traditional testing methods. In a field study numerous different IC systems where used during construction of a number of experimental road sections in South Africa. These sections formed part of a road upgrade project typical of most road construction projects currently being performed in South Africa and the results obtained highlight the advantages and disadvantages of using the systems on such sites. Advantages included those commonly listed in IC literature whereas the disadvantages were related to the depth of measurement and manmade (brownfield) geological complexities. From this project potential advances in IC systems as well as additional uses of IC as a shallow subsurface characterization tool have been proposed.

Keywords

Intelligent compaction • South Africa • Pavements • Recycling

39.1 Introduction

Compaction is one of the more important processes in roadway construction, needed to achieve the required shear strength and reduced permeability and in turn ensure design performance. Current procedures (in South Africa) using conventional compaction machines and a predefined number of passes (determined by trial section or experience) are believed to result in inadequate and/or non-uniform material densities, which can be one of the major factors that result in premature pavement failure. Limited Quality Control/Quality Assurance (QC/QA) procedures test, at a limited number of random points, the modulus or density of the newly compacted material and the results are assumed to represent the distribution of compaction achieved across the entire area. Results are also generally only obtained sometime after compaction is complete (typically after the compaction window of stabilized layers), potentially resulting in expensive equipment remobilization.

Intelligent Compaction (IC) can be defined as the compaction of road materials using rollers equipped with in situ measurement, feedback control and global position systems

R. Leyland (✉)
CSIR Built Environment, PO Box 395,
Pretoria 0001, South Africa
e-mail: rleyland@csir.co.za

G. Lollino et al. (eds.), *Engineering Geology for Society and Territory – Volume 7*,
DOI: 10.1007/978-3-319-09303-1_39, © Springer International Publishing Switzerland 2014

that in combination allow immediate measurements of the material's response to the compaction process and therefore real-time adjustments to the process. IC can therefore help to overcome some of the potential inadequacies of conventional methods by identifying "bad" spots almost immediately, preventing wasteful use of construction equipment during compacting operations and improve QC/QA operations.

Formal specifications for the use of IC technology exist in Europe (especially Germany) and were more recently introduced to the United States (NCHRP 2010). In South Africa IC systems have been available for some time but the routine use thereof is scarce and no IC specifications have been introduced. An experimental study to investigate the feasibility of using IC systems available in South Africa was performed and this paper summarizes the process followed and the results obtained.

39.2 Intelligent Compaction Research

A relatively new area of research, intelligent compaction publications only appeared in about 1980 (e.g. Forssblad (1980) in Briaud and Seo (2003, p. 22) but since then there has been a considerable amount of work published on the subject. Terminology used for "Intelligent compaction" systems varies between system producers and even between different user groups. There are, however, two main kinds of systems, *relative* compaction testing systems that measure an index value that can be compared to the reading taken at the same point on previous passes and *absolute* compaction testing systems which measure the materials modulus (Briaud and Seo 2003). Some definitions of intelligent compaction systems do not include relative systems, some specify automatic operation adjustment systems and some also include GPS based recording and on board reporting systems (E.g. Briaud and Seo 2003; NCHRP 2010; FHWA 2011). In this paper "intelligent compaction" (or "IC") refers to any system that measures an index or absolute value at the time of operation and records the spatial location of each measurement.

Although a modulus value is more useful in pavement design, most traditional specifications and QA/QC measures are based on soil density. Soil modulus is a true factor but is dependent on many factors and the reported modulus is associated with a certain stress level, strain level, rate of loading and water content (Briaud and Seo 2003). There is no evidence for a direct correlation between soil modulus and density (Briaud and Seo 2003) and therefore a conversion of specifications to a modulus value would therefore be problematic.

Numerous studies have attempted to correlate roller measurement values (MVs) and traditional measurements (for example Floss et al. (1991), Rahman et al. (2008), Petersen (2005), White et al. (2008), FHWA (2011) and NCHRP (2010). These have shown that many variables effect the MV's including, stress levels, drum behaviour and moisture contents. Additionally the measurement depth is strongly dependent on the static roller mass (Floss et al. 1991 in NCHRP 2010, p. 20) and the soil type, amplitude and frequency of operation (Brandl and Adam 2000 in NCHRP 2010, p. 20). One important finding was that "roller based stiffness values are insensitive to "thin" layers of stiff soil over softer material and that "QA of thin base layers directly atop softer subgrade might be unreliable as roller-measured stiffness can be insensitive to these layers" (NCHRP 2010, p. 65). The NCHRP (2010) study also found that after MV values approach the maximum values they can reveal what is interpreted to be repeated compaction and de-compaction (i.e. alternating high and low values) as the surface structure is broken and compacted by alternating passes.

39.3 Experiment in South Africa

During the experiment the use of a dedicated test site was not encouraged as such a site would not be representative of current common South African road building activities which are mostly rehabilitation and upgrading projects rather than "greenfield" road construction projects. An actual road upgrade project, being performed under typical conditions and under contract by an established contractor, was therefore used. Three different intelligent compaction systems were utilized during the experiment and these included systems that measured index values and systems that measured actual modulus values. Additionally the systems included one that only stored the information for later analysis (i.e. no on-the-fly calculations performed), one that utilized a real time display of the difference between current and previous readings and one that automatically adjusted the machine operation according to the changing index values.

Systems were supplied by the manufacturers or local agents. Each system was then used on multiple different, but similar, pavement sections over a total length of 4.5 km. Roller operators were instructed to compact the material to the best possible condition based on the feedback obtained from the respective IC systems. The base layer consisted of both crushed stone layers and full depth recycled layers (either cement or bitumen stabilized).

Data collection on each section consisted of the full IC record obtained from the operators and layer density and modulus readings (using a light weight deflectometer or LWD) at a minimum of 8 randomly distributed points after compaction was completed. Deflection data was also obtained in the outer wheel track, at a 5 m interval, with the use of a falling weight deflectometer (FWD). The deflection

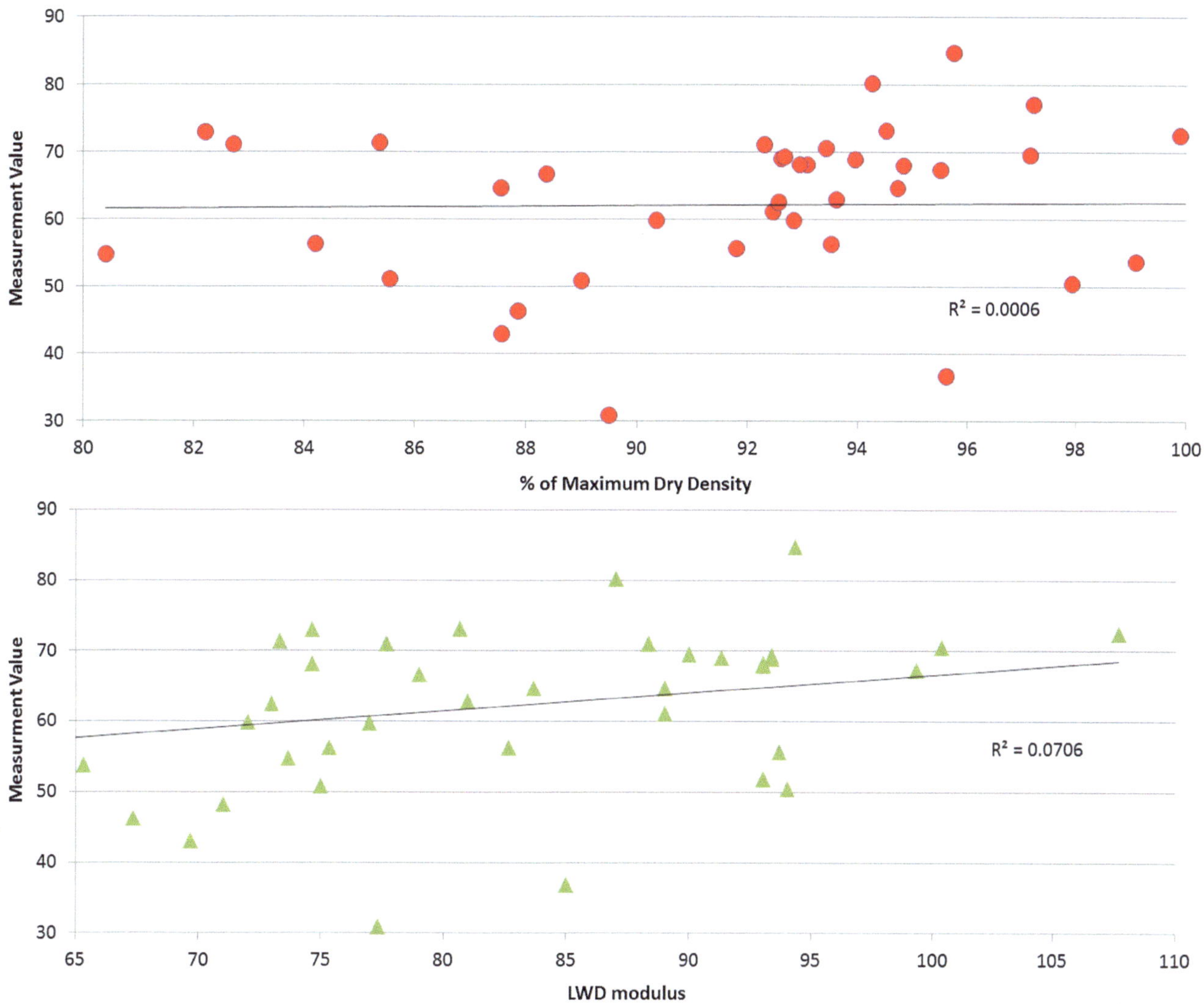

Fig. 39.1 Example of poor correlation between density and MVs (*top*) and modulus and MV's (*bottom*)

data was then used to back calculate moduli for the base, subbase and a semi-infinite subgrade. Initial, field analysis of data, performed directly after data was obtained, included a correlation study between the MVs and other variables. Based on the results dynamic cone penetrometer (DCP) testing was performed at selected testing points. When all sections had been constructed an analysis of the entire database for each IC system was performed.

39.4 Results

39.4.1 Comprehensive Database

An initial correlation study between all density and LWD measurements obtained was performed to investigate if LWD modulus could potentially replace density measurements in the current specifications. The results showed no recognisable relationship between the variables and are therefore in agreement with the statement by Briaud and Seo (2003).

Correlations between each system's MVs and corresponding density (absolute or as a percentage of maximum dry density) measurements produced a best (linear) correlation coefficient of only 0.39 (Fig. 39.1). Similarly poor relationships were obtained when the MV and the LWD modulus were correlated (Fig. 39.1).

When the layer moduli obtained from the back analysis of FWD data were considered attempts were made to correlate each of the layer moduli to not only the final MV recorded but also with the second to last MV and maximum MV recorded. Again no significant relationships were found to exist even when all data that may have been recorded under suspect drum behaviour was omitted. An example of the typical spread of data (for a base modulus correlation) is shown in Fig. 39.2.

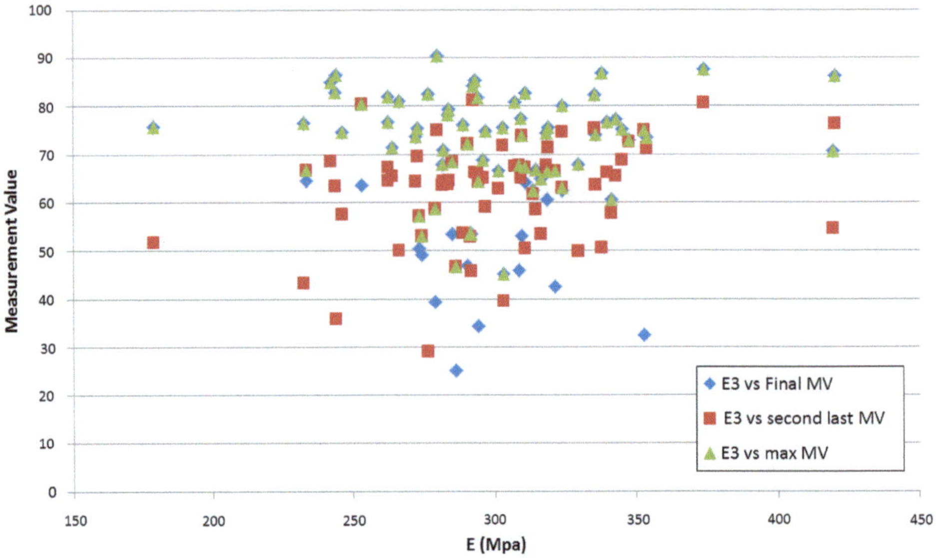

Fig. 39.2 Example of correlation plots between layer moduli and MVs from CAT system

39.4.2 Individual Sections

Similar poor correlations were generally found when each individual section's data was considered. When considering the development history of MVs at the measurement points, repeated compaction and de-compaction as described by NCHRP (2010) was observed on all sections, but not at all measurement points. However, as will be discussed, not all such observations were truly due to compaction/de-compaction. An interesting feature noticed on many of the sections was the difference in MVs between the outer and inner half width. This was observed on sections constructed using all three systems (Fig. 39.3) and the inner half width always had higher MVs than the outer half width. The LWD modulus and density values did not follow the same trends and as such additional DCP testing was performed in both half widths.

The DCP results revealed that the strength of the support layers in the inner half width was generally higher. This was due to the newly constructed lane being wider than the original

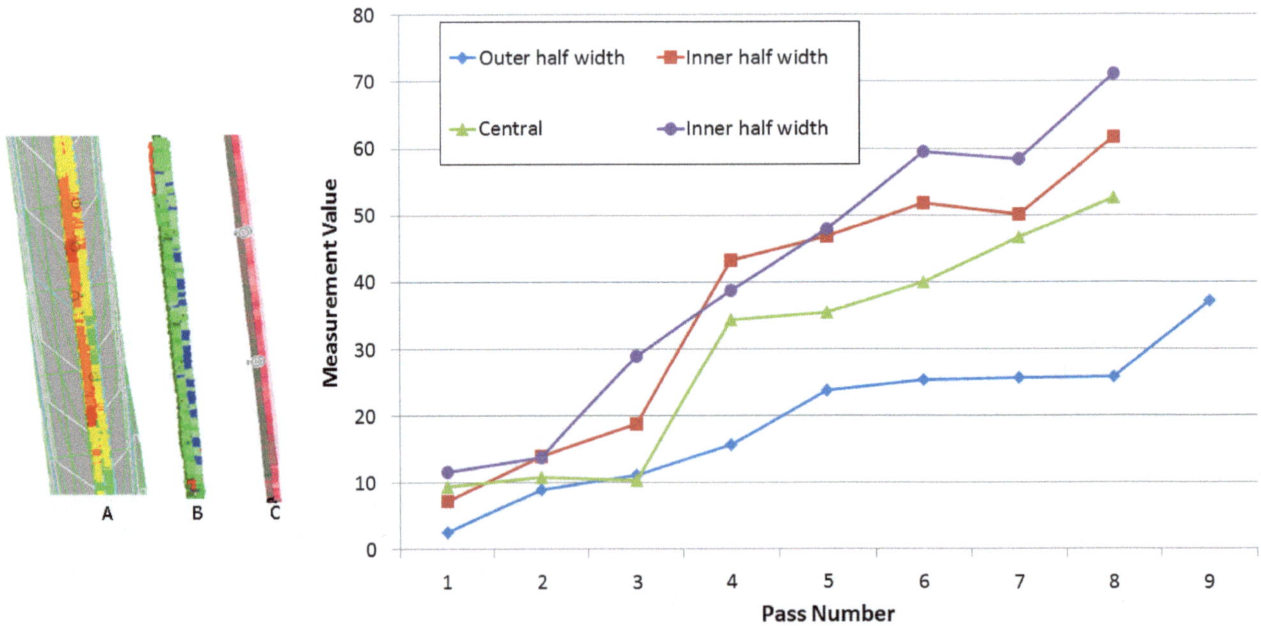

Fig. 39.3 Example of the consistent variation in MVs (*noticeably different colors*) measured on the different half widths of sections and typical MV developments of points in different parts of the sections

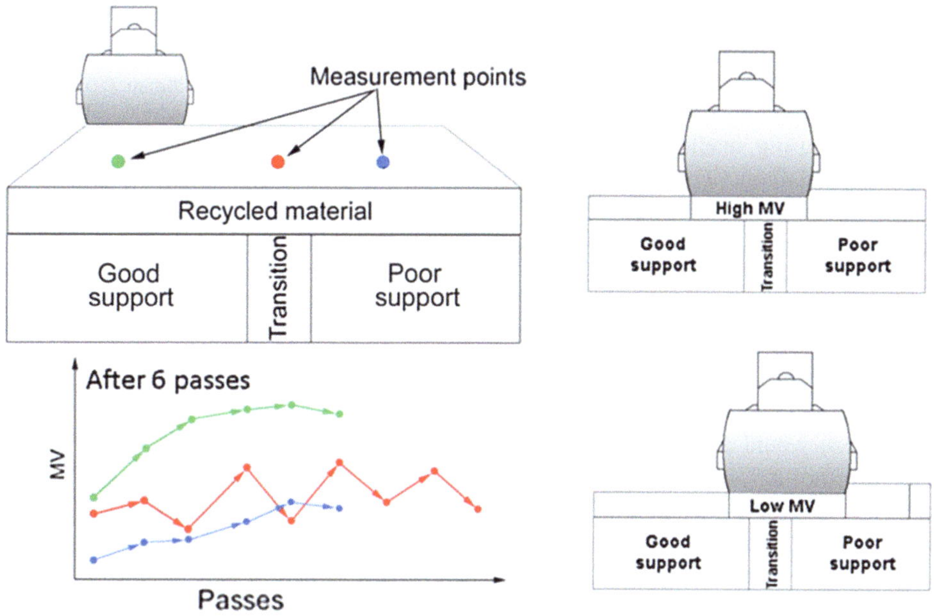

Fig. 39.4 Hypothetical variable support layers and the effect of these on MVs at different points

lane and as such the outer half was generally not underlain by the stronger original pavement layers. The development of MVs at points in the inner half is consistently higher than those at points with intermediate support strength (Fig. 39.3) and the very outer point where support layer strengths were the weakest. Inspection of many such differences in MV's did however reveal that in some cases the differences were not due to support layer differences but rather due to the (apparent) compaction/de-compaction curves resulting in some parts always having final MVs recorded on a de-compaction pass and thus having a lower MV.

39.5 Discussion

The influence of variable support layer strength on the MVs of IC systems has, due to the depth of influence of the systems, been shown to be significant. In addition to causing the obvious differences in MVs between the inner and outer half widths, the variable support layer strengths have also resulted in apparent compaction/de-compaction trends in MVs at many points studied. Such apparent trends occur because the roller passes are almost never located on exactly the same line as the previous passes and they often overlap. A point underlain by good support and located such that the roller will only pass over that point when fully underlain by good support (for example the left hand point in Fig. 39.4) will be attributed only high MVs.

Similarly, low MVs will always be attributed to points in large poor support areas. A point located in the transition zone between good and poor support areas (for example the middle point in Fig. 39.4), or a point located close to the edge of either type of support, will potentially be crossed by the roller both when the majority of the drum is on good and when the majority of the drum is on the poor support areas. Depending on the situation such a point may be attributed two very different MVs. When roller passes overlap such a transition zone is likely to receive alternating high and low MVs as the roller alternates between the inner and outer passes. This explains why some points with very different MVs were seen to have similar support layer strengths. Additionally the central point also records more passes compared to the other points (Fig. 39.4). Such limitation of IC systems in areas of varying support strength are recognized in some existing specifications which limit the use of IC data to subgrades and embankment layers or areas of homogeneous underlying conditions or which require subgrade readings to be taken before construction and used as correction factors.

39.6 Conclusions and Recommendations

The use of IC systems on road construction projects involving the upgrading of existing roads (a common situation in South Africa that is likely to increase in prevalence and usually incorporates some widening of the existing road) was shown to not be feasible due to the extreme differences common in subsurface conditions both laterally and longitudinally. These variations do not only result in the parallel zones of different readings but can also result in fluctuating readings at some points and therefore a misleading representation of compaction. This disadvantage may however be an advantage if used to identify isolated areas of poor support strength that, using traditional centreline surveys, would not be noted.

The results obtained during this study confirm the conclusions by previous studies and highlight the practical limitations of using intelligent compaction technology as a routine quality control/quality acceptance tool. The primary problems associated with such technology are the lack of sensitivity when used on thin and or significantly stiffer layers and the inconsistent relationships between established soil properties and intelligent compaction measurement values.

Although IC data may be useful as a QC/QA tool in greenfield projects, it is considered that the underlying pavement variability in typical South African road upgrading projects would result in inconsistent and unreliable data. The record of pass counts and locations may still be useful but such data can be obtained with other simpler systems.

Additional research and development in the field of intelligent compaction will in all likelihood address the limitations of the systems. Some research and development has been performed on an alternative intelligent compaction measurement known as "Machine Driver Power" which is based on the power required to move the roller forward and not the changing reaction of the drum to vibration impacts. Such technology could provide better estimates of the soil compaction state but were not available in South Africa at the time of this study.

Acknowledgements The author would like to thank the South African National Roads Agency LTD for allowing this paper to be written and for the opportunity to work on the experimental sections.

References

Brandl H, Adam D (2000) Flachendeckende Dynamische Verdichtungskontrolle (FDVK) mit Vibrationswalzen—Grundlagenforschung und praktische Anwendung (Continuous compaction control with vibratory rollers—basic research and practical application). Schriftenreihe der Strasenforschung Heft 506, Forschungsvorhaben Nr. 3.147, Bundesministerium fur Wirtschaftliche Angelegenheiten, Wien

Briaud JL, Seo J (2003) Intelligent compaction: overview and research needs, Texas A&M University, College Station

FHWA (2011) Accelerated implementation of intelligent compaction technology for embankment subgrade soils, aggregate base, and asphalt pavement materials—final report. Federal Highway Administration, Washington, DC. Report No.FHWA-IF-12-002

Floss R, Brau G, Gahbauer M, Gruber N, Obermayer J (1991) Dynamische Verdichtungsprufung bei Erd-und Strasenbauten (Dynamic compaction testing in earth and road construction). Prufamt fur Grundbau, Boden-und Felsmechanik Technische Universitat Munchen, Heft 612, Munchen

Forssblad L (1980) Dynamische Verdichtungsprüfung bei Erd- und Straβenbauten. Bundesminister für Verkehr, Forschung und Straβenverkehrstechnik, Heft 612, p 284

NCHRP (2010) Report 676: intelligent soil compaction systems. Transportation Research Board, Washington, D.C.

Petersen L (2005) Continuous compaction control MnROAD demonstration. Final report submitted to Mn/DOT, Report No. MN/RC-2005-07

Rahman F, Hossain M, Hunt M, Romanoschi S (2008) Soil stiffness evaluation for compaction control of cohesionless embankments. Geotech Test J 31(5):1–10

White D, Thompson M, Vennapusa P (2008) Field validation of intelligent compaction monitoring technology for unbound material. Minnesota Department of Transportation, St. Paul, pp 123–164

Unified Qualification Requirements for Ground Engineering and Engineering Geology Professionals

40

Helmut Bock, Markus Herten, Reinhard Schwerter and Kurosch Thuro

Abstract

Eurocode 7 (EN 1997) is the fundamental geotechnical standard in Europe. It states that the design and construction of geotechnical structures requires *"appropriately qualified and experienced personnel"*. Neither EN 1997 nor the various national annexes and/or Building Codes specify precisely what these "appropriate" qualification and experience requirements might be. For our profession, this is a serious situation as it opens the door for all sorts of self-declared "experts" to carry out ground investigations and to submit Ground Investigation and Geotechnical Design Reports on a low-bid basis, all too often to the detriment of safety. Against this background, the German Geotechnical Society DGGT has issued a document entitled *"Geotechnical Professionals—Requirements on qualification and experience"*. The document was developed in consensus between geotechnical engineers and engineering geologists. The requirements incorporate the following three general areas: (1) appropriate studies at tertiary level; (2) professional experience in ground engineering, and (3) continuous professional development. The DGGT aims to register qualified geotechnical professionals with the Chamber of Engineers (*Ingenieurkammer*). This appears to be the most attractive approach to registration as it covers the geo-engineering profession at large, including engineering geology.

Keywords

Consultants • Registration • Geotechnical engineer • Engineering geologist

40.1 Introduction

Eurocode 7 (EN 1997) is the fundamental geotechnical standard which, since 2010, is mandatory in all EU- and EFTA member states. It states that geotechnical designs and supervision of geotechnical constructions require *"appropriately qualified and experienced personnel"*(EN 1997-1: 2004, Sect. 1.3(2)). Neither the EN 1997 nor (to the best of our knowledge) the various national annexes and/or Building Codes specify precisely what exactly these "appropriate" qualification and experience requirements might be. For our profession, this is quite a serious situation as it opens the door for all sorts of self-declared "experts" to carry out ground investigations and submit Ground Investigation and/or Geotechnical Design Reports on a low-bid basis,

H. Bock (✉)
Q+S Consult, Bad Bentheim, Germany
e-mail: qs-consult@t-online.de

M. Herten
Federal Waterways Engineering and Research Institute, Karlsruhe, Germany

R. Schwerter
University of Applied Sciences, Görlitz-Zittau, Germany

K. Thuro
Technical University Munich, Munich, Germany

G. Lollino et al. (eds.), *Engineering Geology for Society and Territory – Volume 7*,
DOI: 10.1007/978-3-319-09303-1_40, © Springer International Publishing Switzerland 2014

Table 40.1 Qualification requirements for tertiary studies

Study stream study extension (SE)	Number of semesters (total)[c]	Academic degree	ECTS-credit points[a]			
			Study total	Foundational[b]	Geotechnics[b]	
					Core	Suppl.
Civil engineering	6 to 8	Bachelor	180 to 240	60	15	25
SE—Geotechnics[e]	8	Dipl.-Ing. (FH)[d]	240			
	9 or 10	Dipl.-Ing.	270 or 300			
	10	Master	300			
Geosciences	6 to 8	Bachelor	180 to 240			
SE Eng.-Geol.	10	Dipl.-Geol.	300			
Master in Eng.-Geol.	10	Master	300			

[a] ECTS = European credit transfer and accumulation system
[b] See Table 40.2 for details
[c] Minimum number; semester = section of half-year studies
[d] Traditional German degree in engineering from a University College
[e] Geotechnics may also be a study stream in its own right

all too often to the detriment of the safety of geo-engineered structures, or their cost effectiveness.

This unsatisfying, and potentially dangerous, situation has been similarly identified in various European countries. For instance, in the UK there is a Register of Ground Engineering Professionals which is jointly sponsored by the Institution of Civil Engineers, the Geological Society of London and the Institute of Materials, Minerals and Mining (ICE 2011). That Register aims at providing external stakeholders, including clients and other professionals, with a means of identifying individuals who are suitably qualified and competent in ground engineering—be they from consultancy, contractors, public bodies or academia.

A similar development has taken place in Germany. The German Geotechnical Society DGGT, the national umbrella organisation for both ground engineers and engineering geologists, has issued a document entitled "*Geotechnical Professionals—Requirements on qualification and experience*" (DGGT 2013). In the following, an outline is given of the relevant DGGT requirements (Sect. 40.2). Registration of qualified geotechnical professionals, as pursued by the DGGT, is considered in Sect. 40.3. Based on a comparison between the two approaches undertaken in Germany and the U.K., it is concluded that a unified European scheme for registration of geotechnical professionals is highly desirable (Sect. 40.4).

40.2 Requirements for Geotechnical Professionals in Germany

According to the DGGT document, qualifications are required in all of the following three general areas:

1. Successful completion of appropriate studies at tertiary level (see Sect. 40.2.1),
2. professional experience in ground engineering (see Sect. 40.2.2), and
3. continuous professional development (see Sect. 40.2.3).

40.2.1 Studies at Tertiary Level

The successful completion of appropriate studies at tertiary level shall be evidenced by a bachelor or master degree (or by the traditional German diploma degrees Dipl.-Ing. or Dipl.-Geol.) (see Table 40.1). Foundational and geotechnical core subjects are identified and quantified in terms of ECTS (European Credit Transfer and Accumulation system) credit points. Core subjects include mathematics, applied mechanics, IT and/or geo-information systems, soil mechanics, foundation engineering and engineering geology (see Table 40.2).

It should be noted that, in accordance with Table 40.2, individuals can satisfy the tertiary study requirements by a degree in one of the following two streams: either by a degree in civil engineering or by a degree in geosciences. In order to keep the geo-engineering profession principally open to any qualified person, there is a provision for individuals with a degree in other tertiary study streams. In such cases evidence must be lodged that the foundational and geotechnical core topics, as specified in Table 40.2, are appropriately covered.

40.2.2 Professional Experience

The minimum number of years required for geo-engineering project experience varies between 2 and 7 years (see Table 40.3), depending on the academic degree (Bachelor vs.

Table 40.2 Curriculum and required ECTS credit points as one of the pre-conditions for geotechnical professional recognition

Study category		Topics		Required ECTS		
		Compulsory	Electives	Compulsory	Electives	Total
Foundational		Mathematics Applied mechanics IT /CAD /GIS	Physics Chemistry Analytical geometry Fluid mechanics	20	10	30
Study stream	Civil engineering	Design Structural engineering Bulk structures Construction methods	Construction materials Steel /timber structures Hydraulic engineering Traffic route construction	15	15	30
	Geosciences	Geology Mineralogy/petrology Structural geology	Hydrogeology Regional/historic geology Quaternary geology Natural hazards			
Geotechnics core topics		Soil mechanics Foundation engineering Engineering geology	Rock mechanics Rock engineering Tunnelling Constitutive laws numerical modelling	10	5	15
Geotechnics supplementary topics		Practical training in geotechnics Thesis in geotechnics or engineering geology repositories Environmental geotechnics, geothermal topic Technical petrology Geophysics, soil dynamics		25		25

Table 40.3 Requirements on geo-engineering project experience

Academic degree	Project experience (Years)	Geotechnical category (EN 1997-1)
Master (M.Sc., M.Eng.) Dipl.-Ing. Dipl.-Geol.	2	1 and 2
	5	3
Dipl.-Ing. (FH) (see Legend Table 40.1)	3	1 and 2
	5	3
Bachelor	4	1 and 2
	7	3

Master) and on the "Geotechnical Category" of the associated project as specified in EN 1997-1: 2004, Sect. 2.1.

More specifically, project experience in the form of on-site work in line with methodological competence is required as follows:

(a) **Project experience** (evidence to be lodged for at least three of these fields):

- Foundation of structures
- Excavations in soil or rock
- Ground improvement
- Rock construction
- Underground construction
- Traffic and conveyance lines
- Waterways
- Earth or rock fill construction
- Deposits and refuse dumps

(b) **Methodological competence** (evidence to be lodged for at least three of these fields):

- Delineation and quality control of soil and rock parameters on the basis of laboratory and field tests
- Setting up of ground models and evaluation and consideration of geologic and geotechnical risks
- Proof of ultimate and serviceability limit states of geotechnical structures
- Numerical modelling in geo-engineering
- Monitoring and documentation of geotechnical construction procedures
- Observational design method, including geotechnical monitoring by field instrumentation and interpretation of monitored data
- Analysis and rehabilitation of ground failures

40.2.3 Continuous Professional Development

Geotechnical professionals have to lodge evidence on their continuing professional development (CPD) following the completion of tertiary studies. The minimum requirement is 24 h of further education on geotechnically related topics,

Table 40.4 Comparison of the relevant schemes for qualified geotechnical professionals developed in the U.K. and Germany

	U.K.	Germany
Issued by	BGA—British Geotechnical Association, sponsored by: • The Geological Society • IoM^3-The Institute of Materials, Minerals and Mining • ICE, Institution of Civil Engineers	DGGT—German Geotechnical Society, The national umbrella organisation for geotechnical engineers and engineering geologists
Targeted to individuals	In consulting, contracting, public bodies or academia	In consulting
Grades and level of competency	"Professional" = *Ability* to ... "Specialist" = *Management* of ... "Advisor" = *Responsibility* for ...	"Geotechnical Professionals", in their capacity to take responsibility for Ground Investigation and Geotechnical Design Reports in the sense of EN (1997)
Requirements for recognition	1. Sound knowledge and understanding of scientific/engineering /technical principles 2. Experience of ground engineering 3. Continuous professional development (CPD)	1. Degree in civil engineering or in geosciences together with quantified ECTS scores in foundational and geotechnical subjects 2. Experience of geo-engineering, quantified in years of experience 3. CPD; quantified
Required attributes with regard to:	• Innovation • Technical solution • Integration • Risk management • Sustainability • Management	Not specified
Sponsor	Required	Not required
Registration	Existing since 2011 with the ICE (under the auspices of civil law)	Intended: chamber of engineers (under the auspices of public law)

accumulated over a time span of 3 years. CPD can be achieved by participation in courses, seminars, symposia and congresses that satisfy DGGT criteria.

40.3 Registration Pursued by the DGGT

At the time of writing, recognition and registration of geotechnical professionals, as outlined in the previous Sect. 40.2, remain pending. For the time being, individuals can use the DGGT requirements to evaluate their own competencies and for promoting themselves as geotechnical professionals to clients and, if necessary, to defend themselves before the courts. In the long term, the DGGT aims to register qualified professionals with the Chamber of Engineers (*Ingenieurkammer*) under Public Law. Due to the federal structure of the Chamber of Engineers in Germany, talks are cumbersome and results cannot be expected by year end 2014.

From an engineering geological viewpoint, registration with the Chamber of Engineers appears to be the most attractive approach towards much needed professional recognition. Alternative certification and registration schemes or titles such as the EurGeol title, promoted by the European Federation of Geologists (EFG) under Civil Law, appear to us to be inferior. These are confined to geo-science and are thus unable to cover the geo-engineering profession at large.

40.4 Comparison with the UK Scheme and Outlook

When comparing the DGGT scheme with that of the U.K. (see Table 40.4), it becomes apparent that there are significant similarities as well as differences.

Most significantly, within both countries there is a joint effort between geotechnical engineers and engineering geologists in developing their respective national schemes. Agreement also exists with regard to principal entry requirements. In detail, however, the German entry requirements are significantly more specific and quantified than the U.K. ones, with the consequence that, in Germany, no sponsorship is required for entry. Beyond that, it is thought that the German entry requirements are a fair and practicable compromise between the extraordinarily detailed requirements as formulated by Turner and Rengers (2010) and those of the U.K. scheme. The U.K. scheme addresses the entire spectrum of geotechnical professionals irrespective of their function or affiliation, whilst the

German scheme focuses on geotechnical consultants in their capacity to take responsibility for Ground Investigation and Geotechnical Design Reports as per EN 1997.

A further difference exists with regard to registration. In the U.K. registration of geotechnical professionals has been established since 2011. It is carried out by the ICE under the auspices of Private Law. In Germany, registration of geotechnical professionals as per the DGGT document has not as yet been established. Efforts are underway to introduce such registration by the States Chamber of Engineers under the auspices of Public Law.

In reflecting on the similarities and differences of the above schemes, and also in considering the urgent need for improved quality in geo-engineering, it is concluded that a unified European scheme for registration of geotechnical professionals is highly desirable. The schemes developed in the U.K. and in Germany could serve as the base for such unified scheme.

References

DGGT (2013) EASV Sachverständige für Geotechnik: Anforderungen an Sachkunde und Erfahrung. - Geotechnik, 36:51–57, Berlin (Ernst & Sohn)

EN 1997 Eurocode 7. Geotechnical design, EN 1997-1. Part 1: General rules (2004), EN 1997-2. Part 2: Ground investigation and testing (2007)

ICE (2011) UK register of ground engineering professionals. ICE, London, p 12

Turner AK, Rengers N (2010) Adaptation of the ASCE body of knowledge—competency-based approach to the assessment of education and training needs in geo-Engineering. Report to JTC-3, p 37

Guidelines for Site Investigation and Analysis of Nuclear Facilities: A Consultants Perspective

41

William Godwin, Ramon Secanell and Christophe Martin

Abstract

Planning site investigations for siting, licensing, or constructing nuclear power plants require implementation of appropriate technical guidelines. Guideline documents have been developed in the USA and internationally to provide standards for how geologic, geotechnical, hydrologic and seismic data are collected, analyzed and presented in safety analysis reports. The authors discuss how guidelines established by both the U.S. Nuclear Regulatory Commission (USNRC) and the International Atomic Energy Agency (IAEA) are used in the regulatory framework of nuclear power plants in France, the United Arab Emirates, Japan and the United States. The importance of the choice of methodologies engineering geologists use to successfully perform nuclear safety-related site characterization studies and analysis is also discussed. Lastly, new guidelines and approaches developed to address seismic and flooding risks in response to the 2011 Fukishima Daiichi Power Plant incident in Japan are presented.

Keywords

Guidelines • USNRC • IAEA • Nuclear • Safety

41.1 Introduction

Over the last few years, many new nuclear plant sites have been characterized using regulatory established procedures. The most commonly used procedures are those defined in the USA by the U.S. Nuclear Regulatory Commission (USNRC), described in Regulatory Guides, and the methodology described by the International Atomic Energy Agency (IAEA) in their Safety Guides. The recommendations described by the IAEA have been accepted by almost 200 members of the IAEA. It implies that recommendations of the IAEA guides are accepted by USA regulators. However, sometimes, the procedures and methodologies imposed by NRC are stricter than the IAEA recommendations (and they are not accepted by all IAEA members).

Some countries have adopted all or part of the USA regulations. Other countries and mainly the "newcomers" (countries desiring to become nuclear operator) adopt some IAEA requirements and safety guides as national regulatory documents. France and Japan, with established nuclear regulatory bodies, have their own regulations. A recent shift has been to adopt (and to respect) the general procedures recommended by the IAEA.

In this article, we describe the methodologies and procedures generally followed by USA, France and other countries (mainly the "newcomers") for seismic site characterization.

W. Godwin (✉)
Fugro Consultants, Inc., 1777 Botelho Drive, Suite 262, Walnut Creek, CA 94596 USA
e-mail: w.godwin@fugro.com

R. Secanell · C. Martin
GEOTER-SAS, Fugro Group, 3 Rue Jean Monnet, 34830 Clapiers, France

G. Lollino et al. (eds.), *Engineering Geology for Society and Territory – Volume 7*,
DOI: 10.1007/978-3-319-09303-1_41, © Springer International Publishing Switzerland 2014

41.2 Regulatory Framework

41.2.1 USA, France and Other Experienced Nuclear Countries

The USNRC has prepared a series of guidelines used extensively by consultants and utilities to collect and analyze of geologic, geotechnical and seismological data (Table 41.1). The most recently used guideline for seismic hazard studies is RG.1.208. The Senior Seismic Hazard Analysis Committee (SSHAC) methodology developed in USA in the last decade is the most accepted methodology to reduce uncertainty in seismic hazard analysis (USNRC 1997). The probabilistic approach is the preferred one used for seismic design purposes for describing response spectra and associated time histories.

In France, the reference law for seismic hazard studies is the RFS 2001-01. It provides the details describing the methodology specifically for seismic hazard analysis. The RFS 2001-01 is based on the deterministic approach, the most commonly used methodology in the seventies and eighties. The rule is based on a definition of the characteristics of "Maximum Historically Probable Earthquakes" considered to be the most penalizing earthquakes liable to occur over a period comparable to the historical period, or about 1,000 years. Secondly, it defines the "Safe Shutdown Earthquakes". In the last few years the probabilistic approach has been used and accepted for the reevaluations of the seismic hazard of existing sites.

Japan is another country with an established regulatory body, the Nuclear Regulation Authority, (NRA) that produces its own regulations. Following the Kashiwazaki-Kariwa event in 2007 and Tohoku earthquake in 2011, Japan has allowed some recommendations given by IAEA.

Table 41.1 Key USNRC guidelines for seismic and geologic hazards

Regulatory guides[a, b]
NUREG/CR 6372 Uncertainty and use of experts ("SSHAC" 1997)
NUREG/CR 5503 Techniques identifying faults and origins (1999)
NUREG/CR 5562 Dating and earthquakes, geochronology (1998)
RG.1.132 Site investigations for foundations (1979)
RG.1.138 Laboratory investigations of soils and rocks (2003)
RG.1.198 Procedures and criteria for assessing seismic soil liquefaction (2003)
RG.1.208 Performance-based approach earthquake motions (2006)

Notes
[a] NUREG references URL: http://www.nrc.gov/reading-rm/doc-collections/nuregs/
[b] RG references URL: http://www.nrc.gov/reading-rm/doc-collections/reg-guides/power-reactors/rg/

41.2.2 IAEA Guidelines

Originally, the IAEA published documents that were used by countries as a basis for national regulations and seismic studies. The NS-R-3 (IAEA 2003) contains global requirements for site evaluation. They are intended to ensure adequate protection of site personnel, the public and the environment from the effects of ionizing radiation arising from nuclear installations.

The Safety Guides on site evaluation listed in Table 41.2 provide recommendations on how to meet the requirements established in the Safety Requirements publication NS-R-3. The Safety Guide associated to seismic hazard is the SSG-9. This Safety Guide allows the use of probabilistic and deterministic approaches. Nevertheless, the probabilistic approach is preferred and the deterministic case could be used as a deterministic control. The reference safety guide for geotechnical aspects is the NS-G-3.6 (the geotechnical conditions are needed for site response analysis).

Table 41.2 Main IAEA guidelines related to seismic and geologic hazards

IAEA Safety guides
NS-G-1.6 Seismic design & qualification (2003)[a]
NS-G-2.13 Seismic safety existing installations (2009)[b]
NS-G-3.6 Geotechnical site evaluation (2005)[c]
IAEA Specific safety guides
SSG-9 Seismic hazards in site evaluation for nuclear installations
IAEA Guides under development
DS433/DPP423 Safety aspects in siting

Notes
[a] http://www-pub.iaea.org/MTCD/publications/PDF/Pub1158_web.pdf
[b] http://www-pub.iaea.org/MTCD/publications/PDF/Pub1379_web.pdf
[c] http://www-pub.iaea.org/mtcd/publications/pdf/pub1195_web.pdf

In the siting process, the reference guide is the DS433 which now supersedes IAEA Safety Guide 50-SG-S9 Site Survey for Nuclear Power Plants. It will take account of the Safety Requirements on Site Evaluation for Nuclear Installations NS-R-3, especially, in relation to exclusion criteria for the site selection of nuclear power plants.

41.3 Contents and Methodologies Used to Perform a Site Evaluation Report

The procedures recommended by IAEA Safety Guides and the procedures imposed by NRC Regulatory Guides are similar in many aspects. However, in some aspects, the NRC methodology is more strict and detailed. In this chapter we will briefly describe the main steps that consultants follow to complete site characterization.

41.3.1 IAEA Procedure Described in SSG-9

Currently, many countries are in the process of selection and/or characterization of sites for a new nuclear power plants (Jordan, Turkey, Indonesia, Vietnam, Lithuania, Poland, etc.). In the majority of these countries, the basic procedures, recommendations and methodologies used to perform seismic hazard studies are those published by IAEA. Following the recommendations of the SSG-9 (IAEA 2010), a Probabilistic Seismic Hazard Assessment (PSHA) follows the following work elements:

- Compilation of a geological, geophysical, geotechnical and seismological database using 4 scales: regional, near regional, site vicinity and site area. The database should be introduced and structured in a GIS.
- Development of a Seismic Source model that includes the description and justification of the geometry of the seismogenic sources and its seismic characterization as well as a selection of a set of appropriate Ground Motion Prediction Equations (GMPEs)
- Seismic hazard calculation to obtain the seismic hazard curves (at free field conditions). Two seismic levels SL-1 and SL-2 should be defined by the Regulator (usually 475 and 10,000 year return periods). They are characterized by response spectra and appropriate time histories.
- The Uniform Hazard Response Spectra is derived from the seismic hazard curves. Depending on soil conditions, a specific site response analysis may be required.

The site characterization is performed at four different scales:

(1) Regional scale: takes into account the geological information contained within a radius of 300 km. (2) Near regional: typically a radius of 25–40 km. In this region some new detailed data should be generated with emphasis to geohazards. (3) Site vicinity: evaluation of date contained in a radius of 5 km. At this scale, it is very important to assure the absence of capable faults in this region. (4) Site area: the typical radius is 1 km. It covers the area of the NPP. The generation of new data at this scale corresponds basically to the geotechnical information.

Typically, for these types of studies, the selection and approval of a new nuclear site has 2 phases:

- Site selection. In this phase only limited studies are performed (Probabilistic Seismic Hazard Analysis [PSHA], geohazards studies, etc.). They are based on existing data. The main references to define the seismic characteristics of the site are DS433 and SSG-9.
- Site characterization. This phase starts when the site is selected. New geological, geophysical and geotechnical data are generated in this phase. The new data are used to refine the preliminary studies performed in the previous phase. The main references to define the seismic characteristics of the site are SSG-9 and NS-G-3.6 for geotechnical aspects (site area scale).

41.3.2 NRC Procedure for Safety Analysis Report in USA

Site investigations, analysis and preparation of Safety Analysis Reports in the USA follows a process using US-NRC guidelines presented in Table 41.1.

41.3.2.1 Guidelines for Phases of New NPP

Choosing a site suitable for a new nuclear power plant in the US generally begins with siting study. The study follows a 4-step methodology as described in of the Electric Power Research Institute (EPRI) Siting Guide (2002).

The phase of determining the suitability of a site, even before a technology is chosen is termed by a screening/feasibility study. It typically will include a field and lab testing program, ground motion model and site response calculation following RG 1.208. This analysis is input towards preparation of Safety Analysis Report (SAR) to support an Early Site Permit (ESP).

Following a selection of a particular plant technology, a much more detailed site investigation is conducted to either update the ESP SAR or prepare a new SAR for a Combined Operating License Application (COLA).

Upon receipt of a construction license, the owner will proceed with construction, implementing additional field and laboratory analysis.

41.3.2.2 SSHAC Process for PSHA

Two basic principles underlie the SSHAC approach to PSHAs: (a) all the inputs should represent the composite distribution of the informed technical community (ITC) and (b) the analyst must establish ownership of these inputs.

The goal of the SSHAC process is to "represent the center, the body, and the range of technical interpretations that the larger informed technical community would have if they were to conduct the study" (USNRC 1997). The SSHAC process also identifies a clear definition of "ownership" of the input parameters into the PSHA, and hence ownership of the PSHA results.

Ownership means "intellectual responsibility" such that the regulator will know the individuals who are responsible for developing the PSHA. Four levels of effort (1–4) are defined for capturing the range of uncertainty by the ITC. With each increasing level, there is increasing direct involvement of the ITC and, thus, increasing confidence and documentation that the center, body, and range of uncertainty in the ITC have been captured.

41.4 Impact of the Tohoku Earthquake

New guidelines have been developed to address seismic and flooding risks in response to the 2011 Fukishima Daiichi Power Plant incident in Japan. In response Japanese regulators have their own documents (NRA 2013).

Samaddar (2013) reports on actions undertaken at IAEA's International Seismic Safety Centre (ISSC) to develop necessary technical guidance on issues relating to the safety assessment of a site hosting multiple units under the impact of single or multiple hazards as highlighted by the impact of the Great East Japan Earthquake and Tsunami.

The USNRC recently has taken many actions to ensure the continued safe operation of U.S. nuclear power plants (Miller 2013). These actions include: (1) the performance of inspection activities at all U.S. nuclear power plants to evaluate licensee implementation of procedures and equipment which could mitigate beyond design basis events; (2) the establishment of a Task Force which identified lessons learned which could be implemented to further enhance the safety of U.S. nuclear power plants; and (3) the commencement of a program to identify and take specific near-term and long-term regulatory actions related to these lessons learned.

In France, after the Fukushima event, and as per the L'Autorité de sûreté nucléaire (ASN) or the French Nuclear Safety Authority requirement, Electricité de France (EDF) carried out a complementary safety assessment of its 58 units (Petre-Lazar et al. 2013).

References

Electric Power Research Institute (2002) Siting guide: site selection and evaluation criteria for an early site permit application, EPRI, Palo Alto, CA, 1006878, pp 23–28

Petre-Lazar I, Jadot H, Turpin F, Guisard S, Chantron J, Labbé P (2013) French response to Fukushima event–seismic safety issues. In: Proceedings of the 8th symposium nuclear plants current issues: challenges & opportunities, III-4/1–4/8

International Atomic Energy Agency (2010) Seismic hazards in site evaluation for nuclear installations. Specific Safety Guide No. SSG-9. IAEA, Vienna, pp 26–28

International Atomic Energy Agency (2003) Site evaluation for nuclear installations. Safety Standards Series No. NS-R-3. IAEA, Vienna, 1–26

Miller BW (2013) United States nuclear regulatory commission actions following the Fukushima Dai-Ichi accident. In: Proceedings of the 8th symposium nuclear plants current issues: challenges & opportunities, III-1/1–1/11

Nuclear Regulation Authority (2013) Nuclear regulation authority: nra library, regulatory guide, outline of new regulatory requirements (earthquakes and tsunamis) http://www.nsr.go.jp/english/data/new_regulatory_requirements2.pdf, pp 1–31

Samaddar SK (2013) ISSC activities highlighted by Fukushima event. In: Proceedings of the 8th symposium nuclear plants current issues: challenges & opportunities, pp 111-3/1-3/8

U.S. Nuclear Regulatory Commission (1997) Recommendations for probabilistic seismic hazard analysis—guidance on uncertainty and use of experts: Washington, D.C. NUREG/CR-6372, p 171 with Appendices

Guidelines to Identify and Quantify Expansive Soils in Civil Infrastructure

42

Marcos Musso

Abstract

Expansive soils are geotechnical natural hazards such as landslides and earthquakes, with a widespread area distribution in the word. Different researchers have used simple tests as Atterberg limits and clay content to identify them, but these tests don't quantify the mechanical behavior which is a more complex process involving soil properties as dry unit weight, suction and clay mineralogy. Engineering geological practice would have a useful tool to prevent or minimize problems generated by cyclic volume change- by swelling and shrinking- if a systematic identification procedure is applied to quantify expansive soils. In Uruguay, expansive soils have been investigated during the last decade. Some expansive soils were developed in continental sediments in Libertad formation, composed mainly by loess and mud flow events due to Quaternary semi-arid climate conditions. The soils have variable clay contents but smectite is the dominant clay mineral. The soils developed from low to very high pressure expansion. The aim of this paper is to introduce an useful methodological approach to identify expansive soil and apply it in different sites including field observation and laboratory tests (clay content, cation exchange capacity with methylene blue) and to identify and quantify mechanical behavior as swelling pressure.

Keywords

Expansive soil • Clay • Guidelines • Methylene blue • Uruguay

42.1 Introduction

Expansive soils frequently cause severe damages (crack and fractures) in civil infrastructure by cyclic volume changes as their moisture modifies. The problem is very common in light structures when pressure expansion exceeds load foundation. This is a hidden problem which costs multi millions of dollars per year all over the world to repair the breaks.

It is important to identify expansive soils using indirect features before analyzing samples in laboratory. One kind of expansive soil is vertisol, which has characteristic feature landforms in aerial photographs (gilgai) whereas in the field work it is common to observe surface cracks.

Several researchers have developed charts or summary tables to identify and characterize expansive soil to predict heave. Clay particles content and Atterberg limits were used to distinguish between low to high expansive soil. Seed et al. (1962) used clay size fraction and Atterbergs limits together with swelling potential test to distinguish between low to high expansive soil. They created a useful diagram. Williams and Donalson (1980) developed similar research. Meisina (2000) and Musso (2001) used mineralogical identification with X ray diffraction (XRD), cation exchange capacity (CEC) associated with swelling pressure test. Yitagesu et al. (2009) applied remote sensing techniques associated to soil parameters with reflectance spectra and obtained good correlation with soil properties.

M. Musso (✉)
Dto. Ingeniería Geotécnica, Facultad de Ingeniería, Universidad de la República, Montevideo, Uruguay
e-mail: marcosmusso@gmail.com

G. Lollino et al. (eds.), *Engineering Geology for Society and Territory – Volume 7*,
DOI: 10.1007/978-3-319-09303-1_42, © Springer International Publishing Switzerland 2014

In Uruguay, vertisols are widespread soils that are basaltic, silty clayey rock and loess weathered products. In the south, Quaternary clayey and silty clayey sediments as Libertad Fm. cover Precambrian (igneous and metamorphic) rocks, sedimentary rocks and Cenozoic sediments. Clayey soils developed in Libertad Fm. are very often found and when clay dominant there are smectite (Musso 2001). These soils have more than 90 % of silt and clay, low and medium Atterberg limits and its bearing capacity of foundation is from 100 to 200 kPa (Departmento Geotécnica 1997).

Musso (2001) analyzed more than fifty soil samples taken from the Libertad Fm. and suggested a guide to identify and quantify different expansive potential clayey soil. It contains aerial photo interpretation, field work observation and laboratory testing. This guide was used in some sites with clayey soil in Uruguay with great success. This methodology is an approach that will be improved in the future.

The aim is to show practice methodological approach to identify and quantify expansive soil used in Uruguayan soil that could be used or adapted to other countries with expansive soil.

42.2 Material and Methods

Vertisols have feature landforms in aerial photographs named gilgai. In field work, observation include of pop corn structure in slope cut soil, cracking in houses or civil infrastructure. Soil samples will be collected in this site for laboratory test.

Typical soil tests as grain size and Atterberg limits will be done together for mineralogical identification. The methylene blue test is an easy and quick test to determine CEC and it could identify clay minerals in soil when associated with XRD.

Disturbed soil samples of Libertad Fm. were collected in cut slopes in the subgrade depth of the highway or in borehole from house foundations in the south of Uruguay (Fig. 42.1).

A step by step guide proposed is:

1. Bibliography revision of geological and pedological maps to identify clayey sediments and soils, remote sensing images and air photo interpretation;
2. Field wok: a check list to verify photo interpretation, watch damage buildings, gilgai (micro-relief) and/or surface crack soil, pop corn appearance in cut slope, carbonate concretion, plastic and sticky appearances;
3. Initial laboratory test: soil identification and classification Unified Soil Classification System (USCS), mineralogical identification (CEC, XRD); Proctor test;
4. Heave measure: oedometer expansive test (swelling potential and swelling pressure), or swell in California Bearing Ratio (CBR);
5. Qualifying expansibility (very low, low, medium, high, very high) considering 200 kPa as a limit between low to medium swelling pressure, because it is the bearing capacity value to foundations in Libertad Fm.

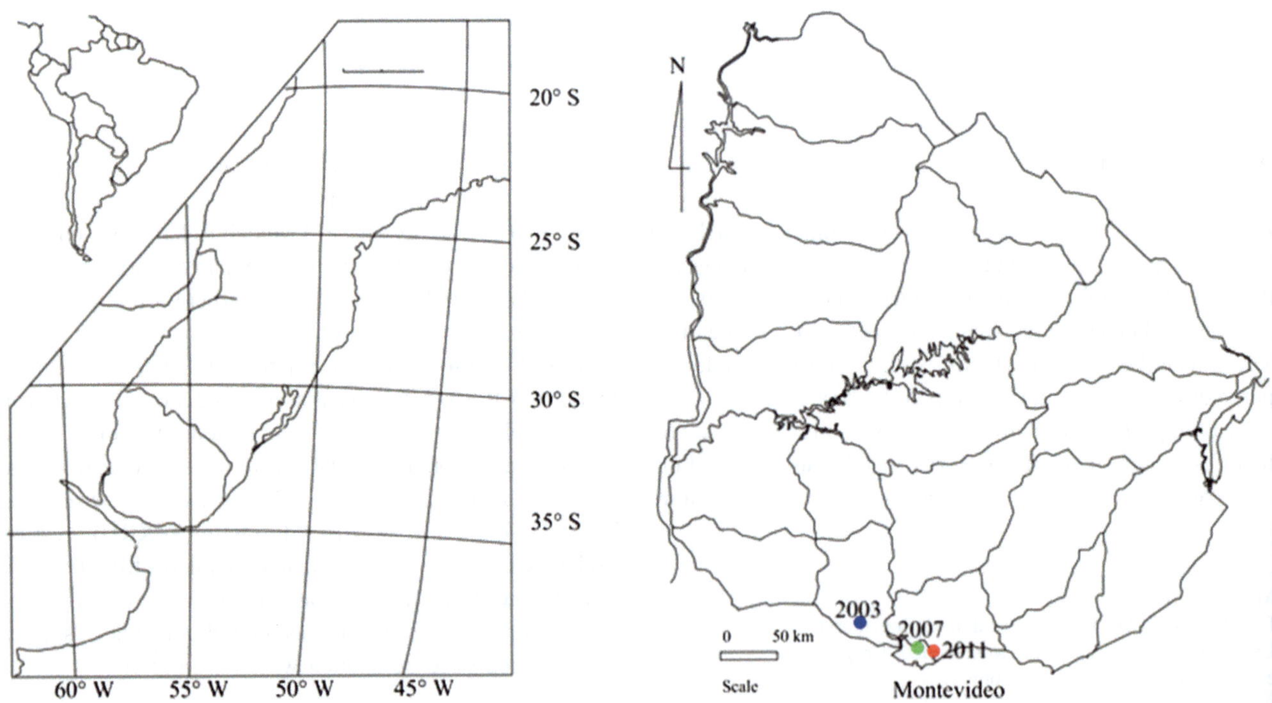

Fig. 42.1 Map location clayey soil samples

Fig. 42.2 **a** (site 2003 Fig. 42.1), **b** (site 2011 Fig. 42.1) Cut slopes show pop corn feature and carbonate concretions

Fig. 42.3 **a** Damage in rigid pavement in highway place of Figure 42.2a, **b** damage in wall in the place of Fig. 42.2b

Table 42.1 Summary of soils properties measured

Soil sample	Liquid limit (%)	Plastic index (%)	Passing #200 (%)	Clay (%)	CEC-clay (cmol/kg)	SUCS	Swelling pressure (kPa)	moisture (%)
[a]2003-R1-1	101	69	99.8	69	108	CH	500–1000	20–17
[a]2003-R1-3	78	56	98.5	48	107	CH	250–500	20–18
[a]2007-4k-111	75	39	98.5	52	84	MH	75–300	24–16
[a]2007-10k400	52	32	90	40	—	CH	50–350	20–14
[b]2011 4-1	54	32	95	33	86	CH	120–250	21–16
[b]2011 5-2	41	24	90	26	84	CL	55–87	22–18

[a] Sample compacted in proctor dry unit weight, [b] sample compacted in dry unit weight 15 kN/m3, moisture variable (Musso et al. 2003, 2011 samples; Rostan et al. 2007 samples)

42.3 Results and Discussion

Clayey Uruguayan soils from different places were analyzed using the step by step guide. Several described items as pop corn features, calcic carbonate, surface cracking, damage to buildings, plastic and sticky appearances were observed during field works. Cut slopes show pop corn feature and carbonate concretions (Fig. 42.2a, b) and the soils in these places have plastic and sticky behavior. Wall and rigid pavement in the same places of cut slope were damaged (Fig. 42.3a, b). Further on, soil samples were obtained in this site from the slope or to a depth of 5 m with manual auger.

All samples have more than 90 % of silt and clay, but clay size fraction content is very different with values between 26 and 69 % (Table 42.1). Clay mineralogy determined with CEC and DRX show esmectite dominant clay in all case (Musso et al. 2003; Musso et al. 2011;

Rostán et al. 2007). Most of the samples were classified as inorganic clay of high plasticity (CH). However, swelling pressure values are different because these values are clay, moisture and unit dry weight dependent.

Swelling pressure in CH soils achieved 350–1000 kPa, higher values than the bearing capacity used in foundations. When the soil has high clay content, the swelling pressure is higher and at the same time, the low clay soil produces the lower swelling pressure.

The limit value between low to medium expansibility could be modified or adapted to other expansive soils weight up it specific bearing capacity.

42.4 Conclusions

Damages in civil infrastructure were observed in several places where soils with macroscopic feature described as superficial cracks, carbonate concretion, plastic and sticky appearances were identified.

The studied soils have smectite as clay mineral dominant and the clay size percent is variable. Then, Uruguayan soils of Libertad Fm. have different swelling potential and this potential depends of mineralogy, clay content, moisture and dry unit weight.

A simple and easy guide that contains field work and laboratory set tests developed, is a useful tool to identify and quantify swelling clayey soil.

References

Departamento de Geotécnica Facultad de Ingeniería UdelaR, Carta Geotécnica de la Región Metropolitana de Montevideo, escala 1:100.000. (Oficina de Publicaciones CEI, Montevideo, 1997)

Meisina C (2000) Prediction swelling/shrinkage potential using the blue methilene method: some samples in Italian clayey soils. In: Paper presented at the Geoeng 2000 an international conference on geotechnical and geological engineering. Melbourne, 19–24 Nov 2000

Musso M (2001) Dissertation, University of São Paulo-Brasil

Musso M, Rostán A, Behak L (2003) Subrasantes expansivas en Rutas Nacionales: ¿mito o realidad?. In: Paper presented at the 4° Congreso de Ingeniería Vial, Asociación Uruguaya de Caminos, Montevideo, 4–6 Nov 2003

Musso M, White E, Behak L, Rottondaro V (2011) Variabilidad de la presión de expansión en suelos arcillosos en el nordeste de Montevideo, Uruguay. Paper presented at the 13 Congreso Brasileiro de Geología de Engenharia e Ambiental, San Pablo, 2–6 Nov 2011

Rostán A, Musso M, Behak L (2007) Zonificación del Potencial Expansivo de los Suelos de la Formación Libertad: Métodos de Identificación y Estabilización. Paper presented at the 6° Congreso de la Vialidad Uruguaya, Montevideo, 7–9 Nov 2007

Seed HB, Woordward RJ, Lundgren R (1962) Prediction of swelling potential for compacted clays. J Soil Mech Found ASCE 88(SB3):53

Williams AB, Donalson GW (1980) Building on expansive soils in South Africa: 1973–1980. Paper presented at the fourth international conference of expansive soils. ASCE Denver, Colorado, 16–18 June 1980

Yitagesu FA, van der Meer F, van der Werff H, Zigterman W (2009) Quantifying engineering parameters of expansive soils from their reflectance spectra. Eng Geol 105:151–160. doi:10.1016/j.enggeo.2009.01.004

Legislation as a Tool to Mitigate Mass Movements in Spain

43

Jesús Garrido, José Delgado, Alejandro L. Grindlay, Ignacio Jiménez Soto and M. Isabel Rodríguez

Abstract

According to the Ministry of Internal Affairs, in the 1990–2012 period, there have been 57 deaths in Spain due to mass movements. The economic losses were estimated at about €180 million each year. Some of these losses have been related to violating the laws and/or the lack of risk-focused land use planning. In accordance with the provisions set out by the 9/2006 Spanish Strategic Environmental Assessment Act, an Environmental Sustainability Report is compulsory and must include a natural hazards map, as it is set out by the national Land Act (2/2008 Royal Legislative Decree). Although there is no natural hazards legislation in Spain, there is some sectoral legislation on several natural hazards but it does not specifically refer to mass movements. Despite most of these codes are mandatory, they are not usually enforced. However, Spanish legislation requires compulsory natural hazards reports in some cases but, unfortunately, there are no guidelines on how to conduct these reports.

Keywords

Mass movements • Prevention • Legislation • Town plannings • Spain

43.1 Introduction

According to the Ministry of Internal Affairs, in the 1990–2012 period, there have been 57 deaths in Spain due to mass movements. The economic losses were estimated at about €180 million each year (Spanish Senate 1998). The total losses have been underestimated because the indirect costs have not been considered. However, the information on economic losses is very scarce since most of the mass movements usually happen in private-owned buildings or civil works and infrastructures under construction; in these cases, mass movements are only known to engineers in charge of these works. Some of these losses are due either to violating the law and/or the lack of risk-focused land use planning and hazard maps.

The importance of legislation and land use planning as a mass movement preventive tool has been pointed out by many authors (Varnes 1984; Gue et al. 2009). They are the most economic and efficient of all possible countermeasures to be taken.

There is no Spanish natural hazards legislation, but there is some sectoral legislation which refers to natural hazards. In spite of this legislation being mandatory, it is not usually enforced. In some countries legislation does not specifically refer to landslides but to natural disasters instead, such as the Salvadoran Civil Protection, Prevention and Mitigation of Disasters Act (2005).

J. Garrido (✉)
Departamento de Ingeniería Civil, Universidad de Granada, Campus de Fuentenueva S/N, 18071 Granada, Spain
e-mail: jega@ugr.es

J. Delgado
Departamento de Ciencias de La Tierra Y Medio Ambiente, Universidad de Alicante, Ap. Correos 99, 03080 Alicante, Spain

A.L. Grindlay · M.I. Rodríguez
Departamento de Urbanística Y Ordenación Del Territorio, Universidad de Granada, Campus de Fuentenueva S/N, 18071 Granada, Spain

I.J. Soto
Departamento de Derecho Administrativo, Universidad de Granada, Plaza de La Universidad S/N, 18071 Granada, Spain

G. Lollino et al. (eds.), *Engineering Geology for Society and Territory – Volume 7*,
DOI: 10.1007/978-3-319-09303-1_43, © Springer International Publishing Switzerland 2014

However, in some other countries there is legislation on landslides such as the "Regulation on Hazard Zoning due to Snow and Landslides, Classification and Utilization of Hazard Zones, and Preparation of Provisional Hazard Zoning" of Iceland (2001) or the "Landslides Prevention Law" (1958) and the "Act for Promoting Prevention Measures against Sediment-related Disasters" (2001, amended in 2005) of Japan. In Australia, building in potential landslide areas is prohibited by law, while some local legislation requires more research to be carried out in questionable areas (Leventhal and Kotze 2008).

The complexity of the natural hazards legal framework in Spain is highlighted in both national and regional legislations. According to the Spanish Constitution and all the regional Statutes of Autonomy, national and regional competences (the latest conferred to Autonomous Communities) are either shared or exclusive competences. Besides, the sectoral legislation and the spatial and urban planning instruments are simultaneously applied.

43.2 Sectoral Legislation

In September 1983 a Special Commission of Inquiry was set up to conduct the investigation on the origin and consequences of floods after the Basque Country and Catalonia floods occurred. Unfortunately this Commission had no specific tasks to carry out.

After the Biescas flash flood (1996), and the Alicante and Badajoz floods (1997) the Spanish Senate created a special commission for the prevention and assistance in catastrophic situations. The commission report (Spanish Senate 1998) suggested including a mass movements risk assessment report on spatial and urban plannings and infrastructures projects. The coordination and cooperation among the national, regional and local administrations was also recommended with the support of the technical and scientific institutions related to them. The competences for natural hazards prevention are carried out by the Spanish Geological Survey, according to the Royal Decree 1953/2000 (amended in 2007 and 2010).

In the nineties, the Civil Protection legislation was the first one to relate both natural hazards and spatial and urban planning. The Civil Protection Basic Standard (1992) establishes that all Spanish regions must have a flood civil protection plan. The phenomena associated to floods, such as mass movements, should be considered too. A seismic civil protection plan must also be done by the Autonomous Communities with intensity-over-6 areas as shown on the Spanish Seismic Hazard Map developed by the National Geographic Institute. The earthquake-induced mass movements should also be considered. The Canary Islands, for instance, should consider mass movements in their volcanic civil protection plan. Besides, the municipal seismic, flood and volcanic plans should also consider mass movements. Almost all of the regional plans, which include a natural hazard map, have already been approved but there are not many municipal plans approved yet.

Moreover, a municipal mass movements civil protection plan might be established by the Autonomous Communities or the municipalities.

The old Basic Building Standard considered studying the bearing capacity but it did not consider the mass movements themselves. However the new Technical Building Code (2006) considered the bearing capacity and mitigating geotechnical hazards, and to a lesser extent, mass movements.

There is also a seismic building code (2002) and a seismic bridge code (2007) to mitigate seismic hazards. Despite the fact that these codes are mandatory, they are not usually enforced in relation with mass movements.

The Water Act (2001) and the Coast Act (1988, amended in 2013) do not specifically take into consideration mass movements.

Furthermore, the Spanish sectoral legislation occasionally establishes that natural hazards reports are compulsory. However, there are no guidelines to carry out a geotechnical or mass movements report in the case it is required by law. Therefore if there are no mandatory guidelines to carry it out, this report is not usually useful because they do not present a minimum content.

In Spain the Ministry of Housing and the Official Spanish Association of Professional Geologists (Ministerio de la Vivienda-Ilustre Colegio Oficial de Geólogos 2008) have published a Methodological Guideline to carry out risk/hazard cartographies, including mass movements. In Portugal there is a Methodological Guide for Risk Mapping and for Development of a GIS at municipal level (Julião et al. 2009). Wang et al. (2012) reviewed 30 landslide guidelines worldwide published in English and French.

Though professional associations or the national, regional and municipal authorities should check out whether these reports are made according to the law, they do not usually do it.

43.3 Spatial Planning and Urban Legislation and Town Planning

According to the provisions set out by Act 9/2006 or the national Strategic Environmental Assessment (SEA), an Environmental Sustainability Report (ESR) is compulsory to be conducted for all urban plans. A natural hazards map must be included in the ESR, as it is established in the

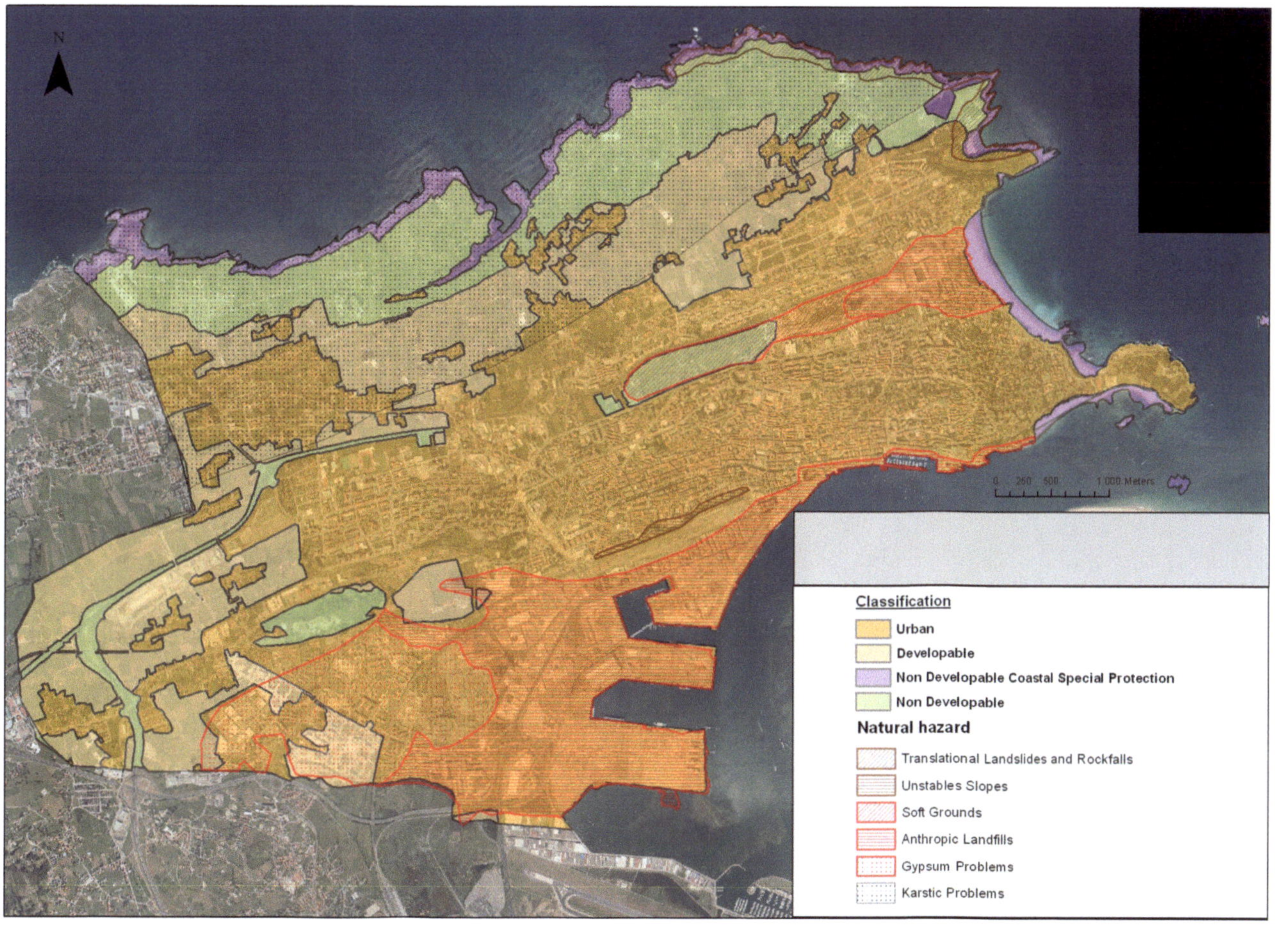

Fig. 43.1 Town planning of the city of Santander. Land use classification and mass movements hazard map

national recast text of the Land Act, the 2/2008 Royal Legislative Decree, which also establishes that land which is exposed to any natural hazard must be classified as non developable. Most of the enacted regional laws about spatial and town planning agree on implementing a natural hazards map. Furthermore, most of the regional land legislation considers that landslides and rock falls prone areas should be classified as non developable or rural land. However, town planning does not consider hazards maps on the land use classification. Despite some mass movements hazard maps consider landslides, rock falls, subsidence and so on, the prone areas are classified as developable land (Fig. 43.1).

Though maps are included in the town plannings they are not usually made in a detailed scale, specifically for those plannings, and/or they are a poor copy of the official maps, which are generally old ones. Unfortunately the hazard maps which are enforced by spatial and urban planning legislation have not been made yet because most of the town plannings are older than the SEA and the Land Act.

Local governments allow building in urban land over high hazard areas because hazard maps are not taken into account in the land use maps (Fig. 43.2).

In the ESR the susceptibility, hazard and risk maps are usually considered synonymous as well the classification of mass movements is not properly used. It might be because professionals who design these maps are sometimes not competent to do so. In spite of these professional competencies are regulated by Professional Associations provisions, some regional SEA legislation establishes that this competence should be regulated to avoid this situation.

43.4 Conclusions

Technical guidelines are a key element to make valuable reports but they should be written by the technical and scientific institutions related to the administration at all levels and should be legally binding. Everything that is not compulsory by law is not usually taken into consideration,

Fig. 43.2 Land use classification (urban land) and geotechnical hazard map included in the Town Planning of the city of Caceres, for El Calerizo quarter, superimposed to *left*) 2005 Google earth; *right*) 2011 Google earth. Orange shaded area means high hazard but with possible corrective measures. Stripped area means high hazard non suitable for building

and being considered, legal mechanisms should be established to enforce the law and to punish its violation. Legislation is the key instrument to natural hazards prevention, either through land use planning, which is the most efficient of all non structural measures, or the more expensive preventive or corrective countermeasures.

Acknowledgements This work was partially funded by the Mapfre Foundation through the Call of "Aid to research 2011".

References

Gue S, Karnawati D, Wong SY (2009) Policy and institutional framework for landslide mitigation and risk reduction. In: Sassa K, Canuti P (eds) Landslides—disaster risk reduction, pp 531–554

Julião RP, Nery F, Ribeiro JL, Castelo Branco M, Zêzere JL (2009) Guia metodológico para a produção de cartografia municipal de risco e para a criação de sistemas de informação geográfica de base municipal. ANPC, DGOTDU, IGP

Leventhal AR, Kotze GP (2008) Landslide susceptibility and hazard mapping in Australia for land-use planning—with reference to challenges in metropolitan suburbia. Eng Geol 102:238–250

Ministerio de la Vivienda-Ilustre Colegio Oficial de Geólogos (2008) Guía metodológica para la elaboración de cartografías de riesgos naturales en España. ICOG. 187 p

Spanish Senate (1998) Conclusiones de la Comisión Especial del Senado sobre la Prevención y Asistencia en Situaciones de Catástrofe, Boletín Oficial de las Cortes, 9 de diciembre, pp 1–40

Varnes DJ (1984) Landslides hazard zonation: a review of principles and practice. Nat Hazards 3 (UNESCO, París), 63 p

Wang B, Reul M, Couture R, Bobrowsky PT, Blais-Stevens A (2012) Review of existing landslide guidelines—national technical guidelines and best practices on landslides. Geological Survey of Canada, Open File 7058, 13 p

Design of a Database for Documentation and Analysis of Lab Data and Rock-Properties Derived in Field Measurements

44

Florian Menschik, Kurosch Thuro, Heiko Käsling and Michael Bayerl

Abstract

Increasing requirements for ground investigation programs in large construction projects are putting pressure on existing data management systems. Today data from a wide range of necessary field and lab investigations are stored in large tables and spreadsheet programs. These tables are often complicated, poorly optimized, or inflexible. To handle large data loads in a more appropriate way, the chair of Engineering Geology at TUM is working on a database concept to optimize easy and fast data management. Its main objective is to make these data accessible to project partners across broadly distributed locations. This database solution will offer easy access, quick output, and fast analysis of field and lab data. Our database system is based on a MySQL database with an HTML-PHP web interface to allow online access. The deployed user management system allows administrators to grant and constrain access to data to guarantee the safety of project data. The base module manages the data of project partners, project data and laboratory data. Simple analysis of the data is integrated in the base module. In addition users can export data for further analysis, thereby allowing integration with a wide range of popular software. This database system is open to add-on modules to develop further analysis tools to investigate the collected data in detail. At the moment, we are developing an add-on to use this database for drill test analysis. Today we are managing over 60 projects with 500 samples and almost 1,200 lab tests with the base module.

Keywords

Data management • Database • Rockmechanical properties

44.1 Introduction

In today's age of bits and bytes and with actual progression of information technology, computers will be used for more and more purposes. This trend is not only observable in private life but also in many lines of business as well.

The possibilities and the accuracy of calculations are much better and higher than 10 years ago. Analysis of data can be obtained on a much higher level than before. But not only are the results getting better, also data volume is rapidly increasing - not only in size, but also in number of files.

To work at the state-of-the-art according to all standards for big projects, many more lab and field tests are needed and much more data are obtained than in previous times—always

F. Menschik (✉) · K. Thuro · H. Käsling
Technische Universität München, Arcisstr. 21, 80333 Munich, Germany
e-mail: menschik@tum.de

M. Bayerl
Hilti Entwicklungsgesellschaft MbH, Hiltistr. 6, 86916 Kaufering, Germany

G. Lollino et al. (eds.), *Engineering Geology for Society and Territory – Volume 7*,
DOI: 10.1007/978-3-319-09303-1_44,

with the aim of reaching a better and more precise understanding of the processes between building and underground during construction work.

With increasing amounts of data we have to develop new methods to get to a better data management with the goal of making data easy accessible, short and long term, and making everything comprehensible.

44.2 Objectives

The significance of a good filing system will increase (Schumacher 2009). For this it is very important to save data only once (without redundancy) and in a clear and structured way. This is necessary to maintain data integrity and keep data available in the long term. In addition, we as engineering geologists and geotechnical engineers, must be able to respond quickly to special requirements in data storage and new analysis methods of properties for new projects.

Today the data of many projects is managed with help of spreadsheet analysis programs (e.g. Microsoft Excel) or in the worst case with word processing programs (e.g. Microsoft Word). If all project partners are using the same software, you can work at least within one file format; but when different programs are used, the normal chaos will be potentially greatly increased. In other mid- and large-sized building projects many different data types are collected in database management systems, specially designed for this purpose. One famous example is the Software 2DOC from Pöyry Infra GmbH. This comprehensive software is used for data management in tunneling and can be found on numerous tunnel and construction sites such as the Brenner Base Tunnel (www-1). Another published positive example of application of this software is the tunnel project Harter Plateau/Linz (Fellinger and Bergler 2009).

A similar concept for the collection, the analysis, and the transfer of rock properties derived in field and lab tests to others would be desirable.

44.3 General Database Structure

Two components are needed to work with a database (Saake et al. 2010, p 8). First, the database itself where all the data are stored. Second you will need a database management system (DBMS). The DBMS manages all incoming data, makes arrangements for data safety, integrity and security and allows, in the ideal case, a parallel multi-user environment.

There are several database types used for different purposes. On one hand simple models like a hierarchic or network model, which are used for simple data structures like a file management system of common computer operating systems. These simple databases are not an improvement to our situation because within these models datasets are stored as single packages with references between these files as it is done in our file system with common documents (.doc, .xls). Object oriented databases are another type of database. This type will work pretty well for smaller datasets with much information. But in general we have many tests with much information. So we need to take the most useful database model—the Entity Relationship (ER) Model. This is one of the most common models used in database systems (Kemper and Eickler 2011, p 71).

In this type of database data is stored in tables (entities) which are related to each other via identification numbers (ID). Due to the relations between datasets and the building of so called master tables, redundant data and mistakes (e.g. spelling mistakes) can be avoided. This model type supports and combines huge data volumes in a fast way and therefore it is suitable for a rock testing database.

44.4 The Rock Database of the TUM

There are several programs and script languages on the market to setup an ER database as shown in Fig. 44.1. Example DB2, SQL, MySQL or Microsoft Access are some of the most common systems. To get an open and cost-efficient database system, the chair for Engineering Geology decided to develop a MySQL-database (at no charge) with an HTML-PHP web interface.

44.4.1 Programming of the Database

The programming of the database system was carried out as part of the master's thesis of Eitschberger (2010) and will be further developed by Menschik during his Ph.D. studies (ongoing).

The data are stored in a MySQL database and can be accessed through a web interface (general usage) and a phpMyAdmin interface (administrative access), which grants easy access to the raw data. phpMyAdmin is also a great possibility for introducing and testing new data queries and expansions for the database before setting up a new part of the web interface.

The web interface itself is built with the software Adobe Dreamweaver as a common HTML web page with php code inside.

44.4.2 Database Structure and Web Interface

The database consists of the following four modules: user management, general project data, lab data and add ons

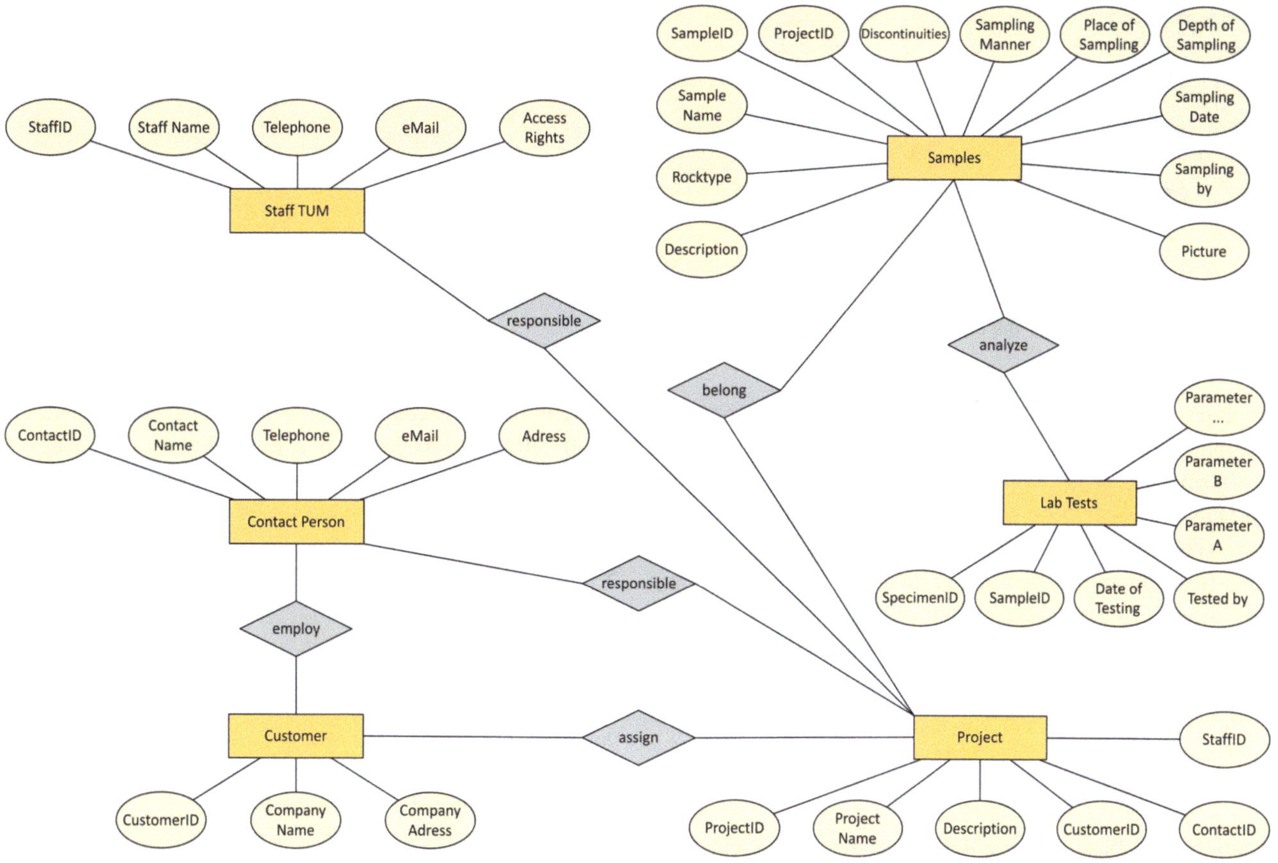

Fig. 44.1 Entity-relationship-model of base module of the rock database of the chair of engineering geology at TUM

The first three modules form the base component of our database (Fig. 44.1) and can be expanded with additional functionality with add ons.

Within the user management moduleall access to single data and the whole database is administered. This module also provides the foundation for granting access to special add ons only to respective customers and users. It allows automatic data sharing only with project partners and not third party members.

The project data module stores all project-relevant data, such as project name and description and customer data. In addition, information about incoming samples, which are processed to test specimen, is collected and related to the project. This information represents the joint of all other data (lab and field) with the project and is essential for further analysis and export of the database.

All data derived in lab tests can be stored in the lab data module. At this moment we are storing data from uniaxial compression, point load, Brazilian tensile, CERCHAR, LCPC and ultrasonic tests.

All data can be entered and queried with the web interface. This interface is accessible worldwide via the internet with any normal web browser (e.g. Mozilla Firefox, Internet Explorer etc.).

First step is the login. The database decides depending on the login data which modules can be accessed and shows the associated view of the interface.

Actually we have three different views: One is administrative (base view); one for project partners; and one is a special view for the first add on.

Only within the administrative view can all modules be accessed. In this view it is possible to add new customers, projects, samples and lab data. All informations can be entered or changed within the web site in simple forms.

In the base view it is also possible to check the status (e.g. receipt of samples, progress of lab tests, issuing of an invoice, etc.) of ongoing projects. This part is also available for project partners.

The web interface also shows all queries of lab data. It is possible to access all raw lab data or calculated mean values of specific samples or projects.

44.4.3 Add Ons

The add ons are further modules that allow for more analysis options and integration of other data (e.g. field test data and their relation to lab data). These data can be connected

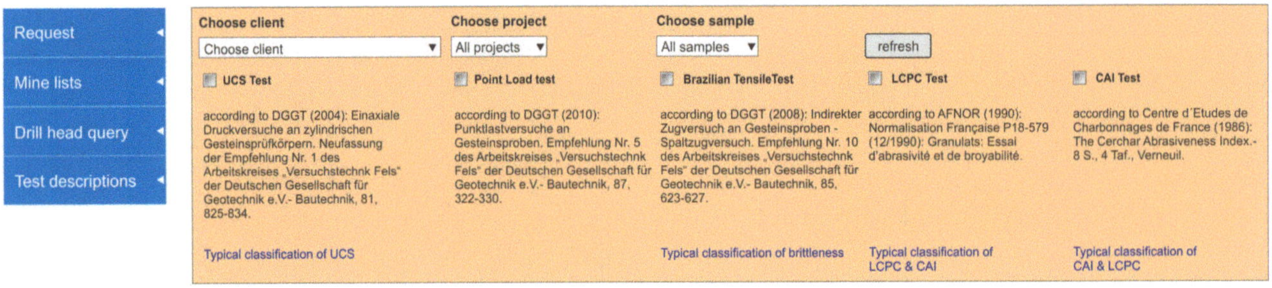

Fig. 44.2 Project view for the database incl. Add-on for drill testing

to a project or to single sample. With this relation the combined analysis of different data is possible.

During the Ph.D. studies of Menschik (ongoing) an add on (Fig. 44.2) for the integration of drill head field testing data (performance and wear) and data of the test sites (quarries and mines distributed over the world) is being developed. This module allows easy correlation of rock mechanical properties with performance and wear data of the drill equipment.

44.5 Pros and Cons of a Database System

With this database we have developed a cost-efficient, easy to handle concept, which allows us to exchange lab and field data with colleagues, project partners and customers easy over long distances. The involved parties don't have to have specialists to work with the database because it's accessible with a normal web browser without any special knowledge about programming. The data are available worldwide in real time as long as there is an internet connection. The database is flexible, easy adjustable and expandable for prospective needs.

In addition, it is possible to export processed and raw data as csv files for further analysis in other spreadsheet calculation or statistical analysis software.

With the phpMyAdmin interface of MySQL we have the possibility to export in several other data interfaces or save the whole database in an archive.

The only disadvantage of the use of such a do-it-yourself data management is that it requires some programming knowledge for the setup and the further development of add ons. But all knowledge which is needed can be learned with small effort and is kept on a low level by the usage of software like Dreamweaver and phpMyAdmin.

44.6 Further Developments and Actual Usage

The next step will be the expansion of the base module with petrographic data. At this moment we are using the rock database for several Ph.D. studies, customer projects and for research projects. It contains over 60 projects with over 500 samples and almost 1200 lab tests.

References

Eitschberger C (2010) Konzeptionierung und Programmierung einer Festgesteinsdatenbank für geotechnische Kennwerte. – 73 S., Masterarbeit, Lehrstuhl für Ingenieurgeologie, Technische Universität München (unveröffentlicht)

Fellinger G, Bergler C (2009) 2DOC in der Praxis. Teil IV: Das Tunnelbauprojekt Harter Plateau/Linz. Felsbau 5(6):284–288

Kemper A, Eickler A (2011) Datenbanksysteme: Eine Einführung. – 792 S., München (Oldenbourg Wissenschaftsverlag)

Menschik F (i. Vorb.) Untersuchung von Leistungs- und Verschleißcharakteristik elektrischer Schlagbohrhämmer. – Dissertation, Lehrstuhl für Ingenierugeologie

Saake G, Sattler K-U, Heuer A (2010) Datenbanken. Konzepte und Sprachen. – 800 S., Heidelberg (mitp-Verlag)

Schumacher (2009) Die Bedeutung einer strukturierten Datenablage. – Infokasten 1 zum Artikel: Schmuck: Datenmanagement im Tunnelbau mit 2DOC. Teil I: 2DOC als Basis zur strukturierten Erfassung und analyse von Daten. - Felsbau, 2:81. http://www.2doc.info/website/2doc/Produktbeschreibung/RefReferen/tabid/654/language/de-DE/Default.aspx, abgerufen am 04.02.2013

Landslide Guidelines and Best Practices for Professional Engineers and Geoscientists

45

Peter Bobrowsky, Doug VanDine and Réjean Couture

Abstract

Documents that summarize "best practices" of a discipline are often widely used to provide bench marks of performance, guidance, expectations and sources of reference to practitioners. Such documents can range from those that are lengthy prescriptive, regulatory and obligatory in the legal sense, to those that are short generalized views of opinion that lack peer consensus. More frequently, "best practice" documents are timely and extensive compilations prepared by a peer community that summarize and illustrate current philosophies and protocols regarding various methods, techniques and procedures. During the past decade a number of countries have published landslide-related best practice documents. Typically these documents are nation specific. They are often written to address landslides in the specific social, cultural, political and natural terrain and territory represented within those nations' borders. As a consequence some of the information they contain is not readily transferable across political borders. Landslide professionals in Canada now have access to their own best practices document. The aim of the document is to provide a Canadian state-of-the-art synthesis of landslide topics including: terminology; socio-economic significance; landslide classification and description; identification and mapping; site investigation, analysis, monitoring and treatment; risk management and evaluation; examples of common landslide types; and professional practice issues. Specialists from the Geological Survey of Canada have coordinated contributions from a number of Canadian landslide specialists representing government, academia and the private sector. Some 60 participants contributed as advisors, editors, authors and reviewers to the 11 open files (chapters) comprising this project. This initiative provides an example of a successful program and strategy that will serve both Canadian landslide professionals and the public on an issue that is important to the health and safety of Canada's population and infrastructure. The document can be readily adopted or adapted by other countries interested in improving professional practices related to landslides within their country.

Keywords

Landslide • Guidelines • Best-practices • Professionals • Canada

45.1 Introduction

Landslides represent a natural hazard that can vary widely in type, size, timing, frequency, geographic location, conditions, cause, trigger, effect, impact, and cost. Our ability to successfully predict, locate, identify, monitor, avoid,

P. Bobrowsky (✉) · R. Couture
NRCan, Geological Survey of Canada, 601 Booth Str., Ottawa, K1A0E8, Canada
e-mail: pbobrows@nrcan.gc.ca

D. VanDine
VanDine Geological Engineering Limited, Victoria, Canada

G. Lollino et al. (eds.), *Engineering Geology for Society and Territory – Volume 7*,
DOI: 10.1007/978-3-319-09303-1_45, © Springer International Publishing Switzerland 2014

control and effectively manage future landslides, however, is quite constrained. After a landslide event, we are in a better position to identify the processes and mechanisms that were involved in causing and triggering the failure, to classify the event, to assess actions that minimize or reduce future impact and to propose responses to control further threats associated with reactivation. Fortunately, prior to an event we can recognize landslide prone terrain and conditions, and can also identify the type of slope instability that is likely to occur, provide a list of probable contributing factors and, under certain circumstances, even forecast relative timing of the event. This applies equally well to landslides on both natural and engineered slopes.

45.2 Guidelines and Best Practices

Many organizations, governments, professional bodies and other groups have long recognized that attention can and should be directed towards better understanding and managing slope instability hazards. Resultant action is often directed towards addressing landslides within the political boundaries of a particular nation. Such action is often focused on reducing the vulnerability of individuals through public awareness and education (e.g. lectures and publications to help citizens) or on improving the resilience of the built environment through proactive means (e.g. legislation, land use plans, engineering regulations).

In 2006 the Government of Canada, through the Geological Survey of Canada (GSC) embarked on a project to inform and educate the non-professional community regarding landslide hazards. The goal was to produce and widely distribute a document that could be easily understood by the average citizen. The document explained the basics of landslides and their behavior including why they occur, where they tend to occur and what activities tend to contribute to their occurrence. Moreover, the document provided advice as to options for avoidance, mitigation and management of unstable slopes. The resulting English version of 'The Landslide Handbook', jointly produced by the GSC and United States Geological Survey (Highland and Bobrowsky 2008; http://pubs.usgs.gov/circ/1325/) was subsequently translated and published in Mandarin, Spanish, Portuguese and Japanese.

In 2009 the GSC initiated an effort to produce a document to serve the technical community of landslide professionals. This community appreciates that sharing collective knowledge and experiences with its peers provides one of the best strategies for addressing this hazard.

The intention of our initiative was to produce a document that would provide guidance and best practice examples for Canadian engineers, geoscientists and other landslide practitioners. The goal was not to prepare a legislated document or a compendium of standards, nor was it to limit the practice of Canadian landslide professionals. Rather it was to provide professionals with relevant information that could be successfully incorporated into their regular professional practice.

The role of the GSC was to facilitate contributions from approximately 60 qualified Canadian professionals. These professionals contributed as advisors, editors, authors and reviewers. Landslide expertise in Canada is distributed across the country in all provinces and territories and in all sectors; local, regional and national governments as well as academia and the private sector.

As they were completed, all contributions (11 separate chapters) were released as individual GSC Open File publications. Comments and edits on the drafts were, and still are, invited and encouraged from both the national the international landslide communities. Such input is essential to ensure a high quality, and is timely because each open file will be revised, updated and eventually compiled as chapters into a single publication: a GSC Bulletin.

45.2.1 Terminology

One of the first open files released was a glossary of landslide terminology (ftp://ftp2.cits.rncan.gc.ca/pub/geott/ess_pubs/288/288066/of_6824.pdf). Open File 6824 (Couture 2011) provides a useful compilation of terms commonly used by the landslide community in Canada, and elsewhere, and the volume set the standard for common usage of terminology by all authors contributing to the subsequent open files. This open file includes an extensive glossary of landslide related terms and formal definitions that have been extracted from many available popular references. The revised version of this open file will also reference the source of each term to its appropriate citation.

45.2.2 Existing International Guidelines

To ensure that the Canadian technical guidelines addressed key issues, a review of existing national and international guidelines, standards, and best practices-type publications was undertaken. Thirty-five readily accessible English and French documents were evaluated and summarized in Open File 7058 (Wang et al. 2012; ftp://ftp2.cits.rncan.gc.ca/pub/geott/ess_pubs/289/289863/of_6996.pdf). This open file assesses each contribution as it pertains to the following key topics: landslide risk management; landslide zoning; geotechnical assessment; land use planning; mitigation and remediation; and codes of responsibilities.

45.2.3 Landslide Classification and Description

An accepted and widely used landslide classification system assists professionals to appropriately observe and describe a landslide. Such a classification system also provides a basic platform to facilitate common dialogue, study and communication amongst professionals. Open File 7359 (Cruden and VanDine 2013; ftp://ftp2.cits.rncan.gc.ca/pub/geott/ess_pubs/292/292505/of_7359.pdf) provides a method to classify and describe landslides and includes a discussion of related topics such as landslide size, intensity, travel angles, causes and indirect effects.

45.2.4 Identification and Mapping

Open File 7059 (Jackson et al. 2012) focuses on Canadian concepts, approaches and practices involved in landslide identification, mapping, map elements and components, as well as types of landslide maps and field mapping methods (ftp://ftp2.cits.rncan.gc.ca/pub/geott/ess_pubs/292/292122/of_7059.pdf). Field description and other related aspects are discussed and where applicable suggestions for successful methods and techniques are given.

45.2.5 Socio-Economic Significance

Open File 7311 (Guthrie 2013) summarizes the socio-economic significance of landslides and considers both direct (e.g. injury, loss of life, damage to infrastructure and property, loss of resources) and indirect costs (lost wages, costs for traffic redirection, etc.). This contribution provides a balanced review of the socio-economic significance of Canadian landslides in the context of other natural hazards and landslides world-wide. The open file includes a detailed review of the economic and other losses associated with 56 notable landslides in Canada (ftp://ftp2.cits.rncan.gc.ca/pub/geott/ess_pubs/292/292241/of_7311.pdf).

45.2.6 Risk Management

Open File 6996 (VanDine 2012) reviews the elements and steps of an effective landslide risk management process: initiation, risk assessment (identification, analysis, evaluation) and risk treatment. The associated aspects of communication and consultation as well as monitoring and review as they relate to landslide risk management are also addressed and it provides the basis for Open File 7312 (ftp://ftp2.cits.rncan.gc.ca/pub/geott/ess_pubs/289/289863/of_6996.pdf).

45.2.7 Risk Evaluation

Open File 7312 (Porter and Morgenstern 2013) examines landslide risk evaluation (ftp://ftp2.cits.rncan.gc.ca/pub/geott/ess_pubs/292/292234/of_7312.pdf). It assesses individual vs. societal risk, voluntary and involuntary risk, tolerable vs. acceptable risk, mortality rates, economic risks, qualitative and quantitative risk methods, partial risk and several other related topics. The authors include known regional examples of safety criteria for landslide and earthquake studies from across Canada.

45.2.8 Professional Practice

Although not technical in content, Open File 6981 (VanDine 2011; ftp://ftp2.cits.rncan.gc.ca/pub/geott/ess_pubs/289/289423/of_6981.pdf) examines what is meant by professional practice, reviews the current requirements for professionals involved in landslide studies, the qualification of professionals, responsibilities of professionals and general quality management in the Canadian context. This open file also addresses the related issues of professional liability and insurance including property owner's landslide insurance as applicable in Canada.

45.2.9 Other Documents

Two other key open files are almost ready for publication (as of September 2014) and they represent the final contributions to the collective effort. Hungr and Locat (in prep) provide a compilation and review of common landslide types from across Canada. Bobrowsky et al. (in prep) provide a review of the various methods and techniques commonly used to investigate, analyze, monitor and treat landslides in Canada and elsewhere.

45.3 Summary

With the above referenced open files, the GSC is now providing the Canadian professional landslide community with a series of readily accessible, online technical publications that provide guidance and best practice. These documents build upon and complement other existing international examples of landslide guidelines and best practices. The aim of this Canadian endeavor is to proactively contribute to a strong landslide risk reduction philosophy that is now shared across the globe.

Acknowledgements We are grateful for the enthusiastic support provided by Natural Resources Canada (Public Safety Geoscience Program).

References

Bobrowsky PT, VanDine D, Bean S, Powell S, Lato M (in prep) Site investigation, analysis, monitoring and treatment—Canadian technical guidelines and best practices related to landslides: a national initiative for loss reduction. Geological survey of Canada, Open file XXX

Couture R (2011) Landslide terminology—national technical guidelines and best practices on landslides. Geological survey of Canada, Open file 6824

Cruden D, VanDine D (2013) Classification, description, causes and indirect effects—canadian technical guidelines and best practices on landslides: a national initiative for loss reduction. Geological survey of Canada, Open file 7359

Guthrie R (2013) Socio-economic significance—Canadian technical guidelines and best practices related to landslides: a national initiative for loss reduction. Geological survey of Canada, Open file 7311

Highland L, Bobrowsky PT (2008) The landslide handbook—a guide to understanding landslides. USGS circular 1325

Hungr O, Locat J (in prep) Common landslide types in Canada—Canadian technical guidelines and best practices related to landslides: a national initiative for loss reduction. Geological survey of Canada, Open file XXX

Jackson LE Jr, Bobrowsky PT, Bichler A (2012) Identification, maps and mapping—Canadian technical guidelines and best practices related to landslides: a national initiative for loss reduction. Geological survey of Canada, Open file 7059

Porter M, Morgenstern N (2013) Landslide risk evaluation—Canadian technical guidelines and best practices related to Landslides: a national initiative for loss reduction. Geological survey of Canada, Open file 7312, p 21

VanDine D (2011) Professional practice and insurance issues—Canadian technical guidelines and best practices related to landslides: a national initiative for loss reduction. Geological survey of Canada, Open file 6981

VanDine D (2012) Risk management—Canadian technical guidelines and best practices related to landslides: a national initiative for loss reduction. Geological survey of Canada, Open File 6996

Wang B, Ruel M, Couture R, Bobrowsky PT, Blais-Stevens A (2012) Review of existing landslide guidelines—National technical guidelines and best practices on landslides. Geological survey of Canada, Open File 7058

Engineering Geology at Post-graduated Level at the NOVA University of Lisbon

46

Ana Paula F. da Silva, Pedro C. Lamas, M. Graça Brito and José A. Almeida

Abstract

The Portuguese and international construction companies have a long tradition of regularly employing Geological Engineers for their staff to deal with complex soil and rock problems. Good practices everywhere recommend the contribution of an expert in that area for several specific engineering interventions such as mining or tunnelling, dams, highway and railway constructions specially since they are continuously interfering with adverse geological environments. To ensure an adequate guidance and practice in engineering geology and to prepare the students for future professional challenges in those environments, the University NOVA of Lisbon offers a Master and PhD programs in Geological Engineering under the Bologna framework. The candidates of these programs are mainly undergraduates in geology or mining, geological or civil engineering. In this context, several topics have been highlighted for course's syllabus and for MSc and PhD thesis, such as:

- ground modelling;
- ground improvement and monitoring;
- excavability assessment of open cuts and rehabilitation of slopes;
- waste management and soil rehabilitation;
- special foundations requirements for renewable energy structures;
- risk management.

These topics demand the use of software for data acquisition, processing and visualization of results. This paper discuss the methodology used at the Nova University to comply with good practices in the domain of engineering geology as well as with innovation and quality assurance/quality control requirements of the engineering industry.

Keywords

Engineering geology • Teaching • Geological engineering • Standards • Modelling

46.1 Introduction

In 1986, in consequence of Portugal's admission in the European Economic Community - EEC, presently European Union - EU, the country was involved over the following two decades in governmental programs of public works, most of them urgent given the Portuguese significant delay in road infrastructures, energy demand and so on. This justifies that, until recently, the main employability areas for graduates in Geological Engineering were those related to

A.P.F. da Silva (✉) · P.C. Lamas · M.G. Brito · J.A. Almeida
CICEGe and Dept. Ciencias da Terra, Faculdade de Ciencias e Tecnologia, Universidade NOVA de Lisboa, Lisbon, Portugal
e-mail: apfs@fct.unl.pt

G. Lollino et al. (eds.), *Engineering Geology for Society and Territory – Volume 7*,
DOI: 10.1007/978-3-319-09303-1_46,

engineering design firms and contractors and, in less amount, the management and exploitation of natural resources and elaboration of geoenvironmental rehabilitation studies; this situation has recently overturned, due to the present economic crisis. In any case, as elsewhere, research institutes or universities in Earth Sciences, Geo-engineering, Environmental or Civil Engineering have been also potential employers.

This was the framework, since the mid-eighties, in which the Geological Engineering teaching, a five years course degree at Faculty of Sciences and Technology (FCT) of the Nova University of Lisbon (UNL), has developed and evolved until today (Rodrigues-Carvalho et al. 1994; Oliveira 2008, 2010; Silva and Lamas 2010), with a continuous update of subjects taught and the introduction of new syllabus in the Engineering Geological domain (Oliveira 1997; Silva and Lamas 2010).

In general, throughout those years, graduates quickly found employment in the above mentioned areas. The course has always been the one, in Portugal, with the higher component in Engineering Geology subjects in its syllabus and it has always been recognised by the Ordem dos Engenheiros (OE), the Portuguese Association of Engineers.

More recently, the Bologna process has introduced a new structure (Silva and Lamas 2010): a three cycle program, encompassing an undergraduate bachelor (3 years) followed by two optional Master Sc. courses (2 years), one dealing with Geotechnics, the other with Georesources; and finally, a 3 year PhD program. The syllabus of Geological Engineering (GE) course at FCT - UNL had to comply with a series of criteria embracing several perspectives, including recommendations from 5 year graduates alumni or by the industry demand requirements, the implementation of Eurocodes and European standards or the adoption of new technologies and specific tools, namely software, in the framework of Geo-engineering.

One year ago, a recent update has been introduced at all the FCT's MSc degrees, contemplating the teaching of Entrepreneurship subjects.

This paper discuss the methodology used at the Nova University to comply with good practices in the domain of engineering geology as well as with innovation and quality assurance/quality control requirements of the engineering industry.

46.2 Engineering Industry Demands and Evolution of the Syllabuses

As mentioned above, changes have been introduced in the syllabus of the course in order to progressively adapt it to the needs of civil and mining industry. Recently, this becomes more and more relevant, especially taking into account the mobility needs for newly formed young specialists in GE due to the globalization of all economic activities and, particularly in the case of Portugal, the search for better work conditions abroad. So, with the scope of preparing future professionals with the required levels of quality triggered by the progress of the society needs and the emergency of new technologies and methods (e.g. 3D modelling, ground improvement and monitoring, waste management and soil rehabilitation, use of recycled materials and risk management, just to name a few), several scientific and pedagogic efforts have been made to encompass the new skill and knowledge requirements so that the new specialists would be able to implement the best practices in all engineering or geological situations, anywhere around the world.

Clearly, the course syllabus has undergone several major changes, from its beginnings in mid-eighties, till 2000. Initially, the units of the Geology domain predominated in the curriculum, and they reached almost 50 % of the total credits units in 1996 (Silva and Lamas 2010). Inversely, the units incorporated in the Engineering Sciences domain showed a downward tendency during the same time, reaching less than one half of the Geology credits. Meanwhile, among the Engineering Sciences, there was an improvement in the weight of Engineering Geology contribution to the GE graduation, accounting for at least 50 % of the Geo-engineering weight in the syllabus ever since.

At the beginning of the 21st Century, a substantial increase in the contribution of the curricular units in the Engineering Sciences domain, specially GE, and a reduction of the ones from Geology, reinforced that domain. Meanwhile, particular specialized horizontal subjects where added, such as Occupational Safety and Health, to comply with OE directives.

Since the end of 2006, the Bologna Process and after a first 3 years cycle (bachelor/180 ECTS) of studies, two 2 years long separated second cycles of studies were considered to complete the GE formation in the domains of Georesources and Geotechnics respectively, with 120 ECTS each. In this new framework, Geology and Engineering Sc. domains preserve their relative weight, but with a reduction of the ECTS in the Geology domain. The curricular plan suffered some evolution towards the adjustment of the graduates' skill requirements to new requests triggered by the progress of the society and emergency of new technologies (namely 3D modelling). Former units making part of the new GE (Geotechnics) MSc curricular plan were adapted namely by developing subjects already embraced (geo-hazards, retaining structures, geo-environmental engineering), and new ones were also added, covering topics considered relevant for the new century challenges, namely ground improvement (Rocha et al. 2013, Santos et al. 2013), monitoring (Lopes et al. 2006) and engineering geological mapping (Brito et al. 2006; da Silva and Rodrigues-Carvalho 2006; Almeida and Kullberg 2011) and detailed site investigation methods (Almeida et al. 2005).

Since 2009, following the implementation of a third cycle's adapted to the Bologna Process, a PhD Program in GE was created and officially registered at FCT - UNL and specific investigation in the framework of risk analysis, 3D geological modelling (Ferreira et al. 2010; Quental et al. 2012; Charifo et al. 2013) and engineering geological characterization of natural materials have been undertaken.

46.3 Good Practices Requirements

In order to ensure a multidisciplinary education of the future GE specialists of the Nova University of Lisbon, subjects subsidiary of Engineering Geology have progressively integrated the course syllabuses, allowing a more complete understanding on the consequences of mining or civil engineering interventions upon the geological environment, and its reverse. Therefore, the knowledge of the expert in Engineering Geology for the 21st century must embrace the teaching of new or updated:

(a) technics or methods of site investigation, ground improvement or construction,
(b) standards of testing materials, both from European Union (EN) or American Society for Testing and Materials (ASTM), their revised versions as well as some technical recommendations (e.g. International Society of Rock Mechanics - ISRM),
(c) geological, hydrogeological or geomechanical modelling of ground and dimensioning requirements, namely according to Eurocodes 7 and 8,
(d) stability analysis and risk management,
(e) observational methods, monitoring and quality assessment/quality control methods,
(f) rehabilitation methods, namely of cut and natural slopes or geotechnical structures - e.g. dams,
(g) remediation and environmental requalification of derelict industrial areas.

The purpose of this approach is to allow the future specialist to have an integrated vision of:

- the behaviour of the structure or infrastructure to build/exploit resources,
- the main study tools to use during the design stage, whether for site investigation, lab testing or modelling/dimensioning,
- the selection and characterization of construction materials, including recycling materials,
- the applicability and procedures of the major ground improvement/remediation methods, to mitigate the effects of adverse conditions at a selected site,
- the key construction/exploitation procedures and their main applications and limitations to contribute to a safe and economic engineering intervention, being able to integrate risk assessment teams;
- the different types of monitoring surveys that may allow to complement and control the engineering construction/exploitation behaviour,
- to assess the rehabilitation needs of a previous structure/infrastructure and therefore to contribute to the definition of the best final adopted solution.

One may think it's a very long and bold list of subjects to be taught, but the MSc formation must advance the basic knowledge on them, allowing the future specialist to expand these skills according to their needs, whether autonomously, or in the scope of specific advance courses. The ultimate purpose is to give the post-graduated, from the first moment, the aptitude to have a wide area of intervention.

At the PhD level, the applied research will usually pick one of those themes, and will contribute to an innovative scientific or technical advance in one of the themes listed previously.

46.4 Final Remarks

The post-graduation courses in the framework of engineering geology offered at the FCT - UNL are MSc and PhD degrees in Geological Engineering since its beginning. The underlying rationale for this is that it is considered adequate to prepare the future specialist in areas subsidiary to engineering geology in order to allow the future expert to face challenges in a more interdisciplinary mode and therefore apply knowledge and skills acquired, and to subsequently expand applied thematic knowledge, supported by advanced training.

Recently, the pursuit of sustainable development policies has increased the need for developing new areas of engineering geology studies mainly those related to the rehabilitation of waste landfill sites, requalification of mining areas and tailing dams and the requalification of *brownfield* areas.

The professional development would be facilitated by the basic knowledge learned there, allowing the necessary preparation and confidence to meet the increasing responsibilities that will be faced by the post-graduate.

The knowledge gained over the years of study at FCT - UNL is considered essential for understanding the geotechnical and geo-environmental defiances and to a successful integration at the professional level, both in Portugal and internationally, since presently there are MSc and PhD graduates in countries like Angola, Australia, Colombia, France, Ireland, Mozambique, The Netherlands, Peru, Poland, Spain, Switzerland, United Kingdom or Venezuela.

References

Almeida JA, Kullberg JC (2011) Rockfall hazard and risk analysis for Monte da Lua, Sintra, Portugal. Nat Hazards 58:289–310

Almeida JA, Rocha MS, Teixeira A (2005) Spatial characterization of limestone and marl quality in a quarry for cement manufacturing. In: Leuangthong O, Deutsch CV (eds) Geostatistics Banff 2004, Quantitative Geology and Geostatistics 1. Springer, The Netherlands, pp 399–408

Brito MG, Costa CN, Almeida JA, Vendas D, Verdial PH (2006) Characterization of maximum infiltration areas using GIS tools. Eng Geol 85:14–18

Charifo G, Almeida JA, Ferreira A (2013) Managing borehole samples of unequal lengths to construct a high-resolution mining model of mineral grades zoned by geological units. J Geochem Explor 132:209–223

da Silva AP, Lamas PC (2010) Teaching engineering geology at NOVA university of Lisbonâ€"An update and prospective. In: 11th International IAEG Congress Proceedings, New Zealand. CRC Leiden, Press/Bal-kema, pp 4079-4084

da Silva AP, Rodrigues-Carvalho JA (2006) Engineering geological mapping for the urban planning of Almada County, Lisbon, Portugal. In: Culshaw MG et al (eds) Engineering geology for tomorrow's cities, IAEG paper 165. The Geological Society, Bath

Ferreira A, Charifo G, Almeida JA (2010) 3D geological modelling, the example of the Farim-Saliquinhé phosphates mineralization. Geophysical Research Abstracts, EGU General Assembly 2010, vol 12. Vienna, European Geoscience Union, p 10927

Lopes M, Avillez G, Costa CN, Almeida JA (2006) Groundwater contamination plume monitoring in sealed waste dumps. Eng Geol 85:62–66

Oliveira R (1997) Teaching environmental subjects in engineering geological education. In: Proceedings of the international symposium on engineering geology in the environment, pp 3649-3654

Oliveira R (2008) Geo-engineering education and training: the past and the future. In: First international conference on education and training in geo-engineering, Constantza (Romania), pp 79-86. 2-4 June

Oliveira R (2010) Geologia de Enegnharia. Génese, evolução e situação actual. Revista Luso-Brasileira de Geotecnia 118:3–13 (in Portuguese)

Quental P, Almeida JA, SimÃµes M (2012) Construction of high-resolution stochastic geological models and optimal upscaling to a simplified layer-type hydrogeological model. Adv Water Res 39:18–32

Rocha C, da Silva APF, Santos-Ferreira A (2013) Geotechnical characteristics and design restrictions of Tavira (Portugal) fishing port: soil improvement for vertical quays. In: Indraratna B, Rujikiatkamjorn C, Vinod J (eds) Ground improvement and ground control, transport infrastructure development and natural hazards mitigation, vol 1. Research Publishing, Singapore, pp 568–571

Rodrigues-Carvalho JA, Chaves MA, da Silva AP (1994) Developments on teaching engineering geology for undergraduate students in the New University of Lisbon. In: Oliveira R et al (eds) 7th International IAEG Congress Proceedings, Lisboa, vol 6. Rotterdam, Balkema, pp 4875-4880

Santos C, da Silva APF, Santos-Ferreira A (2013) Jet grouting underpinning of a building on a marl embankment platform. In: Phung DC (ed) Geotechnics for sustainable development. Construction Publishing House, Hanoi, pp 451–456

Dangers of Foundations and Below Ground Construction in the Coastal Zone of the Rhine Delta Sediments, The Netherlands

47

Pieter Michiel Maurenbrecher

Abstract

Visitors to the Netherlands generally comment how neat the land looks. This visual impression starts from coming into land by air and remains on transit at the airport and transfer to one's destination in Holland. A mess does exist and is hidden from view; it's beneath the ground surface. Welcome to the delta on the North Sea consisting as most deltas of weak or soft alluvial soils. The mess reflects the way it is dealt with when attempting to build something below the surface such as foundations or underground spaces, accompanied by lack of comprehensive procedures for dealing with the subsurface when making proposals to identify and deal with risks to the environment. Surprisingly guidelines, statutes and codes of practice are being produced as a consequence of increasingly ambitious projects utilizing and having a bearing on the subsurface. On hindsight many risks could have been reduced considerably by simple monitoring of the reaction of the subsurface during construction and life time of the structures.

Keywords

Delta • Foundations • Tunnels • Risks • Guidelines

47.1 Introduction

A main topic of the Engineering Geology course at Delft was, set up by the late Professor David Price in the late 1970s and 1980s, was Foundations and slopes: soils and rocks, static and dynamic loading. Much of these lectures were delegated to his assistant professors and in due course with changes in curricula was split and updated. On retirement he endeavoured to further edit an update the engineering geology lecture course into a book. Published posthumously the book, in draft, was reviewed and further edited and finally published by close colleagues under the editorship of Dr. Michael de Freitas, (Price 2009).

The bias of the chapters concerning foundations and the extraction of ground fluids increased over the years towards the situation in the Netherlands: to the soft and loose sediments of the predominantly Quaternary-Holocene fluvial-deltaic glacial land deposits (from north to south) Eems, Rhine, Meuse and Scheldt estuaries; of which the coastal provinces of Groningen, Friesland, North Holland, South Holland and Zeeland are the most urbanized and have the worst ground conditions. Designing, constructing and investigating for foundations for new structures or structures showing distress is not a straightforward situation of deciding a choice between shallow and piled foundations based on the bearing capacity of the ground beneath it. The initial site investigation for foundations are fast and simple: a few cone penetration tests are made in one morning to about twenty meres depth. The same day a report is issued recommending the number and size of piles and the depth of penetration (usually to a dense sand layer varying in depth from 10 to 20 m where most of the bearing load is

P.M. Maurenbrecher (✉)
Prinsesmariestraat 19, 2514 KD, Hague, The Netherlands
e-mail: maurenbrecher@yahoo.com

G. Lollino et al. (eds.), *Engineering Geology for Society and Territory – Volume 7*,
DOI: 10.1007/978-3-319-09303-1_47,

transferred to through the base of the pile). Most piles would be precast concrete driven piles.

47.2 History of Subsurface Structures

Subsurface structures besides foundations historically consist of drainage facilities not only as sewers but also buffer basements for temporary storage for heavy (storm) flows, culverts and conduits (usually along the line of ditches, channels and canals which have been in-filled. Tunnels and below surface roadways in trenches were constructed in the 1930s. The Maas tunnel beneath the main water way of Rotterdam harbour followed the design of the New York road tunnels of the Hudson River consisting of floating and sinking reinforced concrete caisson units into a prepared trench and ballasting the caissons with rock fill or by anchor tension piles. The Maas Tunnel north approach is by a dual carriageway below ground level. 30 years later a similar carriageway was built from the east into the centre of the Hague. Since then many office blocks span the carriageway of which one was founded on the trench retaining walls and on a central pier resting on the road way concrete base; the tension piles beneath reverting into compression!

Towards the end of the 1980s a Tunnel Boring Machine (TBM) with earth pressure balance drill face bored a tunnel beneath the Oude Maas river south of Rotterdam as a trial for later use along the North South Line (NZL; Noord Zuid Lijn) metro, Amsterdam. Despite one passive blow-out failure causing tunnel flooding and delays, the over-all experience was adequate to recommend its use as a tunneling method in Amsterdam.

47.3 Plans, Project-Finance, Politics, Public, Permits, Procedures and Perils

Amsterdam's first metro built in the 1980s using traditional trenches supported by retaining sheet piles and diaphragm walls and a concrete floor. The structure was prevented from becoming buoyant by its weight or with the aid of tension piles. The top of the trench was spanned with a reinforced concrete roof and paved usually by brick to revert to the original street scene. The technique was well established and predictable. But in the narrow streets with historic buildings in close proximity and often including shopping streets the disruption was enormous. After this experience the subsequent proposed north-south metro would be fraught with vociferous opposition as Amsterdam and its visitors did not want years of disruption. The budget for its construction using less disruptive methods meant further opposition at municipal levels. On the positive side the line would connect Amsterdam's northern suburbs, the centre and southern suburbs. The line would have to be tunneled below the River IJ separating the two halves of the city, beneath the monumental central station and pass through the heart of the city (Dam square and the Rokin) ending for the time being at the Europa square near to the RAI station to the south. Many public hearings were held accompanied by vociferous objectors consisting of both business and residents/property owners groups. . Existing transport facilities were not coping and the city's northern half, much expanded in post war years threatened to become isolated.

To placate the opposition an elaborate geodetic survey network was set up to continually monitor the slightest movement of buildings next to the metro route. These assurances did not prevent the more influential businesses such as Amsterdam's main department store at the Dam square from attempting to procure a court injunction to delay the TBM from boring beneath its premises. The concern of the Department store was not ill founded as further south the first failure occurred; the Vijzelgracht station sprung a leak through a gap in the diaphragm wall construction joint resulting in removal of sand from a geological horizon known as '1st Sand Layer'. This layer supports many of the wooden driven piles beneath the historical buildings so that despite the monitoring the building alongside the station suddenly subsided to cause its residents to be evacuated for safety. The NZL went into crises as admonitions came from the major and aldermen of Amsterdam city council causing the director of the NZL consortium to be relieved of his job. Construction was stopped at the other stations leaving the site engineers little option but to review the monitored data and was not resumed until the project had been reviewed with regard to possible further failure risks and assess the costs that such failures would entail. Ground freezing was adopted to lessen chances of leakage from defects in the diaphragm walls.

47.4 Other Projects Other Failures and Learning from Them

Preceding the Vijzelgracht station leak by ten years was a major leak in the floor of the tram tunnel in The Hague. The tunnel constructed as a trench sprung a leak in its concrete floor due to some blemish for which no adequate detection method has been devised. Completion was delayed by over a year and divers plugged the leak by grout injection in a sealed high pressure chamber over the inflow. Decompression chambers were required for the divers after working one shift (The method was used in the deepest NZL station, the Ceintuurbaan to inspect the base floor and carryout further grouting where blemishes were found.).

One of the recent trams through The Hague's tram tunnels is the nearest to what The Hague has resembling a Metro. Larger and faster than the municipal trams, the Metro is known as Randstad Rail. One line extending over a combination of tramlines and railway lines is from The Hague South to the satellite town of Zoetermeer, the other line from The Hague Central Station to Rotterdam Central Station using a new overhead railway line in The Hague and a refurbished regional railway line from The Hague to Rotterdam. It connects with stops and stations the towns of Leidschendam–Voorburg (effectively suburbs of The Hague) and the satellite towns of Nootdorp, Pijnacker and Berkel and Rodenrijs (with a shuttle bus to Rotterdam–The Hague airport).

Unlike the NZL, Randstad Rail did not have to contend with much opposition. The line improved fast access to the main line stations of The Hague and Rotterdam and their town centres. The satellite towns had more frequent trains than the regional train that ran before. Stations were relocated, especially that of Pijnacker which was moved towards the main road from Delft to Zoetermeer and connecting with the bus line Delft-Pijnacker-Zoetermeer. The station went below ground level so that vehicles above could move unhindered as the previously a level crossing caused hold ups when the trains passed or stood still at the old station nearby.

The construction of the new station did require a building permit issued by the Province of South Holland. Its granting required satisfying environmental issues with regards to its construction and its operation. The application is supported with documentation showing not only the design but also specifications dealing with the environmental impact during both construction and subsequent operation. Aspects discuss drainage, groundwater regime, disposal of any dewatering, and the quality of the water. Similar documentation would be sent to the local drainage authority for approval for any changes made (diversion works, drainage of rain water and possible seepage into the station) or add further conditions. The drainage authority is responsible for sewerage, main drains, prevention of flooding, canals and lakes, maintaining water levels in open water systems and groundwater. The documentation for the permit request would contain specifications for dewatering of the station excavation: dealing with the flow rates, disposal of the water and monitoring: extraction flows, sampling for contaminants and maintaining groundwater levels. At the start of construction the provincial inspectors issued reprimand letters to the contractor for lack of flow readings and having faulty meters.

Specifications would also be included to predict the levels of disturbance from vibration and noise. Ideally these would also be monitored. The minimum requirements would be photographic damage surveys carried out on buildings, but also paving, utility structures such as culverts before commencement of works before claims should arise for damage.

The building nearest to the sheet pile and pile driving for the viaduct, a villa and its pharmacy extension suffered severe settlement. The owner attributed this to vibrations when driving piles and sheet piles. Piles were driven within 12 m of the property. No monitoring was done prior to settlement occurring; Prorail, the client for the station engaged a reputable geotechnical firm to examine the situation. Tell-tales were placed over cracks and survey level points installed around the perimeter of the building. Measurements were made at regular intervals (on average once a month) for the next two years. The settlement leveled out with time as if a full scale consolidation test had been performed. The villa shallow foundations were examined and calculations showed that the bearing capacity was inadequate for today's standards. The geotechnical report included some sketches but gave no comment. One sketch showed contour lines of settlement within the perimeter of the building indicating a tilt towards the deepest section of the station and viaduct Another sketch showed sheet pile anchors extending to beneath the building, again, without comment. Anchors can cause settlement in weak clays (Kempfert and Gebreselassie 1999). The CPT profiles show soft peaty silty clays extending to 19 m below ground level below which the older sand layers deposited in the last ice age; piles are driven to this layer. Pile driving does not seem to effect the cone resistance from tests performed before and after except in sands. The geotechnical report stated never-the-less that the damage was caused by the works. A case history in (Maurenbrecher 2009) discusses shortcomings with regards to driving of open-ended tubular piles as an only option in a small plot in the centre of The Hague. Suffice to say the contractor was totally unaware of the vibrations and excessive settlement this would cause to the adjoining building. The then emeritus professor of foundations at TU Delft said the cause could be sand infiltration into the base of the pile if too much bailing of water is done to encourage penetration. The paper indicates that densification of the sands could cause substantial settlements based on CPTs cone resistance only in sands.

47.5 Guidelines, Codes of Practice and Statutes

Sufficient activity has taken place the last three decades with the ensuing experience and research to provide better guidelines involving standards, codes of practice and even statutes when embarking on making changes to the subsurface in the urban-deltaic area which constitutes much of the Netherlands. This is because often disturbance and damage to third parties is caused through these activities. Not only construction activities induce changes; they can also occur in non-urban areas such as distress caused by extraction of groundwater or, even, fossil fuels especially that of the Groningen gas field. Much has been published and considered with regard to consequences of fluid (water and/or hydro-carbon) extraction associated with land subsidence in all the major producing field worldwide, even in the Po Delta of Italy (Price 2009). The consequences are no less serious than that of subsurface activities in urban areas, the land-surface area which settles can extend over many kilometres and sink up to two or three metres drastically altering adversely the gradients of the existing drainage infrastructure.

Building activities also cause changes in the subsurface. Two recent publications addressing the influence of pile driving and underground works are by (Meijers 2007) and (Korff 2012) respectively. Significant about these publications is that settlement and subsidence can be predicted. This follows that a good deal of modeling can be done before construction begins to predict and consequently minimize risk to damage caused by sheet piling activity and construction of tunnels and cut and cover metro stations usually involving deep diaphragm walls. Meijer's study was initiated as a result of sheet piling along extensive works in the denser sands associated with upgrading of the A2 motorway by-pass at Eindhoven. Korff's work is based on the NZL construction in Amsterdam and its now extensive database from the geodetic monitoring. The survey monitoring may not have provided sufficient warning for the Vijzelgracht leakage but they have shown that much of the modeling that predicted displacements did, I fact, occur as predicted.

The NZL pointed towards better direct engineering geophysics testing methods to assure quality control of jet-grouted piles for underpinning the monumental Central Station and belatedly diaphragm walls construction joints (Spruit et al. 2014). The construction joints of the new underground route of the railway through Delft were checked in this way. The method resembles cross-hole sonic profiling. Sufficient measurements can be made to produce a tomographic picture of the construction joint indicating the size of defects in addition to their location. Up to submission of this paper the Delft railway cut and cover tunnels and new underground station platforms has been on schedule. Such techniques should be become standard for all subsurface construction. van der Salm et al. (2012) mention that much better measurements of the shear forces causing compaction can be obtained from geophones inserted in the subsurface. The success of geophysical sonic techniques could easily be used to test the integrity of the base-floor of the cut and cover tunnels by lowering transmitters and receivers through existing access probe tubes for the diaphragm wall joints to floor level at laterally opposite walls. Tomographic analysis can be made in the horizontal plane by placing many receivers (geophones) at floor level as there would be sufficient access tubes for these at construction joint probe access tubes.

More practical codes of practice, standards or preferably even statutes should ensure that monitoring should start prior to construction begins especially when monitoring ground water levels which for Pijnacker gave confusing and misleading results (insufficient standpipes and infrequent measurements) so that large ground water fluctuations appeared to be errors with insufficient standpipes nearby to compare with. Weak peaty clays can consolidate considerably by only half a metre change in water levels; the measurements should be performed at the same time as the geodetic measurements. Both measurements should be mandatory.

References

Kempfert HG, Gebreselassie B (1999) Effect of anchor installation on settlement of nearby structures in soft soils. Proceedings international symposium On Geotechnical Aspects of Underground Construction in Soft Ground, IS-Tokyo '99, 19–21 July 1999, pp 665–670

Korff M (2012) Response of piled buildings to the construction of deep excavations. Ph.D. thesis, University of Cambridge, IOS Press BV, Amsterdam, p 280

Maurenbrecher PM (2009) The smaller the plot the larger the problems: inadequacies of site investigation in The Hague. In: Culshaw MG, Reeves HJ, Jefferson I, Spink TW (eds) Engineering geology for tomorrow's cities. Geological Society, London. Engineering Geological Society Special Publication No. 22 (paper 443, on CD-ROM insert)

Meijers P (2007) Settlement during vibratory sheet piling. Ph.D. thesis, TU Delft, p 235

Price DG (2009) In: de Freitas MH (ed) Engineering geology, principles and practice, 1st edn. Springer, Berlin, p 450

Spruit R, van Tol F, Broere W, Slob E, Niederleithinger E (2014) Detection of anomalies in diaphragm walls with crosshole sonic logging Can. Geotech J 51:369–380

van der Salm R, Holscher P, Snethlage A-J (2012) SBR-A Richtlijn niet eenduidig voor trillinggevoelige funderingen (SBR-A Guideline not unambiguous for vibration susceptible foundations). Geotechniek 16(3):8–11

Part V

Fifty-Year-Long History of IAEG in Events and Personalities

Convener: Dr. Olga Eremina—*Co-conveners*: Kiril Anguelov, Julia Stanilovskaya

The 50-year-long IAEG history involves many interesting events, facts (congresses, conferences, symposia), and more so—it has known many outstanding personalities, i.e., scientists of worldwide repute, who were the founders of our association and who made a great contribution to the development of engineering geology on a world scale. It is our duty to keep their names and deeds in the memory of the young generation of geoscientists. This Congress is a high time to arrange such a part, until there still remain many witnesses of the events that took place during this 50-year-long period, as well as still working colleagues and disciples of the remarkable founders of IAEG, who could share their memories. Associate Prof. Dr. Kiril Anguelov (Bulgaria) is among those who participated in the IAEG work since its very early period, and I believe him to be a very appropriate co-convener to this part. Mrs. Julia Khalilova (Russia) is a young researcher in environmental geoscience, a chairperson of YES organization in Russia.

Evgenii Mikhailovich Sergeev (1914–1997) as One of the Founders and Leaders of IAEG

48

Victor Osipov, Olga Eremina and Nadezda Rumyantseva

Abstract

The paper elucidates the role of an outstanding Soviet and Russian scholar, academician of the Russian Academy of Sciences E. M. Sergeev in the development of International Association of Engineering Geology. The short biography of E. M. Sergeev is provided, the fields of his professional activities are described. The main achievements of Prof. Sergeev in the office of Vice-President (1972–1978) and President of IAEG (1978–1982) are analyzed. It is shown that academician Sergeev has made undoubtedly a significant contribution to the foundation and development of international community of engineering geologists.

Keywords

Evgeny M. Sergeev • IAEG president • Environmental problems • Engineering geology of the Earth • Hans Cloos medal

Evegenii Mikhailovich Sergeev (Prof., Dr. Sci. in Geology, Full Member of the Russ. Acad. Sci) was an outstanding scholar in the field of engineering geology; First Vice-Rector of the Moscow State University, Head of the Department for Soil and Rock Engineering and Engineering Geology at the Moscow State University; founder of the Scientific Centre for Engineering Geology and Geoenvironment at the Russian Academy of Sciences, Vice-President (1972–1978), President (1978–1983), and Past President (1983–1997) of the International Association of Engineering Geology (Fig. 48.1).

Prof. E. M. Sergeev was born in Moscow, and graduated from the Moscow topographical college in 1932. He started his professional carrier as a topographer in the Far East of Russia, where he worked for 3 years. In 1935, he entered the Moscow State University (MSU), the Geological and Soil Science Faculty. Since then, his entire life was closely connected with geological science. In 1940, Evgenii Sergeev graduated with honor from the Moscow State University, and took the position of the assistant professor at the department of soil and rock engineering of MSU. The Second World War interrupted his scientific research. In 1941, from the very first days of Nazi invasion to the USSR, he joined the Soviet army as a volunteer; and fought at the War front lines until 1943. E. M. Sergeev was badly wounded in the Stalingrad battle: he lost his leg. After the hospital, he returned to the Moscow State University in the autumn 1943. In 1944, E. M. Sergeev defended his Ph.D. thesis entitled "The moistening heat of soils", and in 1954 he presented and defended his Doctoral thesis on the topic "Genesis and Composition of Soils and Rocks as the Basis for Their Classification and Properties Study". Since 1954 for nearly 35 years Professor Sergeev headed the Department of Soil and Rock Engineering and Engineering Geology at the Geological Faculty of Moscow State University. In the years of his guidance, this department became an undisputable leading center of engineering geology in the entire USSR. E. M. Sergeev was the dean of Geological Faculty of MSU in 1954–1957 and in 1963–1964; and he held the position of Vice Rector of

V. Osipov (✉) · O. Eremina · N. Rumyantseva
Sergeev Institute of Environmental Geoscience, Russian Academy of Sciences, Ulanskii per, 13, bld. 2, Moscow 101000, Russia
e-mail: direct@geoenv.ru

G. Lollino et al. (eds.), *Engineering Geology for Society and Territory – Volume 7*,
DOI: 10.1007/978-3-319-09303-1_48, © Springer International Publishing Switzerland 2014

Fig. 48.1 E. M. Sergeev

Fig. 48.2 E. M. Sergeev surrounded by his former students, young professors of Moscow state university

Moscow State University in 1964–1978. In his last years (1989–1997) academician Sergeev was a MSU rector councilor.

The main fields of his professional interests included the soil and rock engineering, regional engineering geology, the environment protection, theory and methodology of engineering geology. He benefited considerably to the development of genetic principles of soil and rock engineering, genetic classification of soils and rocks, to the study of physico-chemical, and physico-mechanical properties of fine soils (sand, loess, clay) as multi-component systems, etc. Within regional engineering geological studies, he organized and led large-scale field investigations in various regions of the Soviet Union, including the Far East, the Eastern and Western Siberia, the Central Asia, and the European non-chernozemic part of Russia. The results of these works were generalized in an 8-volume fundamental monograph "Engineering Geology of the USSR", which was awarded with the State Prize in 1982.

Evgenii M. Sergeev was elected to the USSR Academy of Sciences in 1966 as a Corresponding Member, and in 1979 as a Full Member. He made strong efforts to promoting engineering geology and the environment protection as a new field of natural science within the Academy of Sciences. He was an organizer and the chairperson of the Scientific Council on engineering geology and soil and rock engineering at the USSR Academy of Sciences (1966–1990); he was an organizer and editor-in-chief of the Russian academic journal "Engineering Geology" (1978–1990). On his initiative, Institute of Environmental Geoscience was arranged within the Russian Academy of Sciences (1996).

Apart from profound research, administrative, and organizational activities, Evgenii Mikhailovich Sergeev was a distinguished teacher of students. At the Moscow State University, he delivered lectures in soil and rock engineering for many years; he published above 500 scientific papers including fundamental monographs and textbooks. Under his guidance, more than 75 researchers prepared and defended their Ph.D dissertations, and 12 scientists defended their Doctoral dissertations (Fig. 48.2).

Prof. Sergeev made a great input to the strengthening and broadening of international scientific relations. He was an Honorable Doctor of Bratislava (1972) and Warsaw (1974) universities, a foreign corresponding member of the Belgian Royal Geological Society (1974), and a council member of the International Association of Universities (1965–1975).

Acad. Prof. E. M. Sergeev was tied up with the IAEG since late 1960s, from the very beginning of its organization. He took an active part in the development of the Association together with his colleague and friend Prof. M. Arnould (Fig. 48.3).

In the framework of IAEG activities he organized in the USSR several International Symposia: "Engineering Geological Properties of Clays and Processes in Them" (1971); and "Genetic Principles of Engineering Geological Study of Soils and Rocks" (1974), Symposium "Engineering geological problems in hydrotechnical construction" (Tbilisi, 1979). As an IAEG Vice-President, he took an active part in preparation and holding of the 2nd and 3rd IAEG Congresses in San Paolo, Brazil (1974), and in Madrid (1978). In 1978 at the 3rd International Congress of IAEG in Madrid, Spain, Prof. E. M.Sergeev assumed the office of IAEG President, for the 4-year-long period (since 1978 till 1982). As the President of IAEG, Prof. Sergeev felt his high responsibility for the future of the entire international community of engineering geologists; he put many efforts for increasing the number of national groups of IAEG, for broadening communications among national groups, and for involving low-developed countries in Africa, South America and South East Asia to IAEG. As the IAEG President, academician Prof. Sergeev was in charge of organization of the IAEG Congress in New Delhi, India (1982), the engineering geological session within the framework of the

Fig. 48.3 E. M. Sergeev welcomes M. Arnould at Moscow state university

International Geological Congress in Paris, France (1980), as well as other IAEG events such as International Engineering Geological Symposium in Istanbul, Turkey (1981), Symposium on Engineering Geological Mapping for Planning, Design and Construction in Civil Engineering (1979, Newcastle-upon-Tyne, UK), Symposium on changes of the geological environment under the influence of man's activity: engineering geological estimate and prognosis (1979, Poland) (Fig. 48.4).

President E. M. Sergeev (Russia) and General Secretary R. Wolters (Germany) despite political and even former military disputes between their countries, being both severely wounded in the World War II, formed a very friendly and highly productive team in IAEG Executive Committee. They exchanged mutual visits and carried out the entire work in Association in good cooperation. As a result of President Sergeev's and General Secretary Wolters's activities, during the decade 1972–1981, the number of IAEG national groups rose from 14 to 42, the number of members rose from 960 to 4200, the number of associated members from 21 to 86, and individual membership appeared in 30 countries. The IAEG Newsletter was first issued in June 1974. The main scientific journal of Association, "the Bulletin of the International Association of Engineering Geology" (R. Wolters, Editor-in-Chief, W. Dearman, English Editor, M. Arnould, French Editor) since 1974 has increased significantly its volume and circulation. The record in Bulletin volume was achieved in 1980, when it amounted to 700 pages covering the proceedings of two major IAEG Symposia held in 1979 (on Engineering Geological Mapping, Newcastle, UK; and on the Engineering Geological Problems in Geotechnical Construction, Tbilisi, USSR). The circulation of IAEG Bulletin rose from 500 copies in 1973 to 3000 copies in 1980 (Fig. 48.5).

The early unexpected death of R. Wolters in 1981 (at the age of only 58 years) was a severe shock for IAEG Executives, and personally to President E. M. Sergeev. According to President's nomination, in 1982 R. Wolters was posthumously awarded with the Hans Cloos medal, the supreme award of IAEG. In 1984, the Council meeting held in Moscow, following the initiative forwarded by E. M. Sergeev to commemorate R. Wolters, suggested to establish a new IAEG award, **the Richard Wolters Prize,** to be presented to young professionals in engineering geology for their outstanding scientific achievements. This award was officially adopted at the IAEG Congress in Buenos Aires in 1986; and its first recipient in 1988 became Dr. Kiril Anguelov (Bulgaria), one of the students and followers of Prof. E. Sergeev.

Fig. 48.4 IAEG officials at the meeting in Hannover, Germany

V. S. Shibakova, then the secretary of the Russian national group of IAEG, recollects that E. M. Sergeev's main idea at the position of IAEG President was to activate the work of Association, to raise its role and importance in the International Union of Geological Sciences, to unite professional efforts of national groups, to increase their membership, and in general, to bring engineering geology to a higher development level. In his President's address (published in IAEG Newsletter No 7, March 1979) he said: "It is very gratifying to note that Engineering Geology is a good example of a science now undergoing extensive development. This is very remarkable for Man has become a prominent geological force and his growing impact on

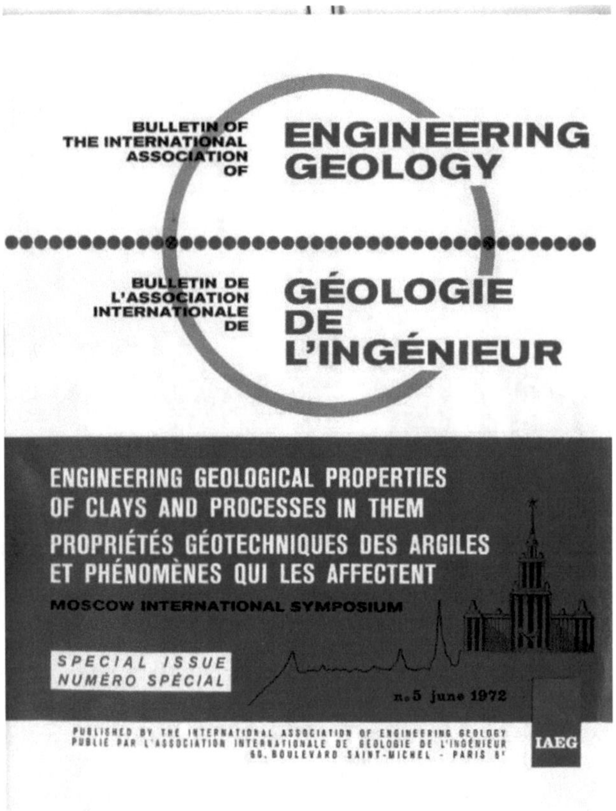

Fig. 48.5 Special issue of IAEG bulletin on Clay symposium in Moscow

Nature may be well compared with the effects of geological processes."

In 1980, at the International Geological Congress in Paris, E. M. Sergeev as the President IAEG, for the first time at the international tribune announced the idea about the importance to involve environmental problems in the field of study by engineering geology. With the permanent concern for engineering geology and its future, he was convinced that the rational use and protection of natural environment should be one of the most important targets of engineering geological investigations. With the permanent concern for engineering geology and its future, he was convinced that the rational use and protection of natural environment should be one of the most important targets of engineering geological investigations. Based on this thesis, the General Assembly of IAEG adopted the Declaration IAEG on its participation in the solution of environment problems (1980). This declaration also reflects President Sergeev's basic scientific ideas: the genetic principle and multi-component approach to the study of soils and rocks for engineering geological purposes.

In order to strengthen communications among engineering geologists all over the world and to promote studies in regional engineering geology, President IAEG Evgenii Mikhailovich Sergeev took an initiative to write an international monograph on Engineering Geology of the World by united efforts of engineering geologists all over the world. The book was published in Moscow, in 1989 by Nauka Publishers, with 1262 number of printed copies. The same year the monograph was presented at the IAEG exhibition stand at the International Geological Congress in Washington, USA.

In 1986, for the outstanding contribution to the development of IAEG, for the outstanding services in the field of engineering geology and promotion of international cooperation academician Prof. E. M. Sergeev was awarded with the highest award of the Association—the Hans Cloos medal.

In June 1997, when E. M. Sergeev has passed already, the IAEG Council meeting in Athens adopted the decision to rename IAEG into the International Association of Engineering Geology and the Environment. The decisive role that influenced the voting results was the reminder that the appeal to geoscientific community for paying attention to environment protection was first announced by IAEG President E. M.Sergeev in 1980.

The 50th Anniversary of the First International Conference on Permafrost

49

Jerry Brown and Julia Stanilovskaya

Abstract

The First International Conference on Permafrost was held at Purdue University, School of Civil Engineering, West Lafayette, Indiana, United States, on 11–15 November 1963. Some 285 registered engineers, researchers, manufacturers and builders participated representing Argentina, Austria, Canada, Germany, Great Britain, Japan, Norway, Poland, Sweden, Switzerland, the USA and the USSR. This was the first post-World War II major contact with a group of senior Soviet frozen ground researchers. The Proceedings is considered to be the first multi-national, English-language collection of papers devoted entirely to permafrost topics. Since 1963, nine additional international conferences have been held: two more in the United States (Fairbanks 1983, 2008), two in the Soviet Union and Russia (Yakutsk 1973, Salekhard 2012), two in Canada (Edmonton 1978, Yellow-knife 1998), and one in Trondheim, Norway (1988), Beijing, China (1993), and Zurich, Switzerland (2003). The International Permafrost Association (IPA) was formed in 1983 and subsequent conferences were under the IPA auspices. A brief review of these 10 conferences is presented. Discussions between IAEG and IPA took place by correspondence and at a meeting in Vail, Colorado, in June 2007 where it was decided to initiate a commission on engineering geology in permafrost regions.

Keywords

Conference on permafrost • 50th anniversary • International permafrost association

49.1 First International Conference on Permafrost

The year 2013 marked the 50th anniversary of the First International Conference on Permafrost (ICOP), held at Purdue University's School of Civil Engineering in West Lafayette, Indiana, United States, 11–15 November 1963 (Woods and Leonard 1964). The conference was an historic event in that it brought together, for the first time, leading researchers and practitioners from North America and other countries with diverse interests and activities in the study and applications of perennially frozen ground, cold regions engineering, and related laboratory investigations. The 285 registered participants represented engineers, researchers, manufacturers and builders from the USA (231), Canada (42), the USSR (5), Sweden (3) and Argentina, Austria, Great Britain, Japan, Norway, Poland, Switzerland, and West Germany. This report is based, in part, on a paper published in 2013 (Brown 2013).

The conference was organized by the Building Research Advisory Board of the U.S. National Academy of Sciences–National Research Council (NAS-NRC). The carefully edited Proceedings, published in 1966 by the NAS (NAS

J. Brown
International Permafrost Association, P.O.Box 7, Woods Hole, MA 02543, USA

J. Stanilovskaya (✉)
Sergeev Institute of Environmental Geoscience, RAS, P.O.Box 145 Ulansky Per. 13, Moscow, Russia 101000,
e-mail: stanilovskaya@geoenv.ru; stanik85@mail.ru

G. Lollino et al. (eds.), *Engineering Geology for Society and Territory – Volume 7*,
DOI: 10.1007/978-3-319-09303-1_49,

Table 49.1 Summary statistics for the ten international conferences on permafrost and publications

Conference	First	Second	Third	Fourth	Fifth	Sixth	Seventh	Eighth	Ninth	Tenth
Year	1963	1973	1978	1983	1988	1993	1998	2003	2008	2012
Location	USA	USSR	Canada	USA	Norway	China	Canada	Switzerland	USA	Russia
Registrants	285	400	452	851	305	275	268	296	685	550
Countries represented	9	16	13	24	19	21	25	24	33	24
Plenary sessions	All	All	8	6	8	3	2	6	5	7
Proceedings volumes	1	2	2	2	3	2	1	2	2	5
Pages	563	1,649	1,202	1,937	1,620	1,360	1,276	1,322	1,740	683
Papers	104	87	147	276	289	189	188	230	358	194
Abstract volume(s)	1	1	1		1	1	1	2	2	2
Extended abstracts	–	–	–	–	–	–	–	96	187	374

1966), is considered to be the first multi-national, English-language collection of papers de-voted entirely to permafrost topics. In addition, the Soviets prepared a special book containing 26 papers organized into nine topical themes.

The closing session included summary reports by the panel moderators. These reports and discussions, published in the Proceedings, reflected the relevant issues of that period. Many of these are pertinent today and are recommended reading for students, young researchers, and established professionals. Although the original printed Proceedings volume may no longer be readily available for purchase, it is accessible in many university and government libraries and it is obtainable via the ICOP DVD (IARC 2008), and on several permafrost and other web sites.

49.2 Subsequent International Conferences on Permafrost

The first conference's resolutions recommended that a second international conference be planned and held with the objectives of further interdisciplinary support and participation. The Purdue conference essentially broke the "ice" between East and West permafrost researchers and set the stage for the Second ICOP. That conference, organized by Melnikov (Director of the Permafrost Institute in Yakutsk), was convened in 1973 and represented the first large international conference held in this restricted area of Siberia, and followed a smaller conference in 1969 (Brown 2012). These accomplishments were the result of Academician Melnikov vision and leadership.

All subsequent permafrost conferences maintained the interdisciplinary principles set forth at the Purdue meeting, and had both plenary and special thematic sessions, but the review papers were not always published in the conference Proceedings. These conferences included two more in the United States (Fairbanks 1983, 2008), two in Canada (Edmonton 1978, Yellowknife 1998), one each in Trondheim, Norway (1988), Beijing, China (1993), and Zurich, Switzerland (2003), and a second conference in Russia (Salekhard 2012) (Table 49.1). Following the formation of the International Permafrost Association (IPA) at the 1983 ICOP, subsequent conferences were organized and administered under the auspices of the IPA. A review of the first eight conferences is available (Brown and Walker 2007) and in a Ninth ICOP (NICOP) brochure published in celebration of the 25th anniversary of IPA and commemoration of the Fourth International Polar Year (Walker and Brown 2008). Starting with the NICOP conference, the Permafrost Young Researchers Network (PYRN) participated in conference activities.

49.3 Publications of International Conferences on Permafrost

Because plenary papers and thematic reviews hold special importance as temporal benchmarks illustrating the state of the science at the time of their appearance, it is worth reviewing approaches employed for each conference and the corresponding topics. Table 49.2 contains a summary of the actual topics presented in plenary, special or topical sessions for all ten conferences. Several topics reoccur in a number of conferences such as ground ice, thermal conditions, mountain permafrost, and select engineering topics including pipelines and other linear construction. More recently reviews related to coastal, subsea and mountain permafrost, carbon, climate change, planetary and Southern Hemisphere permafrost were presented and published. The topic of ground ice remains relevant in discussions of climate change, thermokarst and carbon content of permafrost terrains. For NICOP a special issue of the journal

Table 49.2 Summary of conference topics for plenary, special and/or theme sessions

First (1963-USA)	Second (1973-USSR)	Third (1978-Canada)	Fourth (1983-USA)	Fifth (1988-Norway)	Sixth (1993-China)
Plenary sessions	*Topical sessions*	*Theme papers*	*Panel sessions*	*Special session*	*Special session*
North America permafrost	*North Amer. reviews*	Ground ice	Foundations/embankments	Temperature/climate	Changing climate
Permafrost problems	Thermal	Hydrogeology E. Siberia	Heave/ice segregation	Climate change	Human changes
Soils/vegetation	Distribution	Vegetation/revegetation	Subsea permafrost	Svalbard	Degradation QTH
Ground ice	Ground ice	Disturbance/protection	Pipelines	Coasts North America	Periglacial/mountain
Geomorphology	Physics/mechanics	Geophysics	Environmental protection	Coastal processes	Mountains North America
Phase change	Groundwater	Structures	Climate/geothermal	Railway Canada	Mountains Europe
Thermal aspects	Mapping	Dams		Airfields Alaska	Mountains Asia
Physico-mechanical	Construction	Alyeska pipeline		Yamal development	Linear construction
Exploration					Frost damage China
Sanitary/hydraulic	Soviet reviews				
Earthwork/foundations					
Seventh (1998-Canada)	Eighth (2003-Switzerland)	Ninth (2008-USA)	Tenth (2012-Russia)		
Plenary lectures	*Plenary lectures*	*Plenary sessions*	(Ninth plenary cont'd)	*Plenary lectures*	
Field investigations NA	Lowlands permafrost	*Living in Alaska*	*Processes*	Carbon budget	
Geothermal/engineering	Mountain permafrost	Climate simulations	Hydrology	Engineering guidelines	
Living with frozen ground	Ice age permafrost	Thermal state	Trace gas budget	Coastal/subsea	
	Subsea permafrost	TAPS design	Heat-water transfer	Engineering advances	
	Southern hemisphere	Thermokarst	Subglacial freezing	Map Russia	
	Planetary/astrobiology	*Thermal state*	*Washburn tribute*	Mountain permafrost	
		Russia	Antarctic periglacial	Thermokarst	
	Plenary Reviews (12)	Europe	Alpine perspective	Monitoring (Norway)	
	Exploraton (2)	Active layer	Polar periglacial	Antarctic	
	Interactions (2)	Central Asia	Mass movement		
	Properties	*Engineering*			
	Infrastructure (3)	Design			
	Slopes	Geotechnics			
	Modelling	Russian approaches			
	Warming	Qinghai-Tibet railway			
	Monitoring				

Permafrsot and Periglacial Processes was available at the conference and contained seven review papers (Lewkowicz 2008), in addition to the 17 plenary papers published in the Proceedings.

Throughout the 50-year history of the International Conferences on Permafrost, publication of Proceedings has been the major legacy of each conference. Over the course of the 50 years more than 2,000 papers in English were published in the ICOP Proceedings, involving the efforts of many hundreds of reviewers (Table 49.1). Beginning with the Eighth ICOP, a second form of publication was initiated involving Extended Abstracts. One of the rationales for the Extended Abstracts was to make available timely results that would not have been available in the multi-year, publication process. For the Tenth conference a new publication was developed; the Transaction of the International Permafrost Association (Burn 2013).

49.4 IAEG and IPA

Official relationship between the International Association of Engineering Geologists (IAEG) and IPA began in 2008. The IPA and the IAEG formed the IAEG Commission C21 "Engineering Geology for Development in Permafrost Regions" and the IPA Subgroup with the same title during NICOP in Alaska in 2008. The goal of IPA Subgroup and IAEG C21 is to develop new methodologies for risk assessment and hazard evaluation of permafrost degradation in the context of climate warming by integrating a probabilistic analysis and mapping of permafrost instability. The key objectives of the C21 are: to review and collect examples of risk assessment and hazard evaluation in permafrost regions; to develop new methodologies for risk assessment and hazard evaluation in relation with permafrost change; to document new methodologies and guidelines for permafrost map-ping related to engineering geology goals; to network with practitioners and academics with an interest in permafrost problems world-wide; and to develop and maintain a dedicated area within the IAEG website. In 2012 the Commission met during TICOP in Salekhard, Russia. The ICOPs are the main platforms for continued and productive collaboration.

References

Brown J (2012) Fifty years of permafrost collaboration with the Soviet Union and Russia. In: Hinkel KM (ed) Proceedings of the tenth international conference on permafrost, international contributions, vol 1. The Northern Publishers, Salekhard, pp 1–14

Brown J (2013) Guest editorial. The 50th anniversary of the first international conference on permafrost and beyond. Permafr Periglac Process 24(4):263–267

Brown J, Walker HJ (2007) Report from the International Permafrost Association: a brief history of the international permafrost conferences (1963–2003). Permafr Periglac Process 18:385–392

Burn CR (2013) Editorial: transactions of the International Permafrost Association. Permafr Periglac Process 24:96–98. doi:10.1002/ppp.1780

International Arctic Research Center (2008) Compendium of the proceedings of the first nine international conferences on permafrost 1963–2008. Version 1, June 2008, DVD. http://research.iarc.uaf.edu/presentations/NICOP/NICOP-DVD-ftp-files/NICOP-files-1023am-6_20/

Lewkowicz AG (ed) (2008) Special issue: recent advances in permafrost and periglacial research. In: Ninth international conference on permafrost, Fairbanks, Alaska, USA, June 29–July 3, vol 19. Permafrost and Periglacial Processes pp 105–260

National Academy of Sciences (1966) Permafrost: international conference proceedings, Washington, D.C. Publication No. 1287, 563 p

Walker HJ, Brown J (2008) Part 1: international permafrost conferences, 1963–2008. In: Ninth international conference on permafrost, University of Alaska Fair-banks, pp 3–19

Woods KB, Leonard GA (1964) Permafrost. Science 143:1060–1061

Presidents and Leaders of the IAEG Visited the Medeo Dam in the Soviet Union

50

Shibakova Valentina and Koshkin Andrey

Abstract

The mud-flow protection dam Medeo near Alma-Ata city (constructed in 1972) sustained a powerful mud-flow blow (15 July 1973) stopped it and prevented from the disaster the Alma-Ata city. Five Presidents of the IAEG: E.M. Sergeev, M. Arnould, M. Langer, O. White, R. Oliveira as well as secretary general R. Wolters and H. Vidal (German NG) visited Medeo Dam in the Alatau region of Kazakh republic of the Soviet Union in the period of time 1974–1991. V. Shibakova accompanied the distinguished guests and took part in all technical and visiting programmes as the secretary of the Soviet NG. All gests had the possibility to be acquainted with historical materials of Medeo 1973 mud-flow occurence. Collection of old photos which illustrated those visits and the recent photo of Medeo Dam are presented in the paper.

Keywords

Mud-flow • Protection dam • Reservoir

Medeo Dam attracted great attention of specialists from all over the world-it was the first case of the history in the world hydrotechnical construction the resistance of the dam to the huge mud-flow. Five Presidents of the IAEG: E.M. Sergeev, M. Arnould, M. Langer, O. White, R. Oliveira as well as secretary general R. Wolters and H. Vidal (German NG) visited Medeo Dam in the Mountainous region of Kazakh republic of the Soviet Union in the period of time 1974–1991. Dr. Valentina Shibakova accompanied the distinguished guests and took part in all technical and visiting programmes as the secretary of the USSR NG. All programmes in Alma-Ata were perfectly organized by A.J. Hegay, the Head of Department "Kazglavselezaschita" and the member of the Soviet NG IAEG. Medeo dam parameters: rockfill dam 110 m in height, 530 m at length, 630 m a thick, the volume of the body dam was 5.3 mln · m^3 the capacity of mud-flow reservoir was 6.2 mln · m^3.

The mud-flow (1973) parameters: mud-flow water discharge reached 5,200 m^3/s, the velocity reached 10–12 m/s, dynamical power reached 1,000 t/m^2 the volume of mud-flow mass was equal to 5.5–4 mln · m^3 of debris, 1.5 mln · m^3 of water. Medeo dam survived, but it was in terrible state: dam reservoir was filled to three quarters of its capacity through the mud-flow reservoir capacity was designed for 100 years. This case shows that intensity of the processes occurring and their temporal scale cannot be reliably forecast. In some years after mud-flow 1973 occurence Medeo dam was under reconstruction. After reconstruction: the dam riched 150 m and the capacity of its reservoir—12.6 mln · m^3. New system of water outlets was constructed (Fig. 50.1).

S. Valentina (✉)
Geological Institute of the Russian Academy of Sciences, Pyzhevsky, 7, 119017 Moscow, Russia
e-mail: valentina-shibakova@yandex.ru

K. Andrey (✉)
General Physics Institute of the Russian Academy of Sciences, Vavilova 38, 119991 Moscow, Russia
e-mail: koshkin@fpl.gpi.ru

G. Lollino et al. (eds.), *Engineering Geology for Society and Territory – Volume 7*,
DOI: 10.1007/978-3-319-09303-1_50, © Springer International Publishing Switzerland 2014

Fig. 50.1 Medeo dam and skating ring Medeo. 1973 dam level shown by *arrow*/*Photo* by V.S. Shibakova, April 1977

At the same time different technical projects were fulfilled in high mountain area of the Tujuksu gletcher's field and the marine lakes. The honorary gests of the IAEG participated technical excursions to Medeo dam located in Small-Almaatinka river valley as well as they witnessed the works on the protection of the environment in the Big-Almaatinka river valley and the construction of new mud-flow protection dam there (Fig. 50.2).

The visitors had the possibility to be acquainted with historical materials of Medeo mud-flow 1973 presented in chronicle TV film. The chronology of those events was the following: July 15–18 h 17 m—mud-flow reached Medeo dam reservoir and was stopped by the dam (Fig. 50.3).

The reservoir was filled by deposits and by water almost up to the crest of the dam. Water outlets were blocked. Water level of reservoir was growing up (Fig. 50.4).

July 16—strong rains in high mountain areas of Almaatinka river basin provided new portions of water and the increase of water level. The first springs appeared in the upper part of the dam (Fig. 50.5).

Fig. 50.2 In Moscow before the trip to Medeo dam: Prof. E.M. Sergeev, Dr. V.S. Shibakova, Prof. M. Arnould. April 1977, *photo* by Prof. E.M. Kolomensky

Fig. 50.3 Prof. M. Arnould at the technical trip to Medeo dam from right to left: Prof. M. Arnould, Dr. A.V. Anishenko, Dr. A.S. Degavez, Prof. E.M. Kolomensky, Dr. A.J. Hegay, Dr. H.V. Novikov. *Photo* by V.S. Shibakova. Medeo. April 1977

Fig. 50.4 Dr. R.Walters (second from right) and V. Shibakova at the crest Medeo dam among specialists of "Kazglavselezaschita", the view to the South. *Photo* by N. Kochnev. May 1978

July 17-18-19: the water level was growing up. A lot of springs in many parts of dam have appeared. The road at the crest of the dam was broken (Fig. 50.6).

July 20 Three pumps (2 m^3/sec power of each one) started the working and three pipelines operated the pumping water from the reservoir. Water level has stopped growing for the first time and then began to decrease slowly (Fig. 50.7).

Medeo dam failure was prevented by the strong efforts of people mobilized to save the Alma-Ata city from the disastrous mudflow. The programmes of visits included the meeting and discussion with scientists and engineers in Alma-Ata city. Dr. R. Wolters, Prof. R. Oliveira, Prof. O. White presented lectures at the Institute of Geology of Kazakh Academy of sciences. Medeo dam and mud-flow 1973 indicated to engineering geologists that the main

Fig. 50.5 Prof. M. Langer at the Big Almaatinka dam (second from the *left*), Sept. 1983. *Photo* by N. Kochnev

Fig. 50.6 Prof. O. White, Dr. V.S. Shibakova and Dr. A.J. Hegay. View to Medeo dam. May 1986

Fig. 50.7 Prof. R. Oliveira, Dr. V. Shibakova (master of scating sport) at the Medeo skating ring with the Kazakh collegues and friends, Sept. 1991. *Photo* by N. Kochnev

Fig. 50.8 Medeo dam mud-flow reservoir. The system of water outlets, young forest at the dam's slope, July 2013. *Photo* by A. Koshkin

efforts should be directed towards the creation of an intensity and time-scale of occurrence of natural and man-induced geological processes (Fig. 50.8).

Reference

Shibakova VS (2012) The Medeo dam and mud flow protection of Alma-Ata. Ingenernaya Geologiya N1:6–15 (in russian)

A Short History of Engineering Geology and Geophysics at the British Geological Survey

51

M.G. Culshaw, K.J. Northmore and D.M. McCann

Abstract

Engineering geology in the British Geological Survey (BGS) began, in a formal sense, with the creation of the Engineering Geology Unit in 1967. Virtually since its inception, despite changing research priorities and economic drivers, the survey and research work carried out by BGS engineering geologists can be conveniently divided into four broad research areas: engineering geological mapping and urban geoscience, geotechnical properties of soils and rocks, engineering geophysics and geohazards. Since the late 1960s engineering geologists have undertaken innovative research initiatives and continue to play an important role in ensuring the delivery of BGS research.

Keywords

History of engineering geology • British geological survey

51.1 In the Beginning

Engineering geology in the British Geological Survey began, formally, in 1967 with the creation of the Engineering Geology Unit (EGU) (a name that changed many times over the next 45 years) under the leadership of Roger Cratchley (a geophysicist with a civil engineering degree). It followed the formation of the Natural Environment Research Council (NERC) in 1965 and the merging of the Geological Survey of Great Britain (GSGB) with the Overseas Geological Survey (OGS) to form the Institute of Geological Sciences (IGS), which was a research institute within the NERC, as well as the national Geological Survey (later to be renamed the British Geological Survey [BGS] in 1984). Based in the IGS London office, the Unit was formed at a time when engineering geology in the UK was gaining increasing recognition as an important sub-discipline of geology. This was reflected by the establishment of five Master of Science postgraduate courses in engineering geology in the 1960 and 1970s and by the formation in 1964 of the Engineering Group of the Geological Society (later incorporating the International Association of Engineering Geology UK National Group).

The Chief Geophysicist of the IGS at the time was instrumental in setting up the Engineering Geology Unit, almost certainly taking the view that it would act as a physical properties laboratory to provide fundamental data for the Geophysics Division. However, Roger Cratchley saw the EGU as developing into a more inclusive research laboratory involving site investigation including the drilling and sampling of boreholes, undertaking a wide range of

M.G. Culshaw (✉)
British Geological Survey, Keyworth, Nottingham, NG12 5GG, UK
e-mail: martin.culshaw2@ntlworld.com

M.G. Culshaw
School of Civil Engineering, University of Birmingham, Edgbaston, Birmingham, B15 2TT, UK

K.J. Northmore · D.M. McCann
Formerly British Geological Survey, Keyworth, Nottingham, NG12 5GG, UK
e-mail: kevin.northmore@tiscali.co.uk

D.M. McCann
e-mail: dmccann2@btinternet.com

G. Lollino et al. (eds.), *Engineering Geology for Society and Territory – Volume 7*,
DOI: 10.1007/978-3-319-09303-1_51,

soils and rock mechanics tests and geohazard (particularly landslide) studies. He was also interested in rock mass assessment using geophysical methods and the stability of major rock formations. As a consequence, Roger Cratchley was responsible for broadening the initial remit of the Unit to encompass both engineering geology and what came to be known as engineering geophysics—a remit that carries through to the present day.

The EGU was formed at a time when the traditional geological field mapping units were viewed as the 'flagship' of the IGS. Almost without exception these units were opposed to the formation of an engineering geological group that they perceived would impinge on their areas of work. Indeed, this opposition resulted in the EGU being excluded from the general IGS national mapping programme which, with the singular exception of a geological and geotechnical study of south-east Essex in the early-mid 1970s, lasted until a programme of government-funded applied mapping projects were commissioned in the 1980s.

From a complement of three staff in 1967 the Unit expanded rapidly, recruiting both experienced researchers and a number of young geologists, geophysicists and engineering geologists (including graduates from the relatively new MSc courses). By the mid 1970s the Unit's complement stood at around 20. Today, the number of engineering geologists and geophysicists is still around this number. A key factor in the role of the group was that from its creation the EGU provided a unique capability that combined geology and soil and rock mechanics with geophysics—a multidisciplinary approach that continues to be the hallmark of the Unit's research programme.

51.2 Early Research (1967–1970s)

Early progress in setting up the fledgling EGU was impressive, and included the commissioning, in 1971, of a rock and soil mechanics, geophysical properties and electronics laboratory complex to underpin research programmes. Early EGU research initiatives were essentially developed by the senior Unit staff, largely reflecting their basic research interests. These included four main areas: engineering geological mapping, geotechnical/geophysical property determinations, slope stability studies and research into the application of geophysical methods to engineering.

In 1968/1969, following consultations with John Knill at Imperial College London, an initial research programme was begun that included the engineering geological mapping of an area designated for the new town of Milton Keynes, some 80 km north west of London, mapping and process studies of active landslide areas on the Dorset coast (200 km south west of London) and investigating the geotechnical and geophysical properties of Jurassic strata based on sampling and testing a range of lithologies from the Dorset to Yorkshire outcrop that trends north-east—south-west across central England. Further innovative studies involved near-shore lithological/engineering geological mapping of the western part of Lyme Bay, Dorset to about 3 km offshore. This latter study resulted in a marine geological map linking the onshore and offshore geology—the only one of its kind produced by the BGS. Such sea floor studies also spurred early work into developing geophysical seabed tools to acquire geophysical property data of offshore sediments.

The Milton Keynes engineering geological mapping project provided valuable experience and lessons for future similar studies. Importantly, the study enabled EGU to develop the principles on which later engineering geological maps were produced and highlighted the need for good subsurface data.

51.3 The Rothschild Years (1970–1980s)

In 1972, publication of the Government's Rothschild Report into the funding of science in British research centres and surveys was of great significance to BGS. This report recommended that approximately 30 % of the core funds be transferred to central government departments, which would then 'commission' work back from them. Previously, the amount of applied geological mapping carried out by the BGS had been minimal with most mapping being of the traditional litho-stratigraphical type. However, in 1972 the Department of the Environment (DoE) commissioned the geological and geotechnical study of the site and 450 km^2 hinterland of the then proposed third London airport in south east Essex (around 70 km east of London). This was a significant milestone for engineering geology mapping and the provision of applied geological data in general. The south east Essex work was completed in 1977 and published as a twelve part report with a wallet of over 50 maps. This was IGS's first truly multidisciplinary project involving input from a wide spectrum of IGS expertise. The study utilized a range of techniques to acquire subsurface data, including continuous seismic profiling along river channels to ascertain the presence and distribution of buried channels. It provided the opportunity to experiment with different ways of presenting the three dimensional geology, with emphasis on the variations in lithological and physical/geotechnical characteristics of the sediments and sedimentary rocks of the area. Also, it allowed novel presentation of geohazard data (particularly related to collapsible loessic deposits and landslides). Significantly, it gave rise to the setting up of a major borehole and geotechnical property database from existing and new boreholes.

Further government contracts included a major study to map and classify landslides in the South Wales Coalfield region (about 250 km west of London) resulting in a landslide database and a series of maps locating and classifying nearly 580 landslides, a regional assessment of foundation conditions for industrial development in the Forth and Cromarty Firth Estuaries in eastern Scotland resulting in geotechnical profile maps classified in terms of bearing capacity and settlement data and geophysical research into indirect methods of assessing ground conditions for civil engineering purposes—particularly cavity location.

Other commissioned research during this period included a major geotechnicalstudy to assess soft, seafloor sediments in the North Sea and identify potential hazards (for example, gas-containing soft clay-silts, 'pockmarks', soft infilled channel deposits, mobile sand waves and foundered strata) that may affect the stability of oil production platforms and pipeline trenches. The work highlighted the technical difficulty and cost of acquiring sediment data and suitable 'undisturbed' samples and led to a further research programme, in collaboration with others such as the Marine Science Laboratory, University College of North Wales, to find methods of indirectly determining properties of rocks and sediments by geophysical methods.

A key aspect of the EGU work at this time was the provision of reports, maps and data aimed to meet the needs of planners. This was enhanced by the increased communication between geotechnical engineers and engineering geologists that was fostered, in no small part, by the Engineering Group of the Geological Society and its journal (the Quarterly Journal of Engineering Geology), particularly its efforts to promote standardization of terminologies, rock and soil classifications and preparation of maps and plans for engineering and planning purposes. Against this background, the EGU tended to concentrate on the strategic approach to regional problems and so concerned itself with regional studies of ground conditions in new urban developments and landslide surveys and stability analyses, both for planning purposes and to guide the subsequent site investigation for design.

51.3.1 The Requirement for National Data Sets (1980s–2005)

In the early 1980s a review of government expenditure and priorities heralded a new era of 'Commissioned Research' in which the contracts were no longer allocated as part of a 'rolling programme' but mainly won competitively. From about 1980 till 1996, the DoE (it had many different names during the period but the key point was that it was responsible for land-use planning policy) funded a series of more than 60 studies of the applied geology/environmental geology of British urban areas (Brook and Marker 1987). For each area a report and a series of maps were produced. Not all the projects were led by BGS but of those that were, the engineering geologists had a significant input to most of them.

A parallel EGU research programme into the engineering geology of major UK formations such as the Gault Clay and Mercia Mudstone Group (later to include Lias and Lambeth Group strata) formed a major underpinning to the overall research programme. This somewhat unglamorous research resulted in a series of technical reports that are freely available on-line. It is interesting to note that the report on the Mercia Mudstone Group is the most downloaded of all the reports and publications in external scientific journals produced by the whole of NERC!

Through the applied mapping projects, a huge amount of geotechnical data was being acquired and entered into project-specific databases on desk-top computers (PC's). However, with improvement in computing power and software, a corporate approach was adopted. In 1992–1993 efforts were directed to bring together and validate the 'local' databases and develop a single geotechnical database design. This was the beginning of the corporate 'National Geotechnical Properties Database,' which currently contains 250,000+ property values from over 63,000 boreholes. A landslide database created for an applied mapping study of Bradford formed the basis for the creation of the corporate National Landslide Database that followed some years later. A karst database has also been developed and populated. These databases formed the basis for the provision of national property and geohazard data sets which, in the late 1990s, were commercially licensed with other digital BGS information and become an important source of BGS income.

Of particular significance was the creation, in 1992, of a digital, national geohazard assessment tool (GHASP—*G*eo*HA*zard *S*usceptibility *P*ackage), for the UK insurance industry. This followed a period of exceptionally dry weather in 1989–1990 in south east of England, which led to some £1bn of insurance claims for damage to house foundations caused by shrinkage of mainly Mesozoic and Tertiary clay formations, often rich in montmorillonite. The development of the assessment tool required the digitisation of Britain's geological maps, initially, as raster scans. In the early 2000s, a tax windfall allowed the BGS to vectorise all its 1:50 000 scale geological maps (DiGMap50) (and other data) and develop a new geohazard susceptibility assessment system called GEOSURE that continues today.

51.4 Overseas Research Collaborations

From its early years the EGU undertook many engineering geological assignments around the world, funded largely through the then Overseas Development Administration's Research and Development and Knowledge Transfer Programmes but, also, through winning commercial contracts, occasionally in collaboration with private consultancies. The work mirrored the EGU's UK research areas.

51.5 Engineering Geology and 3D Modeling (2005—Present)

From about 2005, emphasis became directed towards creating 3D digital models and new research in urban areas is now firmly, but not exclusively, based around 3D modeling. Development of a national 3D digital geological model and the property attribution of that model is now a major thrust of BGS research.

2006 saw the start of a 'ramping down' of BGS's 'National Capability' funding, reducing many of its core activities. Also, emphasis has swung towards a regime of competitively won, externally-funded research, rather than 'relatively routine,' though important, commercial commissions. However, new drivers and research directions notwithstanding, the long-term nature of much of the engineering geological work (not least, the effects of climate change on geotechnical behaviour and geohazard risk) means that many of the intentions of the original EGU remain. Under the 2013 operational structure, the EGU research areas still endure as Urban Geoscience, Geotechnical and Geophysical Properties and Processes, Geophysical Tomography, and Shallow Geohazards and Risks within the BGS Geohazards Programme. Better knowledge of the disposition of geological formations, their properties and how they are influenced by natural/anthropogenic processes remain permanent requirements for users of engineering geological information.

Acknowledgments This paper is published with the permission of the Executive Director of the British Geological Survey (NERC). The comments of Stuart Duncan are gratefully acknowledged.

References

Information, in the form of BGS maps, reports and external publications, is available for all the project studies referred to in this paper through the BGS library, the NERC Open Research Archive (NORA—www.nora.nerc.ac.uk) and the BGS website (www.bgs.ac.uk)

Brook, D, Marker BR (1987) Thematic geological mapping as an essential tool in land-use planning. In: Culshaw MG, Bell FG, Cripps JC, O'Hara M (eds) Planning and engineering geology, Engineering geology special publication No. 4. Geological Society, London, pp 211–214

ASAGAI, the IAEG National Group of Argentina: Its History, Process and Present

52

Norberto Jorge Bejerman and Maria Beatriz Ponce

Abstract

The paper presents the history of the "Asociación Argentina de Geología Aplicada a la Ingeniería", whose initials are ASAGAI, since its creation on December 29th 1975. Since then, this Association has been linked with IAEG (situation defined in article 4, paragraph j of the Statutes) and the main objective opportunely established is to promote the application and research of the Engineering Geology, which has guided the life of the Association along its almost 38 years. Throughout time many circumstances shown that the selected path was the correct and this can be currently seen with an Association strong in accordance with the requirements of the moment and occupying an important place among the technical associations of the Argentina. The more important event that ASAGAI organizes is the Symposium, which has been realized since 2009 every 3 years in different cities of Argentina because of the federal character of the Association, with members of the majority of the 24 provinces that integrate the country. Other relevant activities done are the edition of The Journal of Engineering Geology and the Environment, published twice a year, the renewed Website, courses and working groups related with topics of interest for the members.

Keywords

ASAGAI • IAEG • History • Present

52.1 The First Steps

In the evening of December 29th 1975, 28 professionals related with the engineering geology, in a meeting convened with the objective of gather professionals and institutions of the country related with technical and scientific aspects of this field of knowledge decided to constitute the "Asociación Argentina de Geología Aplicada a la Ingeniería" and defined that it will be linked with the International Association of Engineering Geology as the National Group and that all of its members will work in themes related with the geology applied to the engineering. Also they nominated a commission that was in charge of elaborate the statute in order to present it and consolidate legally the Association.

The Statutes were approved on June 26th 1976. The first President of the Executive Committee was Dr. Horacio Rimoldi, who, from that moment until his death, in 2005, was the leader of the Association and participated in the different IAEG Congresses.

N.J. Bejerman (✉)
Córdoba Highway Administration, Figueroa Alcorta 445, 5000 Córdoba, Argentina
e-mail: jorgebejerman@arnet.com.ar

M.B. Ponce
Argentine Mining Geological Service- Institute of Mining Technology, Av. General Paz 5445, Building 14, 1650 San Martín, Buenos Aires, Argentina

G. Lollino et al. (eds.), *Engineering Geology for Society and Territory – Volume 7*,
DOI: 10.1007/978-3-319-09303-1_52, © Springer International Publishing Switzerland 2014

52.2 The History

ASAGAI was the first association in Argentina created with relation to the engineering geology and, currently, is the only organization of Argentina dedicated to these topics, including, since its creation, professionals specialized in them.

Throughout its nearly 38 years the Association has organized many activities. The first Symposium of Geology Applied to Engineering was held in 1981 in San Luis, Argentina, as part of the 8th Argentinian Geological Congress. Up to 1993 these Symposiums were part of the Argentinian Geological Congress and since 1996 the Symposium is organized as an independent activity in different places of the country. Since 2009 the name of the event is Symposium of Geology Applied to Engineering and the Environment because it also includes themes that previously were part of other activities of the Association.

In 1986 ASAGAI organized the Fifth IAEG Congress in Buenos Aires city, which is remembered because of different aspects, for example, the quality of its organization, the field trip and its meals.

The Association has also organized, for example, in 2002, a Geoindicators Workshop in Cordoba city, with the support of the Geological Society of America, and many conferences related with hydroelectric power, such as Yacyretá and Rio Grande.

In 2008, the Executive Committee created the Horacio V. Rimoldi Award, in honor to who was the organizer of the meeting in which the Association was created, its first president, the representative of ASAGAI in the IAEG activities developed up to 2001 and the great instigator of the Engineering Geology in Argentina.

In 2008, the Executive Committee decided to create the category of Honorary Member for members older than 75 years. At the 9th Symposium of Geology Applied to Engineering, two members that assisted to it received a diploma in the Opening Ceremony because of reaching that category (Fig. 52.1).

In 2009, in occasion of the 9th Symposium of Geology Applied to Engineering, it was granted an Special Mention to the paper "The rocks used as materials for concrete" presented by Pedro J. Maiza and Silvina A. Marfil. In 2012, in occasion of the 10th Symposium of Geology Applied to Engineering (Fig. 52.2), the Horacio V. Rimoldi Award was granted to the paper "Reclamation study of degraded lands by erosion process: Franca, São Paulo, Brazil" presented by Cláudia Marisse dos Santos Rotta and Lázaro Valentin Zuquette.

The quantity of members along the life of the Association is related with the economy of the country and the possibility of employment for the interested professionals in becoming part of ASAGAI. The number of members has been changing depending on the international trends about the importance to become part of this type of organization.

Fig. 1 Awarding of Honorary Membership Certificate to Carlos Bessone and Oscar Albert. From *left* to *right* Carlos Bessone, Oscar Albert, Norberto Jorge Bejerman and Maria Beatriz Ponce

52.3 The Present

Currently ASAGAI is discussing its Strategic Plan, whose Vision is focused on the growth of the Association related with the development of the Engineering Geology.

The Mission considers to promote this field of knowledge with the active participation of its members, to provide training for the formation of society, focusing on students who need to go deeper into the knowledge of these disciplines, and, as specific objective, to stimulate the participation of the members in order to increase the value of the engineering geology in the society.

Some actions were taken in order to accomplish the mission and vision of the Association, such as the modernize the Statutes, improve the communication with the members, update the Website with more contents for the members and give specific courses related with issues of interest for the members and the society and, as a result of all those actions, the economy of the Association has stabilized.

The number of members is one hundred and twenty three. Thirty eight of them are members since 2012 and 2013 as a result of the activities and the interest of them to be part of an association with the characteristics of ASAGAI. The members include professionals from seventeen of twenty four provinces that integrate the country.

The present and the future can be synthesized with some phrases:

Fig. 2 Participants of the Field trip in the 10th Symposium of Geology Applied to Engineering (Villa Carlos Paz, Córdoba Province 2012)

- Strengthening of the association internal structure by means of commissions created to advance in technical issues of interest for the members.
- Improvement of the administrative procedure in order to accelerate it.
- Development of actions in the academic community in order to improve the engineering geology at universities.
- Identification of issues of the engineering geology that require an specific training.
- Development of alliances with other societies of the country and the region, i.e. South America, in order to create working groups about issues of common interest.

A Brief History of the Past 65 Years of Romanian Engineering Geology Experience

53

Mihaela Stănciucu, Cristian Mărunţeanu, Emilia Elena Milutinovici, Paul Nedelcu and Leonin Bogăţeanu

Abstract

The Romanian engineering geology experience is a complex and intricate picture that has to be described in terms of contributions to design and execution of engineering works realized during the last century and of individual or collective personalities which significantly contributed at the birth or development of this domain. The history described below refers to engineering geology roots which intertwine with hydrogeology and geophysics in Romania. Six decades of professional activity in the field of engineering geology represent a huge amount of work, effort and often sacrifice of many generations. The quality of their works is demonstrated by the fact that most part of their achievements is still in function and their successors spread all over the world and successfully apply the professional principles inherited. The complete itemization of all persons related to this activity is an impossible task, and authors apology for eventually overlook of persons or contributions and underline that there were mentioned only personalities that passed away.

Keywords

Collective personalities • Professional principles

53.1 The Birth of Romanian School of Engineering Geology and National Group

The origin of the Romanian school of engineering geology rises in the first decade after the World War II, from the old geological and mining educational institutes. In 1948 the Institute of Geology and Mining Technology detached from Polytechnic Institute of Bucharest (the most important technical university in Romania founded in 1818 http://www.upb.ro/en/short-history.html), and became the first national educational center dedicated to applied fields of mining and metallurgy, oil and gas exploitation, economic engineering and engineering geology, under the direction of Professor Nicolae Petrulian (1902–1983), (Băncilă 1989; Bucur 1998; Florea et al. 2010).

From this point, the further evolution of this high education school of engineering geology followed in the same tonality the turbulent evolution of the country. It was changed for short time (1952–1957) in a faculty beside Mining Institute and further transferred to Oil and Gas Institute (1957–1974), which becomes The Institute of Petroleum, Gas and Geology (Florea et al. 2010). Since 1974, the Faculty of Geology and Geophysics merges with the Geology Department of the Geography and Geology Faculty from the University of Bucharest (founded 1864

M. Stănciucu (✉) · C. Mărunţeanu
University of Bucharest, Bucharest, Romania
e-mail: stanciucumihaela@yahoo.com

E.E. Milutinovici
Search Corporation, Bucharest, Romania

P. Nedelcu
I.S.C.E, Bucharest, Romania

L. Bogăţeanu
I.S.P.C.F, Bucharest, Romania

G. Lollino et al. (eds.), *Engineering Geology for Society and Territory – Volume 7*,
DOI: 10.1007/978-3-319-09303-1_53,

http://www.unibuc.ro/e/n/despre/History.php), and supported several and radical transformation by joining with Geography or Biology Faculties.

Starting from 1989, an Engineering Geology Department was created in the frame of the Faculty of Geology and Geophysics, beside University of Bucharest, and few years later, in 1992, the Romanian Association for Engineering Geology is founded and joins I.A.E.G. as national group.

53.2 First Romanian Engineering Geology Experiences

One of the first citations regarding the applied geology for engineering purposes is dated 1921–1922 when the Romanian Railways demand Professor Gheorghe Macovei (1880–1969) to realize a complex geological study on Bistrita Valley, in order to extend and update the railway works. The objective was realized later, between 1949 and 1953, with the expertise in East Carpathians geology of Professor Ion Băncilă (1901–2001) which helped to the best route choice and design of Piatra Neamţ-Bicaz railway (Bancilă 2000).

Preliminary engineering geology studies for dams are realized starting from 1950 by Gh. Cernea and Th. Joja for Siret River, Şt. Ghika Budeşti for Argeş r., N. Gherasi for Sebeş r. and I. Băncilă for Bistriţa r. (Bancilă 2000) in the frame of Romanian Geological Committee. Regarding those pioneering times, the prospecting geological field trips where described by I. Băncilă as an adventure in which the compass and the photo camera where considered subversive tools and results of the studies were always supervised and approved by Soviet cancellers (Bancilă 2000). Readers must be informed that during the communist regime, in spite of scientific and technical interest, for professors or practitioners it was forbidden to adhere to any international professional association and the participation to international conferences was strictly controlled and surveyed by the authorities. Representative and, unfortunately not unique, was the case of one of the greatest personality of Romanian engineering geologists, Professor Ion Băncilă, who even he was official decorated for the substantial contribution to the design and execution of the main Romanian large dams, did not have the permission to participate to the First I.A.E.G. Congress, Paris, 1970, to present his paper (Bancilă 2000).

Later on, this activity will be extended in the frame of the Institute for Power Studies and Design (I.S.P.E.—founded 1949) and the Institute for Studies, Design and Water Management (I.C.P.E.G.A.—founded 1953) which long time further on will concentrate a main part of engineering geology activity in Romania. An objective summary of this activity counts 237 dams, design and executed between 1949 and 2000 (Cojocar 2005; ROCOLD 2000).

By far, one of the most dominant personalities in the field of engineering geology was Professor Ion Băncilă, "recommended" by his long and highly professional activity. He successfully participate to design and execution of the most important large dams in Romania (Bancilă 2000): Izvorul Muntelui-Bistriţa River (1961), Vidraru-Argeş r. (1965), Iron Gates-Danube r., (1971–1984) and many others, of less importance: Stâmtori-Firiza r. (1964), Rovinari-Jiu r.(1965), Paltinul-Doftana r. (1971), Valea de Peşti r.(1972), Poiana Uzului-Uz r. (1973), Săcele-Târlung r. (1975), Cibin r. (1979), Ştefăneşti-Prut r. (1978), Pecinegu-Dâmboviţa r. (1984), Valea Neagră-Runcu-Mara r. (1987), Siriu-Buzău r. (1994), Maneciu-Teleajen r. (1994). At the time of birth of the first large dam (Izvorul Muntelui-Bistriţa 1950–1953), I. Băncilă was sustained by Czech specialists Quido Pfefermann Zaruba and Vojtech Mencl, but later on, beside him grows up a large group of engineering geology specialists in design and execution of large dams in Romania and in other countries. Beside them, I. Băncilă coordinates the edition of an appreciated book of Engineering Geology (1980, Technical Printing House, two volumes, in Romanian).

In the same time (1949), other large scale engineering works are started (Danube-Black Sea Canal) and involve several teams of Romanian and Soviet specialists. Between Romanian specialists the most significant where Professor Gheorghe Macovei and Professor Radu Ciocârdel (1915–2010) who, beside professor engineer N. Maslov, and the geologist engineers N. Prosciuhan and C. Pestovski, "piled up a broad documentation, materialized in The Technical and Economic Memoir regarding the Construction of the Danube-Black Sea Canal, which contained 17 volumes" (Scurtu and Cojocaru 2013). After an apotheosis debut, the works for the canal has been stopped for two decades, but the large experience of Russians in engineering geology inspired the Romanian leaders to sustain this domain in further years (Bucur 1998). Danube-Black Sea Canal was accomplished later, starting from 1972, in 3 years, based on 144 volumes of studies and researches, plus 358 volumes representing the proper designing part, with schemes, graphs, tables about the Canal and the adjacent works and with more than 70 Romanian architects, civil, hydrotechnical and geological engineers, climatologists, botanists, etc. working in shifts (Scurtu and Cojocaru 2013).

All these engineering geology experiences related to dam construction, drainage and waterproofing, water impoundment, underground or open mining, harbors and canals, tunnels, required solid knowledge of applied hydrogeology, domain born in Romania half a century ago (Zamfirescu 2000). For this reason, many personalities who contributed to the foundation of engineering geology proved also competences in the field of hydrogeology. Starting with Gh. Macovei who realized the first hydrogeologic map of South Dobrogea, later on I. Băncilă (1956, 1958, 1965), P. Bomboe (1963), R. Cădere (1964), R. Ciocârdel (1952, 1957, 1969), V. Harnaj (1963, 1965).

53.3 Romanian Engineering Geology School

At the Institute of Geology and Mining Technology the first course of engineering geology was sustained in early 1950 by Professor Ştefan Ghika Budeşti (1904–1959), succeeded by Professor Petre Bomboe (1924–2011). The broad personality of Professor Petre Bomboe dominated the Romanian School of engineering geology for many decades. He was the youngest and long-lasting dean of the Faculty of Geology and Geophysics from the University of Bucharest (17 years), position which allows him to sustain engineering area of geological sciences. His field of activity covered beside engineering geology and hydrogeology also statistics and applied mathematics for geology purposes. He also participated to the foundation of the Department of Geology at Dar Es Salaam University, Tanzania, and implementation of engineering geology at Antofagasta University, Chile. He coordinates also some of the most appreciated courses of Engineering Geology 1980, or Mathematical Geology, 1979, (University Printing House, in Romanian).

Beside him, following the next decades, many outstanding professors contributed at the strengthening of the Romanian engineering geology school: Alexandru Codarcea (1900–1974), Iulian Gavăt (1900–1978), Gheorghe Murgeanu (1901–1984), Ion Huber-Panu (1904–1974), Radu Cădere (1905–1993), Viaceslav Harnaj (1917–1988), Ştefan Airinei (1920–1989), Radu Botezatu (1921–1988), Vasile Lăzărescu (1926–1989), Mircea N. Florea (1926–2011) et al. The limited space of this publication do not allow us to present the scientific activity of all of them, and we confine to underline the personalities and enthusiastic commitment for the true values of Romanian engineering school of the last two of them.

Professor Vasile Lăzărescu was one of the greatest and most beloved teachers of the Faculty of Geology and Geophysics for more than 40 years. In spite of all political harassments that surrounded him under communist regime, he excels in complexity and profundity of understanding the geologic environment. He contributed to tectonic maps of Romania and Europe in the frame of UNESCO project related to Balkan seismicity and wrote one of the best books of Physical Geology (1980).

Professor Mircea Florea had successfully sustained for four decades courses of Soil and Rock Mechanics, Geotechnics and Stability of slopes. He published over 100 articles, most of them related to the contrast between mechanical properties of bedrock and shallow deposits, piping and unsaturated water flow phenomenon and their implication in the stability of tailing dams. His deep and true generosity sustained over years many generations of engineering geologists, which still preserve for him a great respect and gratitude. He is the author of several important books: Landslides, 1979, Soil and Rock Mechanics, 1983, and Tailing Dams Stability, 1996, (Technical Printing House, in Romanian).

As we mentioned before, the communist regime limited the scientific communication with occidental word, but permitted the change of knowledge with the rest of communist countries (Vietnam, North Korea, Cuba), with African new states or South American countries. In consequence, large numbers of foreign students applied for Romanian school of engineering geology, and many Romanian geologists worked abroad between 1960 and 1980. Between 1974 and 1989, in Romania the needs for engineering geologist grows up and the Faculty of Geology and Geophysics increased considerably the number of graduate students. Most of them contributed to the economic and industrial development of Romania (before 1989): 213 dams design and executed after 1970, (Cojocar 2005; ROCOLD 2000), over 10,000 km of new railways with art works, 82,000 km of roads, 62 airports, over 200 tailing dams and mining waste deposits, 70 km of metro lines with 51 stations, etc.

53.4 Romanian Association for Engineering Geology

In 1992, at the initiative of founding members, Romanian Association for Engineering Geology (ARGI) was founded. Almost as many years, ARGI is a member without interruption in the International Association for Engineering Geology and the Environment (IAEG). Most of the members works in universities, private or multinational companies and offer their expertise for rehabilitation and new construction of subway, airports, motorways, national and county roads, railway sections, bridges overpasses or underpasses providing traffic improvement in major cities, unconventional energy sources (wind turbines and photovoltaic cells), natural disasters hazard and risk maps, hazardous and radioactive waste disposal, geo-environmental protection, cultural heritage preservation etc. During the last two decades several conferences, national and international symposia organized by ARGI, among which we mention:

- International symposium "Engineering Geology and Geoenvironment Protection", Mamaia, May 23–28, 1994; volume published;

- National Symposium "Mining geological engineering", Motru 1996; volume published;
- National Symposium Landslides—the impact on the environment and society, hazard zoning and prevention of effects", Bucharest, 30 September—1 October, 2004;
- National Symposium (with international participation) "Landslides—the impact on the environment and society" October 25–27, 2007 (organized by the Geological Institute of Romania—IGR in collaboration with ARGI), May 21–22, 2010;
- Contributions to annual Sympozia organized in 2005–2012 by the Faculty of Geology and Geophysics;
- Conference held within the framework of the bilateral program "Engineering Geology and Waste Disposal" between Bucharest University and Bundesanstadt fur Geowissenschaften und Rohstoffe from Hannover, in Bucharest, 1997;

In August 1999 ARGI supported organization in Bucharest University of the event The 5th Conference "Mechanical Behavior of Salt—MECASALT 5", sponsored by companies and universities from the United States and Romania. Participation of ARGI members in government committees on laws of expertise: We cite only the Law no. 575 of 22 October 2001 approving the plan for upgrading of the national territory—Section V—natural risk zones, or geotechnics regulatory in specialized Technical Committee No. 6 "Geotechnical engineering and foundations" of the Ministry of Regional Development and Public Administration.

International activity: The ARGI President is member of the IAEG Council, which governs and manages IAEG. In this capacity he represented ARGI at the majority of IAEG congresses every 4 years, for the election of the IAEG President and of the Executive Committee, and at other IAEG annual meetings of the Council. The Romanian Group of engineering geology is part of the core of countries with continuous activity in the IAEG for more than 20 years and had in the past few years, in particular by the efforts of the treasurer of the association, an ever-increasing number of members and associate members, contributors IAEG, with and without subscription to Bulletin of Engineering Geology and the Environment. A part of ARGI achievements was presented to the IAEG Secretariat as contribution to mark in 2014 the anniversary of 50 years from the establishment of the International Association for Engineering Geology and the Environment (IAEG).

References

Băncilă I (1989) Geology of hydrotechnical works. Ed. Tehnică (in Romanian)

Bancilă I (2000) Arch over times. Ed. Istros (in Romanian)

Bucur IC (1998) Pages from the historical geology of. evolution of knowledge. Contributions, people, facts, cases. Petrom, SA. (in Romanian)

Cojocar M (2005) Tradition and state of the art. Hidroconstructia

Florea MN, Gheorghe A, Neguţ A (2010) Romanian school of geological and geophysical—past and future. Monitorul de Petrol şi Gaze 4(98), (in Romanian)

http://www.upb.ro/en/short-history.html

http://www.unibuc.ro/e/n/despre/History.php

http://www.argi.info.ro

ROCOLD—Romanian National Commitee on Large Dams. (2000) Dams in Romania

Scurtu I, Cojocaru I (2013) History of the construction of the Danube-Black Sea Canal. Rivers and lakes in ottoman Empire international symposium 2–4 mai 2013, Kayseri

Zamfirescu Fl (2000) A century of hydrogeology in Romania. Hidrog. din România, Bucureşti, Asoc (in Romanian)

The Activity of Uzbek Scientist in IAEG

54

N.G. Mavlyanova and K.L. Rakhmatullaev

Abstract

Uzbek scientists participated in IAEG activity as a part of the Soviet group. Professor G. Mavlyanov (1910–1988) was the leader in the field of engineering geology, hydrogeology and seismology in Central Asia. He founded two large institutes in Uzbekistan—Institute of Hydrogeology and Engineering Geology (HYDROINGEO) in 1960 and Seismology Institute in 1967. Professor G. Mavlyanov was a participant of the Congress of IAEG in Paris in 1970. He took an active part in IAEG development from 1977 to 1986, and he organized visits of Presidents and members of IAEG to Central Asia.

Keywords

Professor Mavlyanov • Central Asia • History of activity in IAEG • Loess ground

54.1 IAEG Presidents' Visits to Uzbekistan

IAEG Presidents E. M. Sergeev, M. Arnould, M. Langer, O.White, R. Oliveira, Secretary General R. Wolters and H. Vidal (German NG) as well as other members of IAEG visited Uzbek Academy of Science. All of these visits were organized by Gani Mavlyanov. In spring of 1977, IAEG President Professor M.Arnould had a visit Tashkent. G. Mavlyanov organized for him a field excursion to the massifs of loess soil; and showed him some sections of different structures, properties and collapsibility. Professor M. Arnould was very interested in this excursion and he was greatly impressed by the unique geological objects shown to him. M. Arnould apprehended with a great interest the Mavlyanov's theory about the polygenetic origin of loess soils in Uzbekistan. After this visit M. Arnould distributed among other experts the information on unique sections of loess soil in Uzbekistan, and he told about the works by Prof. Mavlyanov and his colleagues (Shibakova 2012).

The Secretary General of IAEG R. Wolters visited Tashkent in 1978. The program of his visit was also wide: he delivered lectures at Seismology Institute, got acquainted with materials of the Tashkent earthquake of 1966 and with problems in seismic-resistant construction. The field excursion to the Charvak water reservoir and to loess sections was organized to him. G. Mavlyanov demonstrated him the biggest loess deposits with a thickness of 80–100 m and explained his view on the genesis and distribution of loess soil, properties and nature of collapsibility. R. Wolters could see the famous Charvak loess cross section, the several layers of buried soils, he was very impressed by this sight. He said that is very important to save these loess deposits such as a geological heritage.

N.G. Mavlyanova (✉)
Sergeev Institute of Environmental Geoscience,
Russian Academy of Sciences, Ulanskii Per, 13, Bld. 2,
Moscow 101000, Russia
e-mail: georisk2012@mail.ru

K.L. Rakhmatullaev
Mavlyanov Institute of Seismology of Uzbek Academy of
Sciences, Zulfiyahonim Str., 3, Tashkent, Uzbekistan 100128
e-mail: mavnad13@rambler.ru

G. Lollino et al. (eds.), *Engineering Geology for Society and Territory – Volume 7*,
DOI: 10.1007/978-3-319-09303-1_54, © Springer International Publishing Switzerland 2014

Fig. 54.1 Visit of E. Sergeev to Samarkand in 1980 (from *right* to *left* G. Mavlyanov, E. Sergeev, Kh. Baimukhamedov et al)

Fig. 54.2 Visit of W. Dearman to Institute of seismology of Uzbek Academy of Sciences (Tashkent, 1986, from *left* to *right* G. Mavlyanov, Kh. Rakhmullaev, V. Shibakova, scientist from Uzbekistan, W. Dearman)

Professor E.Sergeev was in Uzbekistan in 1980. He had a big interest to loess deposits and of their genesis. His students Kh. Rakhmatullaev et al. investigated the microstructures of loess (Figs. 54.1 and 54.2).

Later Prof. G. Mavlyanov together with his colleagues took an active part in preparation of the international monograph "Engineering Geology of the Earth". He was the author of the chapter "Arid regions".

54.2 The Activity of G. Mavlyanov

Professor Gani Mavlyanov was the leader in the field of engineering geology, hydrogeology and seismology in Central Asia. His famous fundamental book "Genetic types, engineering geological properties of loess and loessial soils in Central and South part of Middle Asia" published in 1958 is still very actual now because it includes a lot of

Fig. 54.3 Professor Vladimir Obruchev and Gani Mavlyanov (Moscow 1955)

results of field and laboratory works about loess deposits. His theory about polygenetic genesis of loess was interesting to many scientists even to the famous Prof. V.Obruchev, who developed the theory about aeolian genesis of loess. So, he has several discussions with G. Mavlyanov (Fig. 54.3).

Gani Mavlyanov founded two large institutes in Uzbekistan. The first is Institute of Hydrogeology and Engineering Geology (HYDROINGEO) in 1960. At that time new cities were built, mining industry was developed and intensive irrigation of new loess fields went on, and huge water reservoirs were constructed. G.Mavlyanov solved arisen engineering geology problems in HYDROINGEO institute where he was a director. After the destructive Tashkent earthquake in 1966 he founded Seismology Institute in 1967.

In 1970 G.A. Mavlyanov (Uzbek academy of Sciences) published a paper in Russian in the Uzbek Geological Journal, in which he set out a programme of study for loess researchers working in the general field of engineering geology. It was a valuable paper in that it set out some interesting topics of study and it gave a snapshot of geotechnical needs in the Central Asian environment in 1970 (Jefferson et al. 2004). Still now, this program of study remains acute for loess researches. Mavlyanov (1970) made a commendable attempt; most of his targets are current and important. The main items of this program are the following.

1. We require the creation and acceptance of a uniform standard theory of the formation of loess soils. The absence of such a theory can slow down further development of a science of loess. In my opinion, this theory should be based on a polygenetic principle of the formation of loess soils. The validity of the polygenetic principle is confirmed by widespread researches: in Central Asia, in western Siberia, in the Far East, on Sakhalin Island and the Southern Kuriles, in India, and in the territory of 21 states of the USA from New York in the east and almost up to San Francisco in the west. Loess is associated with river basins and within a river basin various types of loess may be found. For example, in deposits of the river Amu Darya in Central Asia we found loess of aeolian, proluvial and diluvial genesis and loess-like deposits of proluvial, diluvial, alluvial and lacustrine origin. In deposits of the river Ganges in India we found loess soils of eluvial, diluvial, proluvial and alluvial genesis. Obviously the same situation exists in deposits of other rivers with associated loess provinces.
2. Research is required into the structure and physic-mechanical properties of loess soils of various genetic types in horizontal and vertical directions. It is necessary to produce maps showing the distributions of various properties.
3. Study the correlations between genesis, the geological/geographical properties of loess soils and loess-like deposits and their physical—mechanical properties.
4. Development of a standard terminology for concepts connected with the loess problem. There is no unity in the names, for example, of properties of loess soils referred to in the various disciplines: engineering—

geological, geotechnical, physical—mechanical, technical, building, physical—technical. There is confusion about the contents of the terms 'loess', 'loess-like deposits', 'loessial', etc. The border between loess and non loess silts is not well defined. The scientific terminology needs to be neat and precise and definitive—to allow progress to be made.

5. Creation of a conventional engineering—geological classification of loess soils and loess like deposits based on their genesis and emphasizing the connection between genesis and type and physical—mechanical properties.
6. Study of the conditions of formation of loess soils in view of changes in their geographical environment during all of their geological history. The environment has had a major influence on these soils, therefore it is necessary to find out the conditions affecting deposition, structure and properties during all the history of the formation process.
7. Research on the influence of water on the structure and properties of loess soils. The nature and the amount of water in the system may affect many important properties. In saturated systems collapsibility must be influenced, and major structural properties lost.
8. Study of changes of engineering—geological properties of loess soils which have arisen as a result of long-time irrigation of sites with various geological, hydro-geological and engineering—geological conditions.
9. Drawing up a methodical management plan to facilitate efficient addition of water for controlled collapse operations—taking into account the various types of ground and ground conditions encountered.
10. Development of studies of the theoretical bases of collapse phenomena. In general more collapse is observed in the field than in the laboratory. In Uzbekistan this collapse divergence ranges from 32 % upwards.
11. Forecast of collapse phenomena for the irrigation of fields and various kinds of structures.
12. Creation of more modern equipment; devices are required for field and laboratory definitions of engineering—geological properties of loess soils. There should be a widespread introduction of geophysical methods of research for these purposes.
13. Study of engineering—geological properties and also for soils under active dynamic and static loadings.
14. Development for seismic areas of methods for the forecast of collapse phenomena in conditions of dynamic loadings.
15. Research on the properties of loess soils in seismically active areas and on the determination of the increase in seismic intensity (using engineering geological data) for micro-seismic zoning.

The idea of making lists and setting targets has an honourable history; it certainly reached a considerable intellectual level with mathematicians Klein and Hilbert; the world of loess is so complex that it seems worthwhile to attempt to pick out the really significant geotechnical problems that need to be tackled. The key, central problem is that of loess hydrocollapse and subsidence; this is where effort might be concentrated, and the rewards be the greatest (Jefferson at al. 2004). Nowadays, the Uzbek Academy of Sciences tries to support the efforts of Uzbek scientists to continue they activity in IAEG.

References

Jefferson IF, Mavlyanova NG, O'Hara-Dhand K, Smalley IJ (2004) The engineering geology of loess ground: 15 tasks for investigators—the Mavlyanov's programme of loess research. Eng Geol ENG GEOL 4(1):33–37

Mavlyanov GA (1958) Genetic types, engineering geological properties of loess and loessial soil in Central and South part of Middle Asia. Tashkent, p 609 (in Russian)

Mavlyanov GA (1970) On the problems of engineering—geological studies of loessial deposits. Uzbek Geol J 2:53–55 (in Russian)

Shibakova VS (2012) Academician Gani Arifhanovich Mavlyanov (1910–1988). Eng Geol 3:6–18 Moscow (in Russian)

Author Index

A
Acharjee, Swapna, 141
Aldighieri, Barbara, 115
Alessandra, Magagna, 181
Allington, Ruth, 131
Almeida, José A., 233
Amadio, Vittorio, 115
Andrey, Koshkin, 251
Arattano, Massimo, 125
Averkina, Tatyana, 27

B
Bacenetti, Marco, 125
Bayerl, Michael, 225
Baynes, Fred J., 17
Bejerman, Norberto Jorge, 261
Bernardo, Marcello, 109
Bernhardsdóttir, A. E., 77
Bertacchini, Milena, 97
Bobrowsky, Peter, 229
Bocchia, Maurizio, 121
Bock, Helmut, 207
Bohle, Martin, 161
Brito, M. Graça, 233
Brown, Jerry, 247

C
Cabezas, J. A., 93
Camassi, Romano, 69
Capua, Di Giuseppe, 59
Castaldini, Doriano, 97
Cheng, Fergus, 21
Chiarle, Marta, 125
Contrafatto, Concetta, 125
Couture, Réjean, 229
Crescimbene, Massimo, 69
Culshaw, M. G., 17, 257

D
D'Addezio, Giuliana, 103
da Silva, Ana Paula F., 233
Datta, Partha Sarathi, 63
Davie, Andrew, 21
Delgado, José, 221
de Pascale, Francesco, 109
Desikachari, Vasudevan, 135
Di Capua, Giuseppe, 81
Diez, J. Javier, 151, 157
Di Franco, Sabina, 169, 175
Dutto, Furio, 125

E
Eggers, Mark, 17
Eremina, Olga, 243

F
Falsaperla, S., 77
Farabollini, Piero, 115
Federica, Appiotti, 49
Fenton, Clark, 21
Fernández, Elvira, 157
Fernandez-Fuentes, Isabel, 131
Ferreira, M. A., 77
Ferretti, Maurizio, 49

G
Garrido, Jesús, 221
Giardino, Marco, 125, 147, 169, 175, 185
Gioia, Eleonora, 49
Godwin, William, 213
Goldsmith, Robert, 197
Grindlay, Alejandro L., 221

H
Herten, Markus, 207

K
Käsling, Heiko, 225
Keaton, Jeffrey R., 17
Khan, Mohammad Omer Faruk, 33
Kosevich, Natalia, 27

L
La Longa, Federica, 69
Lamas, Pedro C., 233
Lanfranco, Massimo, 185
Lanza, Tiziana, 85
Lentfer, Kori, 17
Leonin, Bogăţeanu, 265

G. Lollino et al. (eds.), *Engineering Geology for Society and Territory – Volume 7*,
DOI: 10.1007/978-3-319-09303-1, © Springer International Publishing Switzerland 2014

Leyland, Robert, 201
Lombardo, Vincenzo, 147
Longo, Fabrizio, 125
Loppas, Andria, 21
Loudon, Scott, 21
Lozar, Francesca, 147
Luca, Ghiraldi, 181
Lugeri, Francesca Romana, 115
Luigi, Perotti, 181

M
Macedo, Luisa, 81
Magagna, Alessandra, 147
Marco, Bacenetti, 181
Marco, Giardino, 181
Marincioni, Fausto, 49
Mariño, Jersy, 81
Marone, Eduardo, 45
Marone, Luis, 45
Marsh, Rebekah, 21
Marsili, Antonella, 103
Martin, Christophe, 213
Mărunţeanu, Cristian, 265
Maurenbrecher, Pieter Michiel, 237
Mauro, Palomba, 181
Mavlyanova, N. G., 269
McCann, D. M., 257
Menschik, Florian, 225
Mihaela, Stănciucu, 265
Milutinovici, Emilia Elena, 265
Musacchio, G., 77, 89, 121
Musso, Marcos, 217
Muto, Francesco, 109

N
Nespereira, J., 93
Northmore, K. J., 257
Novotný, Jan, 11, 17

O
Oboni, Cesar, 191
Oboni, Franco, 191
Oliveira, Ricardo, 1
Osipov, Victor, 243

P
Parry, Steve, 17
Paul, Daren, 17
Paul, Nedelcu, 265
Paz, Iria, 157
Paz, Rosa M., 151, 157
Pendlebury, Matthew, 21
Peppoloni, Silvia, 59
Perotti, Luigi, 125, 147
Peruzza, Laura, 69
Piangiamore, Giovanna Lucia, 121
Pino, Nicola Alessandro, 69, 89
Ponce, Maria Beatriz, 261
Pravikova, Natalia, 27

R
Racca, Filippo, 125
Rakhmatullaev, K. L., 269
Rapisardi, Elena, 169, 175, 185
Rocci, Lorenzo, 125
Rodriguez, Fernando, 151
Rodríguez, M. Isabel, 221
Rubbia, Giuliana, 103
Rumyantseva, Nadezda, 243

S
Schwerter, Reinhard, 207
Secanell, Ramon, 213
Solarino, Stefano, 39, 55
Soto, Ignacio Jiménez, 221
Speranza, Gabriella, 49
Stanilovskaya, Julia, 247
Stefania, Bertotto, 181
Sultanbek, Yeldos, 21

T
Tatarakis, Leonidas, 21
Teo, Michael, 21
Tepel, Robert E., 165
Thuro, Kurosch, 207, 225
Ting, Charlene, 21
Tosatti, Giovanni, 97
Turconi, Laura, 125

V
Valentina, Shibakova, 251
VanDine, Doug, 229
Vasquez, Shianny, 81
Veiga, Efren M., 151
Villacorta, Sandra, 81

Y
Yao, Xue, 21
Yenes, M., 93

MIX
Papier aus verantwortungsvollen Quellen
Paper from responsible sources
FSC® C105338

If you have any concerns about our products,
you can contact us on
ProductSafety@springernature.com

In case Publisher is established outside the EU,
the EU authorized representative is:
Springer Nature Customer Service Center GmbH
Europaplatz 3, 69115 Heidelberg, Germany

Printed by Libri Plureos GmbH
in Hamburg, Germany